T0396203

VOLUME FIFTY SEVEN

ADVANCES IN
APPLIED MECHANICS

Advances in APPLIED MECHANICS

Edited by

STÉPHANE P.A. BORDAS
Institute for Computational Engineering Sciences, Department of Engineering Sciences, Faculté des Sciences, de la Technologie et de Médecine, Université du Luxembourg, Campus Kirchberg, Luxembourg, Luxembourg

Academic Press is an imprint of Elsevier
50 Hampshire Street, 5th Floor, Cambridge, MA 02139, United States
525 B Street, Suite 1650, San Diego, CA 92101, United States
The Boulevard, Langford Lane, Kidlington, Oxford OX5 1GB, United Kingdom
125 London Wall, London, EC2Y 5AS, United Kingdom

First edition 2023

ISBN: 978-0-443-13705-1
ISSN: 0065-2156

For information on all Academic Press publications
visit our website at https://www.elsevier.com/books-and-journals

Publisher: Zoe Kruze
Acquisitions Editor: Jason Mitchell
Editorial Project Manager: Palash Sharma
Production Project Manager: Abdulla Sait
Cover Designer: Victoria Pearson

Typeset by MPS Limited, India

Contents

Contributors

Stéphane Pierre Alain Bordas
Department of Engineering, University of Luxembourg, Luxembourg, France

Fabrizio Daví
DICEA and ICRYS, Ancona, Italy

Tae-Yeon Kim
Civil Infrastructure and Environmental Engineering, Khalifa University of Science and Technology, Abu Dhabi, United Arab Emirates

John Li
Department of Computer Science, Northwestern University, Evanston, IL, United States

Modesar Shakoor
Centre for Materials and Processes, IMT Nord Europe, Institut Mines-Télécom, University of Lille, Lille, France

Shaoping Xiao
Department of Mechanical Engineering, Iowa Technology Institute, University of Iowa, Iowa City, IA, United States

Preface

In this volume of "Advances in Applied Mechanics," we delve into the diverse and impactful contributions of three distinguished authors, each offering unique insights into applied mechanics.

Shaoping Xiao, along with coauthors John Li, Stéphane Pierre Alain Bordas, and Tae-Yeon Kim, presents an exploration of "Artificial Neural Networks (ANNs) and Their Applications in Computational Materials Science." The collaborative effort spans across institutions, including the University of Iowa, University of Southern California, University of Luxembourg, and Khalifa University of Science and Technology. Their chapter reviews the current advancements in artificial intelligence, particularly in the realm of computational materials science, showcasing the effectiveness of ANNs, including convolutional and recurrent neural networks. The emphasis is on their applications in multiscale modeling, understanding microstructure-dependent material properties, and establishing model-free constitutive relationships.

Fabrizio Davi contributes a study on "Coupled Bulk and Lattice Waves in Crystals Seen as Micromorphic Anisotropic Continua, with an Application to Crystals of the Tetragonal Group." Affiliated with DICEA and ICRYS at the Università Politecnica delle Marche in Italy, Davi's work focuses on the intricate dynamics of wave propagation in anisotropic crystals. The study, paying tribute to Gianfranco Capriz, not only provides qualitative insights into frequencies and dispersion relations in general anisotropic crystals but also specializes these findings for crystals of the tetragonal group. Exact representations for acoustic and optic frequencies, as well as coupled vibration modes, are detailed for various propagation directions, including those along and orthogonal to the tetragonal c-axis.

Modesar Shakoor brings forth research from IMT Nord Europe, Institut Mines-Télécom, Univ. Lille, Center for Materials and Processes in France. His work revolves around "Numerical Modeling of Highly Nonlinear Phenomena in Heterogeneous Materials and Domains," with a focus on single-phase or two-phase flows and damage/fracture of structures. Shakoor's simulations, computationally demanding due to the fine-scale interfaces, leverage advanced numerical methods. The discretization of heterogeneity involves finite element mesh generation, voxel meshes, and fast Fourier transform–based numerical methods. The innovative approach extends to modeling interfaces in flows using level-set functions and quadratic finite element interpolation and in fracture problems using a phase-field approach. Model order reduction methods, borrowing from data science and deep learning, add efficiency to these simulations.

These diverse contributions collectively advance our understanding of applied mechanics, bridging theory and practical applications across the domains of computational materials science, crystal dynamics, and numerical modeling of nonlinear phenomena in heterogeneous materials.

As we navigate through the chapters presented in this volume of "Advances in Applied Mechanics," intriguing connections between the diverse topics emerge. Shaoping Xiao's exploration of ANNs in computational materials science lays a foundation for understanding complex material behaviors. The insights gained from this data-driven approach could potentially inform Fabrizio Davi's study on crystal dynamics, especially in predicting and simulating the intricate wave propagation observed in anisotropic crystals. Modesar Shakoor's work on numerical modeling of nonlinear phenomena in heterogeneous materials adds another layer, providing essential computational methods that could enhance the precision of simulations in both materials science and crystal mechanics. These interwoven threads contribute to a comprehensive understanding of applied mechanics, demonstrating the interdisciplinary nature of advancements in this field.

Looking ahead, there is significant potential for further exploration in the realms of data-driven modeling and simulation. Building upon the foundations laid by Xiao, future research could delve deeper into the integration of ANNs with real-time model feedback, control systems, and the development of digital twins. The application of machine learning principles, as showcased in Shakoor's work, could play a crucial role in refining these models, making them adaptive and responsive to real-world conditions. This approach aligns with the growing trend in engineering and materials science towards creating intelligent systems that can dynamically adjust and optimize their behavior. The concept of digital twins, mirroring physical systems in a virtual space, holds promise for predictive maintenance, performance optimization, and gaining deeper insights into the behavior of materials and structures under various conditions. These areas present exciting avenues for future research, highlighting the potential for applied mechanics to shape the landscape of emerging technologies and engineering practices. Those will be showcased in future volumes of the series.

CHAPTER ONE

Artificial neural networks and their applications in computational materials science: A review and a case study

Shaoping Xiao[a,*], John Li[b], Stéphane Pierre Alain Bordas[c], and Tae-Yeon Kim[d]

[a]Department of Mechanical Engineering, Iowa Technology Institute, University of Iowa, Iowa City, IA, United States
[b]Department of Computer Science, Northwestern University, Evanston, IL, United States
[c]Department of Engineering, University of Luxembourg, Luxembourg, France
[d]Civil Infrastructure and Environmental Engineering, Khalifa University of Science and Technology, Abu Dhabi, United Arab Emirates
*Corresponding author. e-mail address: e-mail address: shaoping-xiao@uiowa.edu

Contents

Advances in Applied Mechanics, Volume 57
ISSN 0065-2156, https://doi.org/10.1016/bs.aams.2023.09.001

Abstract

Current advances in artificial intelligence (AI), especially machine learning and deep learning, provide an alternative approach to problem-solving for engineers and scientists in various disciplines, including materials science. Artificial neural networks (ANNs), including their variations as convolutional neural networks (CNNs) and recurrent neural networks (RNNs), have become one of the most effective machine learning approaches. This paper comprehensively reviews ANNs and their applications in different computational materials science research topics, such as multiscale modeling, microstructure-dependent material properties, and model-free constitutive relationships. In addition, we intend to share AI insights in the materials science community and promote the applications of ANNs in our research.

1. Introduction

Considerable effort has been dedicated to developing next-generation materials (Peng et al., 2017; Ray & Cooney, 2018) and structures for use in multiscale and multi-physical problems. In particular, advanced materials promote such platforms by coupling the predominant material properties to create multifunctional composites with enhanced mechanical, thermal, and other material capabilities. However, understanding such complex phenomena is highly dependent on systematic and accurate estimations of the effective physical properties, if possible. Therefore, rapid advancement requires numerical modeling and simulations capable of quickly and accurately determining such properties. Furthermore, computation has assisted the materials science community in various achievements as an important discovery tool, including rapid process development, quick microstructural analysis, fast property evaluation, and significant performance improvement.

Traditional computational methods have been extensively used to study physical phenomena at different length and time scales independently (Attarian & Xiao, 2022). Those methods include finite element methods (FEMs) (Belytschko, Liu, & Moran, 2000), meshfree particle methods (MPM) (Li & Liu, 2002; Rabczuk, Belytschko, & Xiao, 2004), phase-field methods (PFM) (Boettinger et al., 2003), molecular dynamics (MD) (Ghaffari, Zhang, & Xiao, 2017; Samanta et al., 2019), and quantum mechanics (QM) (Griffiths, 1995). In addition, many other advanced numerical methods have been developed recently. Peridynamics, introduced by Silling (Silling & Lehoucq, 2010; Silling, 2000), is a nonlocal integral-type numerical method for continuum mechanics. Notably, the internal forces in the governing equations of peridynamics are calculated via integrations instead of derivatives. As a result, this method can directly handle spatial discontinuities. It has been successfully applied to fracture mechanics (Bobaru & Zhang, 2015; Silling & Askari, 2014), as well as the studies of plastic deformation (Madenci & Oterkus, 2016), composite materials (Yaghoobi & Chorzepa, 2017; Tuhami & Xiao, 2022), and heterogeneous materials (Jung & Seok, 2016). Besides bond-based peridynamics (Ghaffari et al., 2019) mentioned above, state-based peridynamics (Silling et al., 2007; Silling, 2010) has also been developed. Another recently developed method is the lattice element method (Rizvi, Nikolić, & Wuttke, 2019), a numerical method that investigates rock materials' fracture without predefining a crack path.

Multiscale modeling (Tadmor & Miller, 2011) is an efficient approach to studying the physical phenomena of materials when considering the interactive effects between multiple spatial and temporal scales. Early development focused on the architecture of either hierarchical (i.e., sequential) or concurrent multiscale methods. Concurrent multiscale methods (Wagner & Liu, 2003; Xiao & Belytschko, 2004; Xiao & Hou, 2007a, 2007b; Xiao et al., 2008; Miller & Tadmor, 2009; Rahman et al., 2017; Tadmor & Miller, 2017) employ an appropriate model to couple multiple length/time scales so that simulations at different scales are conducted simultaneously. Most of the developed concurrent multiscale methods are atomistic/continuum coupling methods, in which the molecular model is overlapped with the continuum model. However, the scale-coupling or scale-overlapping challenge in concurrent multiscale methods doesn't exist in hierarchical approaches (Tadmor, Phillips, & Ortiz, 2000). Indeed, researchers pay more attention to passing information between scales. Homogenization (Arroyo & Belytschko, 2003; Ericksen, 1984; Xiao & Yang, 2005, 2006; Xiao, Andersen, & Yang, 2008; Yang & Xiao, 2008), including the RVE techniques (Ghaffari, Zhang, & Xiao, 2018; Grabowski et al., 2017; Subramanian, Rai, & Chattopadhyay, 2015), has been commonly employed to obtain effective material properties to bridge various scales. The current state-of-the-art multiscale methodologies can be found in several review papers (Budarapu & Rabczuk, 2017; Gooneie, Schuschnigg, & Holzer, 2017; Kanouté et al., 2009).

Riding the current wave of artificial intelligence (AI), many disciplines, especially robotics and control (Cai, Hasanbeig, et al., 2021; Cai et al., 2021; Zhu et al., 2022), have applied learning-based approaches. Notably, the data-driven approach has become another powerful tool in scientific discoveries and engineering problem-solving (Versino, Tonda, & Bronkhorst, 2017). One of its new paradigms in materials science is discovering new materials

(Wang et al., 2022) or improving material designs (Pollice et al., 2021) based on the knowledge extracted from extensive materials datasets. Himanen et al. (2019) addressed data-driven materials science's status, challenges, and perspectives. Their review focused on materials data infrastructures and discussed several critical challenges in developing a material search tool. Tripathi, Kumar, and Tripathi (2020) presented big data models for material science data management and feature preservation in another survey. They also reported several challenges in big data analysis, such as data privacy, data preprocessing, and predictive algorithms.

Machine learning (ML) (Mitchell, 1997) is an approach using statistical models to analyze data and draw inferences from its pattern. Particularly, supervised learning models learn the relationship between the input features and the output targets without explicit instructions. As a subset of ML, deep learning (DL) (Schulz & Behnke, 2012) employs artificial neural networks (ANNs) to find appropriate representations from data progressively for good performance. Zhang and Friedrich (2003) presented one of the first reviews on predicting specific material properties of polymer composites by using neural networks. According to their review, a few early works have been conducted to predict fatigue life (El Kadi & Al-Assaf, 2002), tribological properties (Rutherford et al., 1996), and some other mechanical behaviors (Zhang, Klein, & Friedrich, 2002). Neural networks were also used for composite processing optimization (Heider, Piovoso, & Gillespie, 2003) and design optimization (Ulmer II et al., 1998). In another work, Kadi (El Kadi, 2006) summarized the implementation of ANNs in the mechanical modeling of fiber-reinforced composite materials, including static deformation and failure behaviors (Olivito, 2003), creep behavior (Al-Haik, Al-Haik, Garmestani, & Savran, 2004), delamination (Valoor & Chandrashekhara, 2000), crack and damage detection (Bar, Bhat, & Murthy, 2004), impact (Chandrashekhara, Okafor, & Jiang, 1998), and vibration control (Smyser & Chandrashekhara, 1997). Kadi also reviewed the applications of fuzzy ANN in studies of damage and failure in composite materials (Jarrah, Al-Assaf, & Kadi, 2002).

Recently, the applications of ML and DL have caught more and more attention from researchers in the materials science community, and quite a few updated reviews and discussions have been reported. Rodrigues et al. (2021) proposed a roadmap for future research focusing on ML-aided discovery of new materials and analysis of chemical sensing compounds. They also elaborated on the conceptual and practical limitations when applying big data and ML to materials science research topics. Morgan and Jacobs (2020) reviewed some common types of ML models in materials science and addressed the breadth of opportunities and the best practices for their usage. Another recent work (Choudhary et al., 2022) reviewed the applications of DL in atomistic simulation and material imaging.

This paper aims to provide an in-depth review of ANNs, including physics-informed neural networks (PINN), convolutional neural networks (CNNs), and recurrent neural networks (RNNs), and their applications in computational materials science and engineering. We will focus on several advanced research topics, such as multiscale modeling, forward and inverse problems, microstructure-dependent material property prediction, and model-free constitutive identification. In addition, we use a case study to demonstrate the applications of neural networks in studying the material failure of metal-ceramic spatially tailored materials. This paper also intends to share AI insights in the materials science community and promote the applications of ANNs in our research.

2. Artificial neural networks

2.1 Basics of artificial neural networks

A typical ANN (Dreiseitl & Ohno-Machado, 2002), shown in Fig. 1, usually consists of an input layer, an output layer, and one or more hidden layers. This kind of neural network is fully-connected because every neuron connects all the neurons on the previous and subsequent layers. For example, we consider a data set of N distinct training samples (x_I, y_I) where $I \in [1, N]$. Each data sample has p input features $(x_I \in R^p)$ and q outputs $(y_I \in R^q)$.

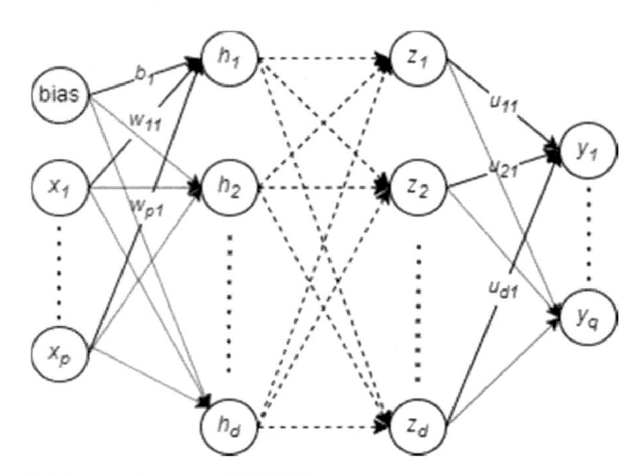

Fig. 1 An artificial neural network (ANN).

Therefore, the corresponding neural network to approximate the relation between the input and the output of the data set has $p + 1$ neurons on the input layer and q neurons on the output layer. The hidden layers can have various numbers of neurons, and it is assumed that each hidden layer has d neurons in Fig. 1.

The neural network training includes a feedforward process and a backpropagation process. During the feedforward process, every neuron in the hidden layers transforms the outputs from the previous layer into a different representation, the input to the next layer. There are two steps in the transformation. For example, the input data is projected into the first hidden layer via the weights, $\boldsymbol{w}$, and biases, $\boldsymbol{b}$. Then, the projected outcome is transformed via the activation function φ, also called the transformation function. Mathematically, this transformation can be expressed as

$$h_j = \varphi_j\left(w_j^T x + b_j\right) = \varphi_j\left(\sum_{i=1}^{p} w_{ij} x_i + b_j\right) j = 1 \dots d \tag{1}$$

It is known that the hidden layer is not limited to having only one type of activation function in neurons. There are a variety of activation functions available, and most of them are nonlinear functions, including sigmoid, hyperbolic tangent, and radial basis functions (RBFs). Particularly, the RBF neurons use distances between samples and centroids as inputs, and L^1, L^2, or L^∞ norms of distances can be used.

If the last hidden layer has the output z, as shown in Fig. 1, the estimated k th output of the Ith training sample is calculated as

$$\tilde{y}_{Ik} = \varphi_o\left(u_k^T z\right) = \varphi_o\left(\sum_{i=1}^{d} u_{ik} z_i\right) \quad k = 1 \dots q \tag{2}$$

where u are weight coefficients. φ_o is the activation function for outputs, and it is usually an identity function for regression problems. A loss function is calculated based on the estimated output targets and the actual outputs to evaluate the neural network's performance. It is also called the data loss function as

$$L_d = L_d\left(y_1, y_2 \cdots y_N, \tilde{y}_1, \tilde{y}_2, \cdots \tilde{y}_N\right) \tag{3}$$

Indeed, training a neural network becomes an optimization problem to find appropriate weight coefficients, including w and u, for minimizing the loss function. This is usually done using the gradient descent method or its variations in the backpropagation process.

2.2 Physics-informed neural networks

Training a PINN (Raissi, Perdikaris, & Karniadakis, 2019) needs a training data set and a physical-based mathematical model, i.e., partial differential equations (PDE), shown in Fig. 2. Without a loss of generality, we assume a system of PDEs with appropriate initial and boundary conditions as below.

$$\frac{\partial y(x,t)}{\partial t} + D[y;\gamma] = 0$$
$$\text{s. t. } y(x,0) = y_0(x),\quad y(0,t) = \underline{y}_0(t),\quad y(L,t) = \underline{y}_L(t) \tag{4}$$

where D is a nonlinear operator parametrized by γ.

The training data can be collected by numerically solving PDEs in Eq. (4). It shall be noted the solutions are on the discretized spatial and temporal grids. Therefore, in addition to the input features, the neural network in Fig. 2 may also take time and spatial coordinates on the input layer. In most existing works of PINN, the fully-connected neural work was employed. Therefore, the output targets $\tilde{y}_l$ can be predicted as described in Eq. (2) via the feedforward process. Such approximations to y result in not only the data loss function but also the residual of PDEs as

$$r = \frac{\partial \tilde{y}(x,t)}{\partial t} + D[\tilde{y};\gamma]. \tag{5}$$

where the derivatives of $\tilde{y}$, i.e., $D[\tilde{y};\gamma]$, can be derived via the automatic differentiation approach. Consequently, the physics loss function is defined

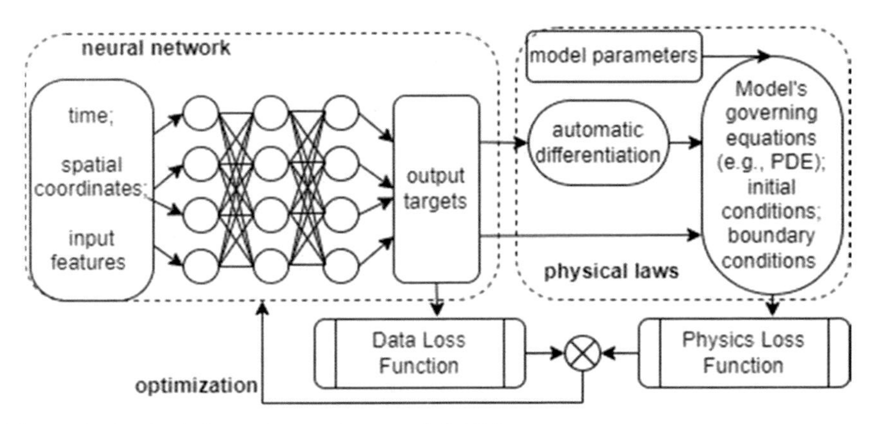

Fig. 2 A physics-informed neural network (PINN).

below, considering residuals of governing equations and initial and boundary conditions.

$$L_p = L_p(|r|, |\tilde{y}(x,0) - y_0(x)|, |\tilde{y}(0,t) - \underline{y}_0(t)|, |\tilde{y}(L,t) - \underline{y}_L(t)|) \tag{6}$$

Then, the total loss function, which is a combination of the data and physics loss functions, i.e., $L_d + L_p$, is implemented in the backpropagation process to optimize the neural network's weight coefficients.

It can be seen that the data loss function measures the difference between the actual outputs and their approximations predicted by the neural network. On the other hand, the physics loss function quantifies how close the input-output relationship approximated by the neural network follows the physical laws. Therefore, this neural network is physics informed. The concept of PINN has been employed in many disciplines, including computational mechanics and materials science, for both forward and inverse problems (Raissi et al., 2019).

2.3 Other neural networks

Other commonly used artificial neural networks include CNNs (Sainath et al., 2015) and RNNs (Schmidhuber, 2015), which mainly aim to handle image and time-series data, respectively. Convolution neural networks have been proven to be very effective and successful in image recognition and classification. Generally, an image can be represented as a matrix of pixel values, and a color image has three channels – red, green, and blue. Therefore, an image sample is indeed a three-dimensional tensor of shape, i.e., (height, width, and channels). To address a classification problem, a CNN usually has four main operations in the feedforward prediction process: convolution, nonlinearity, max pooling, and classification, as shown in Fig. 3.

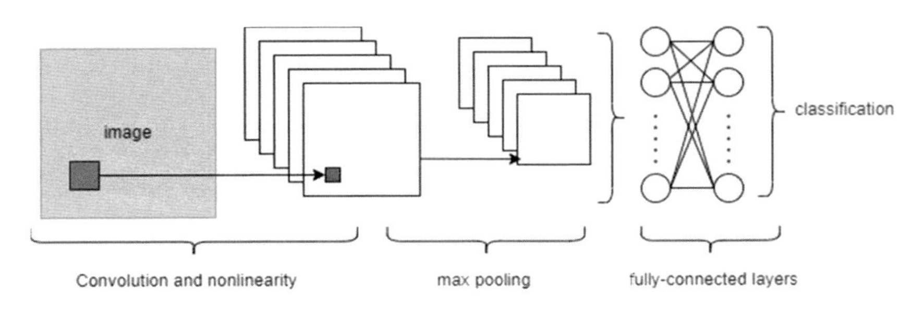

Fig. 3 A convolutional neural network (CNN).

During the convolution and nonlinearity operations (in a convolution layer), the image is decomposed into overlapped image tiles via the sliding window search, i.e., stridden convolution. Each image tile is then fed into a small neural network with nonlinear activation functions. It shall be noted that the same neural network is applied for every single tile individually. The output feature map is a three-dimensional array with a smaller height and width than the original image. If the original image is a color one, it has a depth of three because of three channels. However, the resulting array has a depth the same as the neural network's filter number.

The next operation is max pooling, which is also called downsampling. This operation consists of extracting windows from the input feature maps (the output from the previous convolution layer) and outputting the max value of each filter. The convolution layer and the max pooling can repeat multiple times before reshaping the output feature map as a one-dimensional array to a "fully-connected" network for prediction.

Recurrent neural networks mimic the biological intelligence procedure that processes information incrementally while maintaining an internal memory (i.e., state) for past information. They are often employed for time-series data, and each data sample (i.e., time sequence) is encoded as a 2D tensor of size with time steps and input features. An RNN is a fully-connected neural network that has states. Instead of taking a data sample at one time, an RNN unit loops over time steps, as shown in Fig. 4.

At each time step t, the RNN considers the output from the previous time step t-1 as its current state, takes the tth input entry, and combines them to obtain the output at time t. After applying the activation function, the RNN sets the output as the state for the next time step. Such a recursive update can be written as

$$y^t = \varphi\left(w^T x^t + u^T y^{t-1} + b\right) \tag{7}$$

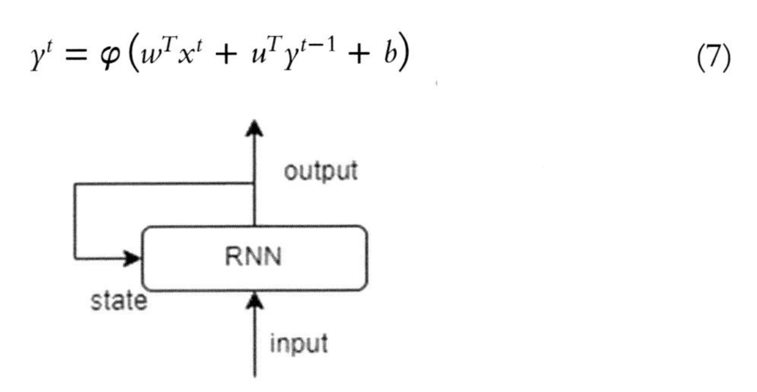

Fig. 4 A classical recurrent neural network (RNN) unit.

where φ is the activation function, w and u are weight coefficients, and b is the bias. A few advanced RNN architectures have been proposed, including long-short term memory (LSTM) (Sutskever, Vinyals, & Le, 2014) and gated recurrent unit (GRU) (Cho et al., 2014).

3. Applications of neural networks

3.1 ANNs in multiscale modeling

Multiscale modeling and simulation have benefited from ML and DL methods (Alber et al., 2019), including fully connected ANNs. Le, Yvonnet, and He (2015) employed ANNs and proposed a decoupled computational homogenization method for nonlinear elastic materials. Their approach computed the training samples' effective potentials through random sampling in the parameter space. Then, ANNs were used to approximate the surface response and derive the macroscopic stress and tangent tensor components. In another work, Unger and Könke (Unger & Könke, 2009) adopted ANNs as material models in a multiscale approach to studying reinforced concrete beams. In another work, Liu, Wu, and Koishi (2019) developed a new data-driven multiscale method, i.e., a deep material network, to describe complex overall material responses of heterogeneously structured composites. They also simulated the macroscale dynamics of gas-solid mixtures by employing information collected from microscale simulations via an ANN model (Lu et al., 2012). White and co-workers (White et al., 2019) used a single-layer feedforward neural network as a surrogate model to predict the elastic response of the microscale metamaterial during the optimization of macro-scale elastic structures. Other achievements include a multiscale multi-permeability poroplasticity model (Wang & Sun, 2018), a 3D architecture of a deep material network (Liu & Wu, 2019), and ANN-assisted multiscale analysis (Balokas, Czichon, & Rolfes, 2018).

In addition, Xiao et al. (2020) proposed an alternative data-driven approach by using neural networks to pass information from the molecular model to the continuum model in a hierarchical multiscale framework. First, intensive molecular dynamics (MD) simulations were conducted to generate the dataset in which the input features included strains and temperature, while the output targets were stress components and material failure mode. Then, the generated data was used to train several DL classification and regression models. Indeed, one of the important emergent ML techniques, extreme learning machines (ELMs) (Huang, Zhu, & Siew,

2006; Guang-Bin Huang et al., 2012) were adopted. It has been shown that an ELM is a fast training method for Single-Layer Feed-forward Networks (SLFNs). An SLFN has three layers of neurons. The term "Single" stands for the only layer of linear/nonlinear neurons in the model and is the hidden layer. Finally, the well-trained learning machines were directly implemented in continuum simulations to predict material failure mode and stress components. In this approach, as shown in Fig. 5, neither constitutive relations nor effective material properties were explicitly derived as achieved in existing hierarchical multiscale methods. The learning machines served as "black boxes" to replace constitutive relations and failure mode decisions in the continuum simulations. Such "black boxes" were trained based on the dataset from molecular simulations; therefore, the proposed scheme is physical-based and data-driven.

Xiao et al. (2021) and Tuhami and Xiao (2022) have extended the above ML-based multiscale framework to study the mechanics of spatially tailored materials (STMs) via FEMs or peridynamics. Spatially tailored materials (Birman et al., 2008), also named functionally graded materials

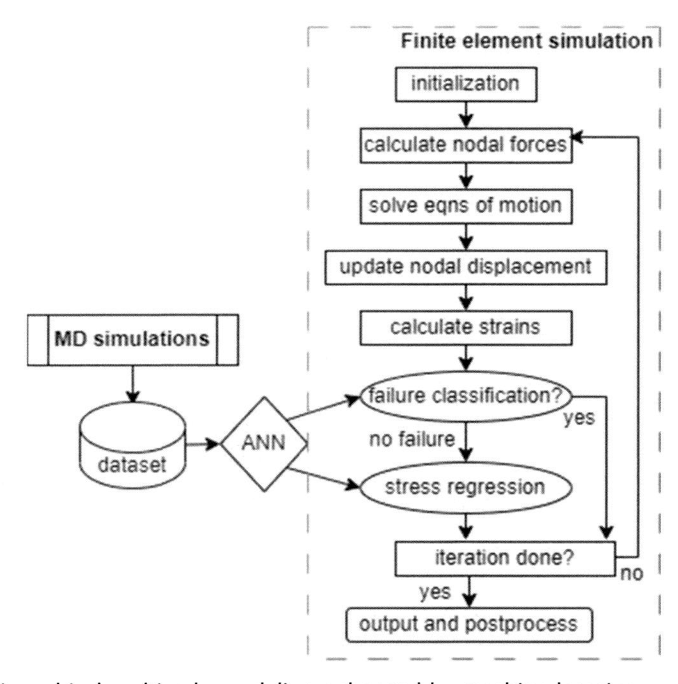

Fig. 5 Hierarchical multiscale modeling enhanced by machine learning.

(FGMs), are one of the next-generation composites for use in multi-physical problems. They are essentially composites consisting of two or more phases of distinct materials in which the volume fractions continuously change in space. This unique class of heterogeneous composites offers advantages over traditional composites due to its ability to leverage the predominant characteristics of the constituent materials and to tailor the effective material properties according to the loading conditions and operating temperatures. A metal-ceramic STM was considered in Ref. Xiao et al. (2021) and studied via a hierarchical multiscale method from micro to macro scales. Microstructure uncertainties, including particle number, size, shape, and location, were considered during data collection via microscale simulations. After being trained via the collected dataset, the ANNs for material property and failure predictions were implemented in macroscale simulations.

3.2 PINNs in forward and inverse problems

Physics-informed neural networks have been adopted in various disciplines, including computational materials science. For example, Haghighat et al. (2020) demonstrated the application of PINN in solving a two-dimensional linear elasticity problem with the following governing equations:

$$\frac{\partial \sigma_{ij}}{\partial x_j} + f_i = 0 \tag{8}$$

where σ is the Cauchy stress tensor, f is the body force vector, and $i, j = 1{,}2$. The constitutive model is

$$\sigma_{ij} = \lambda \delta_{ij} \varepsilon_{kk} + 2\mu \varepsilon_{ij} \tag{9}$$

where δ_{ij} is the Kronecker delta, and λ and μ are material constants. ε_{ij} is the infinitesimal strain tensor and can be calculated as

$$\varepsilon_{ij} = \frac{1}{2}\left(\frac{\partial u_i}{\partial x_j} + \frac{\partial u_j}{\partial x_i}\right) \tag{10}$$

where $u(x)$ is the displacement field.

The FEM solutions were used as the training dataset in their approach. Each data sample had the coordinates (x) as the input features, and the output targets included $u_1(x)$, $u_2(x)$, $\sigma_{11}(x)$, $\sigma_{22}(x)$, and $\sigma_{12}(x)$. Various ANNs were employed/trained to predict each output variable individually. In addition to the data loss, the physics loss function for this specific problem was written below based on the governing equations and the constitutive model.

$$L_p = \left| \frac{\partial \tilde{\sigma}_{11}}{\partial x_1} + \frac{\partial \tilde{\sigma}_{12}}{\partial x_2} + f_1 \right| + \left| \frac{\partial \tilde{\sigma}_{12}}{\partial x_1} + \frac{\partial \tilde{\sigma}_{22}}{\partial x_2} + f_2 \right| + |(\lambda + 2\mu)\varepsilon_{11} + \lambda\varepsilon_{22} - \tilde{\sigma}_{11}| + |(\lambda + 2\mu)\varepsilon_{22} + \lambda\varepsilon_{11} - \tilde{\sigma}_{22}| + |2\mu\varepsilon_{12} - \tilde{\sigma}_{12}| \tag{11}$$

where the output variables with tilde were predicted from ANNs. The other variables, including ε_{11}, ε_{22}, ε_{12}, f_1, and f_2 can be obtained from Eqs. (8) and (10) as below, through automatic graph-based differentiation (Güne et al., 2018).

$$\varepsilon_{ij} = \frac{1}{2}\left(\frac{\partial \tilde{u}_i}{\partial x_j} + \frac{\partial \tilde{u}_j}{\partial x_i} \right) \text{ and } f_i = -\frac{\partial \sigma_{ij}}{\partial x_j} \tag{12}$$

Then, the PINN's weight coefficients can be updated via gradient descent approaches during the backpropagation process. Haghighat et al. (2020) also demonstrated that PINNs could be used to identify the model parameters λ and μ, as solving an inverse problem (Raissi et al., 2019).

In another work, Zhang, Yin, and Karniadakis (2020) extended PINN to solve identification problems of nonhomogeneous materials. In this inverse problem, they sought to identify soft tissue material properties based on the full-field displacement measurements under quasi-static loading. In addition to a PINN employed to approximate the unknown material parameters, another PINN was utilized to solve the corresponding forward problem. Two PINNs were trained simultaneously, as shown in Fig. 6, and the physics loss function was formulated according to the displacement boundary conditions, the incompressibility constraints, traction boundary conditions, and the governing equations (i.e., PDEs).

Recently, Zhang and Gu (2021) extended PINNs into digital material problems, in which a composite, as a 3D-printable material, was considered an assembly of material voxels. They addressed a few challenges, including discontinuous material properties, nonlinear strain due to large deformation, and neural network accuracy. In digital material design, a material configuration is generally a combination of step functions, and its derivatives are often not available. To address this challenge, they adopted the minimum energy criteria as the loss function of a PINN other than the strong governing equations (i.e., PDEs). This energy-based PINN

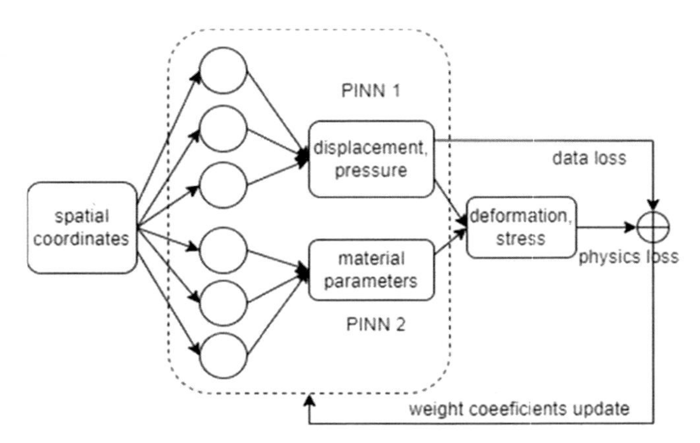

Fig. 6 Two PINNs were used in (Zhang et al., 2020) for an inverse problem.

achieved comparable accuracy to supervised ML models. In addition, adding a hinge loss for the jacobian could enhance the proposed PINN to approximate nonlinear strain properly.

Physics-informed neural networks were also adopted in studying crack propagation of quasi-brittle materials under complex loading (Zheng et al., 2022). In this work, the researchers used PINN to reconstruct the solution of the displacement field. Without labeled data (i.e., a priori information), the PINN based on the energy minimization principle could predict crack propagation with incremental loading pattern and damage evolution model. In addition, they introduced the domain decomposition method and the finite basis algorithm to address complex boundaries and the gradient pathology problem, respectively. Therefore, their approach is robust.

Furthermore, Shukla et al. (2020) applied PINN to identify and characterize a surface-breaking crack in an aluminum alloy substrate material. PINN was supervised with realistic ultrasonic surface acoustic wave data, representing deformation on the top surface of the aluminum plate. It was physically informed by the ultrasonic surface acoustic wave equation as an inverse problem, i.e., the estimation of the unknown wave speed for given acoustic wave data. In other words, the spatially varying surface wave speeds were used as markers of characterizing/identifying the surface-breaking crack in the aluminum alloy substrate. Moreover, they used adaptive activation functions in training to accelerate convergence significantly, even with highly noisy data. Using a small portion (i.e., 10–20% of the total data) of the wave data, PINN accurately predicted the wave speed affected by the crack in the substrate, verifying the efficiency of the PINN by reducing the cost and time of the data acquiring process.

3.3 CNNs in microstructure quantification

Since CNNs have been successfully employed in image processing, they were also recently used in computational materials science to quantify the microstructure of materials (especially composite materials) for predicting material properties. Previous approaches (Torquato & Haslach, 2002) utilized n-point spatial correlations (Kröner, 1977), which could effectively quantify the local neighborhoods in material microstructure as features to measure material properties. However, the number of features could be practically infinite when considering the complete set of possible local neighborhood configurations.

One of the pioneering works (Cecen et al., 2018) aimed to address this core challenge in constructing material process-structure-property linkages for new material design and improvement by using CNNs. In their approach, a 3D CNN was employed to learn the salient features of the material microstructures for material property predictions as a regression problem. Specifically, they collected 5900 microstructure images and conducted finite element simulations to calculate material properties for each microstructure. Each microstructure image consisted of $51 \times 51 \times 51$ cuboidal voxels and was convolved with 32 filters. The filter size was $10 \times 10 \times 10$, which was informed by spatial statistics. The rectifier function was used as the activation function. It shall be noted that average pooling was used instead of the max pooling as in conventional CNNs. After getting 256 features, linear regression was conducted to estimate the effective material property.

In a similar work, Rao and Liu (2020) proposed a three-dimensional CNN, shown in Fig. 7, as a homogenization surrogate model to predict the anisotropic effective material properties for microscale RVEs with random inclusions. A high-fidelity dataset was generated by using FEM simulations, and the trained CNN was capable of capturing the microstructural features of RVEs. Instead of predicting each material property with individual CNNs, a single 3D-CNN was employed to estimate all material stiffness and position components for the studied heterogeneous composites. They also discussed uncertainty quantification and the model's transferability. In another similar work (Mianroodi et al., 2022), a CNN took the nanostructured configurations as input and predicted the corresponding elasticity tensor. Other CNN-enhanced modeling and simulations include stress field

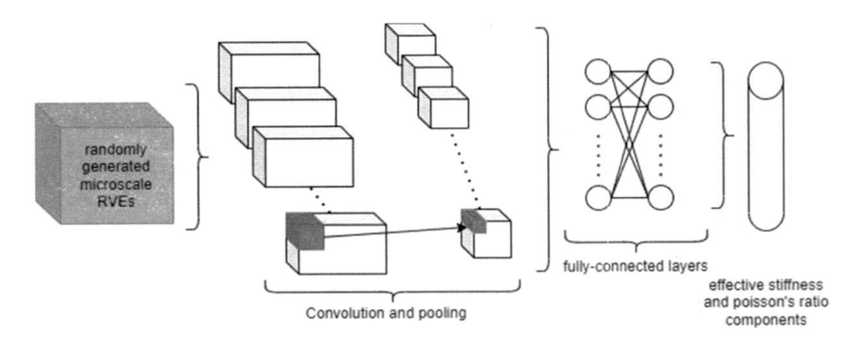

Fig. 7 A 3D-CNN (Rao & Liu, 2020) for estimating effective anisotropic material properties.

prediction in composites (Bhaduri, Gupta, & Graham-Brady, 2022) and optimizing material structures (Yilin, Fuh Ying Hsi, & Wen Feng, 2021).

During the data collection in approaches similar to the above-mentioned works (Cecen et al., 2018), computationally expensive physics-based simulation tools (e.g., FEM) were usually employed to calculate the effective material properties for given microstructures. Such a forward model may not be practical for microstructure design problems in which the target properties are usually achieved via an iterative process. Mann and Kalidindi (2022) combined the microstructure-sensitive design (MSD) framework with the CNN-based surrogate model to reduce the computational cost. They introduced the microstructure hull concept in MSD and used 2-point spatial correlation maps as inputs in the CNN. Such a proposed strategy made exploring the complete search space of possible material properties feasible.

3.4 RNNs in material constitutive identification

Many previous works employed ANNs (Akbari, Mirzadeh, & Cabrera, 2015; Sabokpa et al., 2012; Singh, Rajput, & Mehta, 2016) to approximate constitutive modeling at various strain rates and high temperatures. However, most nonlinear constitutive relationships are stress/strain-history dependent. In other words, the predicted stress depends on the current deformation and the deformation history. More input features that can capture stress/strain history need to be included to address this issue. Furukawa and Hoffman (2004) used an ANN in FEM to conduct cyclic plastic analysis. The inputs in their neural networks included the current and two previous states of the back stress and inelastic strain, while the output was the increment of the back stress. Similar works include a hybrid multilayer perceptron neural network approach to describe the statistical scatter of cyclic stress-strain curves (Janežič, Klemenc, & Fajdiga, 2010).

Since RNNs have advantages over ANNs in analyzing time-series data, they have been a better alternative approach to approximate stress-strain-time relationships. An early work done by Oeser and Freitag (2009) utilized an RNN with the input of previous load (stress) increments to estimate the new displacement (strain) increment for the studies of rheological materials. It shall be noted that the training data was collected from creep simulations. In another work, Freitag, Graf, and Kaliske (2013) employed RNNs as a complete or part of the material model in fuzzy FEM simulations. They assumed that the material parameters in numerical simulations were uncertain. Time-dependent stress and strain were modeled as fuzzy processes (Möller & Beer, 2004), which were discretized via time discretization and α-level discretization. Then, an RNN was developed to map fuzzy strain processes onto fuzzy stress processes, and it was applied to describe time-dependent material behaviors in FEM simulations.

Zopf and Kaliske (2017) coupled an RNN to the micro-sphere approach to approximate the model-free characterization of elastic and inelastic materials, including uncured natural rubber. Such coupling could address the issue that material testing cannot cover the complete stress state space. Therefore, reliable training of the proposed neural network was achieved. Recently, Stöcker et al. (2022) proposed a novel training algorithm for RNN to be more robust. The algorithm could generate adversarial examples based on the neural network prediction errors, i.e., the loss function. Consequently, the neural network could yield reliable material representations even when providing perturbed inputs. Specifically, GRU, an advanced RNN, was used in their approach.

4. A case study

4.1 Metal-ceramic spatially tailored composite materials

In this case study, we consider metal-ceramic STMs to generate the dataset and then employ various neural networks to predict material properties and mechanical behaviors. The STM materials are Ti(Ti-6Al-4V)-TiB_2 composites, and material properties are listed in Table 1 (Wiley, Manning, & Hunter, 1969; American Society for Metals., 1979; Munro, 2000). We only consider the room temperature (20 °C) in this case study.

Table 1 Material properties of Ti (Ti-6Al-4V) and TiB_2 at 20 °C.

	Young's modulus *E (GPa)*	**Poisson's ratio *v***	**Density *ρ (kg/m³)***	**Tensile strength σ_t *(GPa)***
Ti (Ti-6Al-4V)	106.2	0.298	4357	1.17
TiB_2	495.4	0.100	4505	3.73

We use the ceramic volume fraction (CVF) to represent the composition of ceramic in the studied STMs, which can be modeled as continuously variable composition materials. Particularly, the CVF can be expressed as a function of coordinates to indicate the difference between metal and ceramic at a particular spatial location in the STM. For example, in a Ti-TiB_2 plate, if the volume fraction varies along with the plate thickness, it can be written below as a power-law distribution.

$$v_f(z) = v_0 + (v_1 - v_0)\left(\frac{z}{h}\right)^n \tag{13}$$

where h is the total thickness, and z is the thickness coordinate between two surfaces $z = 0$ and $z = h$. In addition, v_0 and v_1 are the CVFs at two surfaces. It shall be noted that n is the control parameter for the ceramic content distribution. If choosing $n = 1$, there is a linear distribution of ceramic volume fraction along the thickness. On the other hand, if $n = 2$, a nonlinear (i.e., quadratic) distribution exists, as shown in Fig. 8. The metal (Ti) is the matrix material when the CVF is less than 50%. After the CVF exceeds 50%, the matrix material switches to ceramic (TiB_2).

There may be more than one directional material variation in STM structures. Generally, Eq. (13) can be revised below for a two-dimensional graded material with a rectangular geometry.

$$v_f(x, y) = v_0 + (v_1 - v_0)\left[\eta_x\left(\frac{x}{w}\right)^{n_x} + \eta_z\left(\frac{y}{h}\right)^{n_y}\right] \tag{14}$$

where x and y are the coordinates of arbitrary material points taking one corner of the rectangle as the origin, w and h are the total width and height, and η_i and n_i are parameters controlling the ceramic content and profile in each direction. v_0 and v_1 are the minimum and maximum CVFs in the material, and they are usually 0 and 1, respectively.

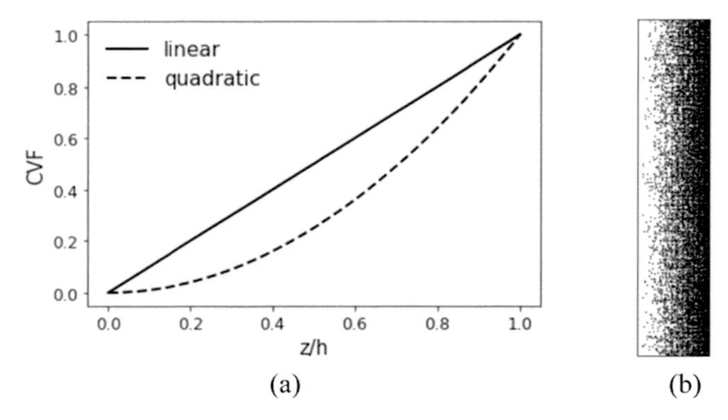

Fig. 8 Metal-ceramics STMs: (A) CVF distributions and (B) an STM plate.

4.2 Peridynamics

In this case study, we will use bond-based peridynamics (Silling & Lehoucq, 2010; Silling, 2000) to model and simulate STMs at the microscale. Peridynamics is a nonlocal model of classical continuum mechanics. However, the governing equations, i.e., the partial differential equations, are reformulated below by replacing the derivative terms with volume integrals of force densities.

$$p\ddot{u}(x,\ t) = \int_{H_x} f(\eta,\ \xi,\ t)\, dV_{x'} + b(x,\ t) \tag{15}$$

where ρ is the material density, $\ddot{u}$ is the second time derivative of the displacement vector, i.e., the acceleration vector, and **b** is the body force density vector. In bond-based peridynamics, the simulation domain is discretized into material points with finite volumes. The pairwise force density vector, f, corresponds to the deformation of bonds between point x and material points in its horizon, i.e., $x' \in H_x$. ξ and η represent the relative position vector and the relative displacement vector, respectively, and they can be defined below, as shown in Fig. 9.

$$\xi = x' - x,\ \cdot\eta = u(x',\ t) - u\cdot(x,\ t) \tag{16}$$

The force density can be calculated from bond strain s and bond rotation γ (Zhu & Ni, 2017) as

$$f(\eta,\ \xi,\ t) = c\ s\ n + \kappa\gamma \tag{17}$$

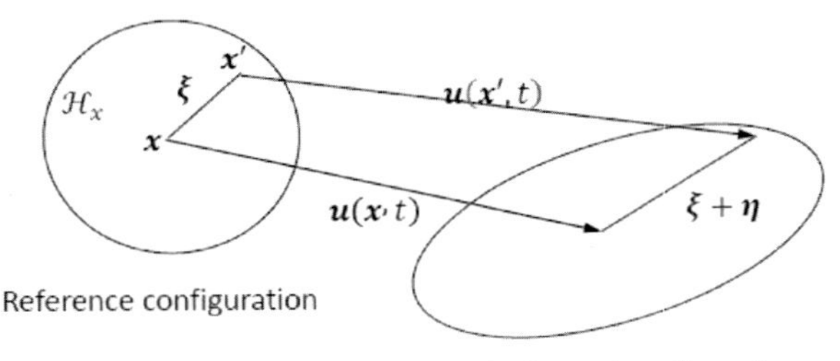

Fig. 9 Reference and deformed configurations in bond-based peridynamics.

where

$$s(\eta, \xi) = \frac{(\|\eta + \xi\| - \|\xi\|)}{\|\xi\|} \tag{18}$$

$$n = (\eta + \xi)/\|\eta + \xi\| \tag{19}$$

$$\gamma(\eta, \xi) = \frac{1}{\|\xi\|}\eta \cdot (I - n \otimes n) \tag{20}$$

c and κ are the first and second micromoduli. Assuming an elastic material has Young's modulus E and Poisson's ratio ν, the micromoduli for plane strain problems can be derived as

$$c = \frac{6E}{\pi\delta^3(1 + \nu)(1 - 2\nu)}, \cdot k = \frac{6E(1 - 4\nu)}{\pi\delta^3(1 + \nu)(1 - 2\nu)} \tag{21}$$

It shall be noted that the metal-ceramic STMs have various CVFs spatially at the macroscale. However, it is assumed that the CVF is a constant at each macroscale material point. In other words, each microscale model of STM has a constant CVF. All microscale models are the two-dimension domain of 30 μm × 30 μm, assuming plane strain. The simulation domain is discretized with 2601 (51 × 51) material points for peridynamics. According to the CVF, the material points are randomly assigned as either metal or ceramic.

In addition, 1–2% porosity (Patil et al., 2019) is added to the microscale configuration by randomly selecting a small number of material points and removing them to generate vacancies. The number of vacancies follows the Poisson point distribution (Xiao et al., 2008). Fig. 10 illustrates two microscale STM configurations with 25% CVF and two with 50% CVF. In

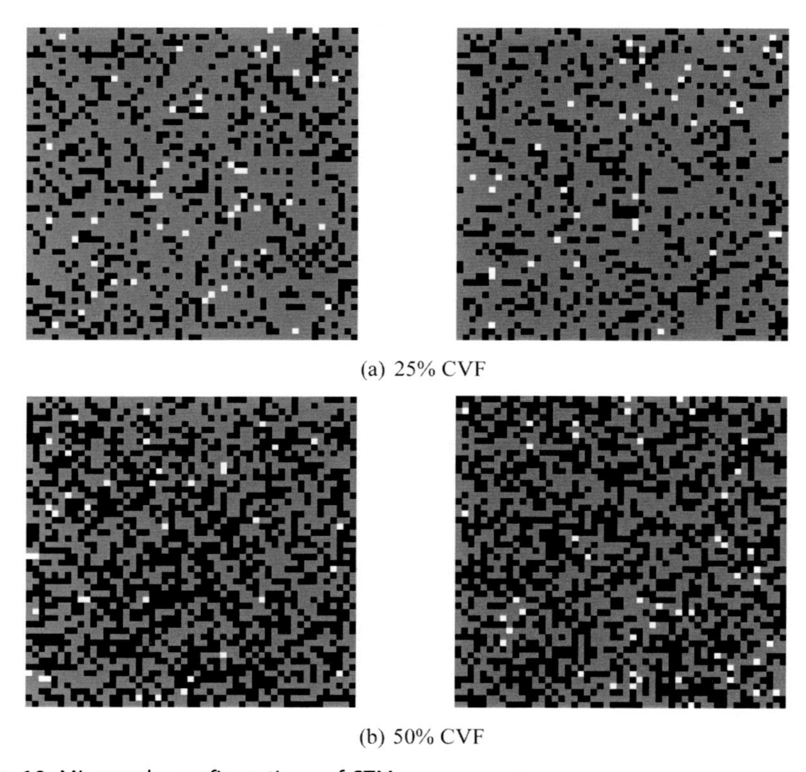

Fig. 10 Microscale configurations of STM.

the microstructure configurations, gray represents metal, black represents ceramic, and white represents vacancy.

In the peridynamic model, we choose $\Delta x = \Delta y = 0.6\,\mu m$ and the horizon radius $\delta = 1.6\Delta x$. Since we consider metal-ceramic STMs in this study, there are three different types of bonds: metal-ceramic, metal-metal, and ceramic-ceramic. According to the material properties listed in Table 1, the first and second micromoduli can be calculated as in Table 2. They are calculated via Eq. (21) for metal-metal and ceramic-ceramic bonds. The prototype micro-elastic brittle (PMB) material model is employed, and the critical stretches, s, are computed based on the material tensile strength. In addition, the corresponding parameters for metal-ceramic bonds are derived via the combining rule. For example, the first micromodulus of metal-ceramic bonds is $c_{mc} = \sqrt{c_m c_c} = 622.55\frac{GPa}{\mu m^3}$.

Table 2 First and second micromoduli in peridynamics models.

	Metal-metal bond	Ceramic-ceramic bond	Metal-ceramic bond
$c\left(\frac{GPa}{\mu m^3}\right)$	326.01	1188.83	622.55
$\kappa\left(\frac{GPa}{\mu m^3}\right)$	47.94	756.53	190.44
s	0.011	0.008	0.0094

4.3 Data collection

The microscale configurations vary at the same CVF. Therefore, 25 simulations are conducted at each CVF between 1% and 99% at a 1% increment. The model is subject to the uniaxial tension in each simulation by applying the prescribed displacement on the top and fixing the bottom. Since the enforcement of boundary conditions cannot be directly applied to the boundary material points in peridynamics, fictitious walls (Ghaffari et al., 2019) are applied to eliminate such "edge softening" phenomena (Nishawala & Ostoja-Starzewski, 2017). A low strain rate is maintained, so the simulation results (e.g., strain-stress relations) can be used for quasi-static analyses. The stress is calculated by dividing the vertical component of the total bond force by the area of the middle cross-section. The Young's modulus is calculated from the slope of a strain-stress curve at 0.005% strain. The highest stress at the stress-strain curve is the tensile strength.

It is observed that the failure strength follows the Gaussian distribution at a particular CVF. Table 3 illustrates the statistical characteristics of failure strength at three different CVFs: 25%, 50%, and 75%. Obviously, the composite with a higher CVF has a higher mean failure strength and a larger standard deviation.

4.4 Deep learning predictive models

In the work of (Tuhami and Xiao, 2022), peridynamics was employed at the microscale to study the mechanical behaviors of metal-ceramic composites at various CVFs. The generated data were used to train multiple machine learning models to predict material properties, including Young's modulus, Poisson's ratio, and tensile strength, taking the CVF as the input. The well-trained predictive models were then implemented in continuum mechanics simulations at the macroscale. In those machine learning

Table 3 Statistics of failure strength (GPa) at different CVFs.

Ceramic volume fracture	25%	50%	75%
Mean	0.956	1.290	1.880
Standard deviation	0.048	0.053	0.076

models, fully-connected neural networks were employed. They took the CVF as the input feature and the material properties as the output target. Therefore, they were regression models. In this case study, we utilize the same way to generate the data. However, not only the CVF but also the image of the microstructure is used as the input feature. Consequently, in addition to ANN as a material failure classification model, CNN is utilized as a regression model to predict tensile strength.

4.4.1 Failure probability prediction via ANN

During the data collection, 25 tensile simulations are conducted at each CVF, and various tensile strengths are obtained due to the microstructure uncertainties. In other words, although the composites have the same CVF, they fail at different strains. According to the strain-stress histories collected from microscale simulations, we can approximate the likelihood of failure occurrence at a particular strain. Consequently, we use a binary map to represent the failure probability in the strain-CVF space, as plotted in Fig. 11. The white domain indicates a 100% probability of material failure in the figure, while the black domain indicates non-failure. However, there is no clear decision boundary to distinguish failure/non-failure domains but rather a fuzzy interface, representing material failure with a probability between 0% and 100%.

Indeed, the corresponding dataset has 154,025 data samples in which the strain and CVF are in the input features while the output is material failure or non-failure, represented by "1" or "0", respectively. Instead of developing a binary classification model, we employ and train a fully connected ANN to predict the failure probability. The neural network comprises two hidden layers (128 and 64 neurons) with the "relu" activation function. In addition, the "sigmoid" activation function is employed in the output layer. Fig. 12 shows the material failure probability map that the ANN predictive model generated in the strain-CVF space. The gray area at the interface of failure/non-failure domains represents the change in the failure probability.

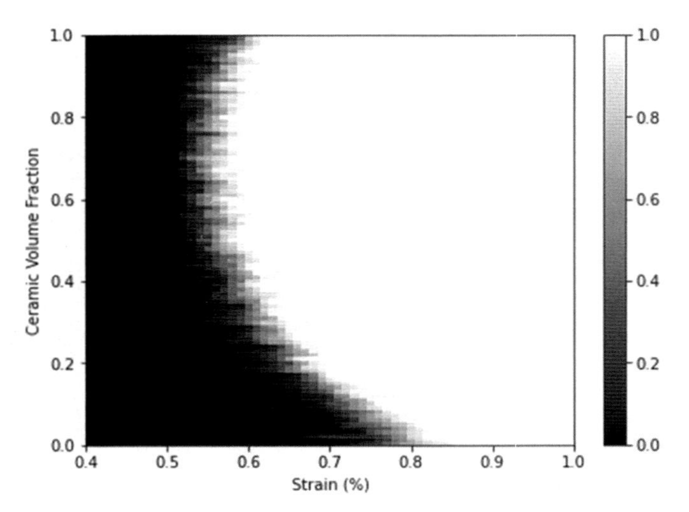

Fig. 11 A binary map representing the failure likelihood approximated from the collected data.

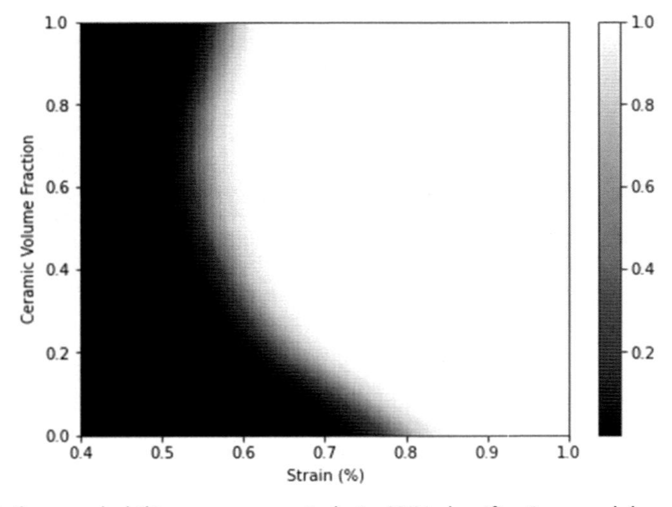

Fig. 12 Failure probability map generated via ANN classification model.

4.4.2 Tensile strength prediction via CNN

Table 3 shows that failure strength can vary at the same CVF due to microstructure uncertainties. Indeed, CNN may extract the microstructure features so that the failure strength can be deterministically predicted. There are 2525 microscale configurations generated during the data collection.

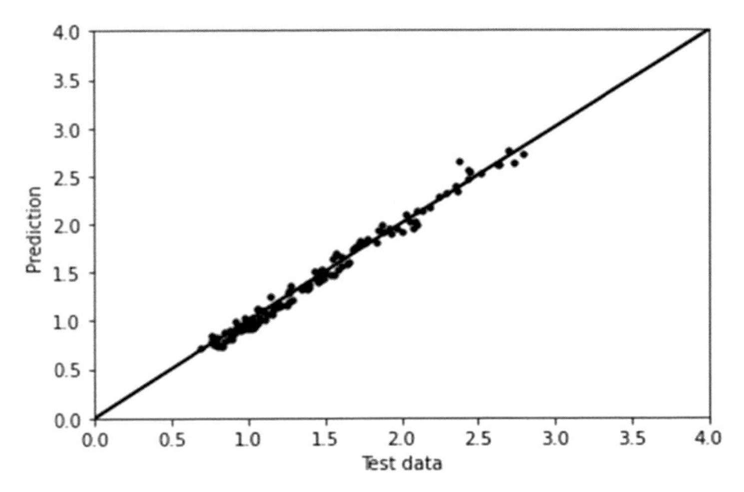

Fig. 13 Strength (GPa) predictions of the test data.

We use those images as the inputs, and the corresponding failure strengths are the outputs. 5% of the dataset is randomly selected as the test set. The others are split into training and validation sets with a ratio of 85:15.

The CNN consists of three convolutional and max pooling layers. Each convolutional layer employs a 3 by 3 kernel. The numbers of filters are 32, 64, and 128, respectively, for convolutional layers. The max pooling uses a 2 by 2 max filter in each layer. After flattening, the fully connected neural network has two hidden layers with 64 and 32 neurons before the output layer. The "relu" activation function is utilized except for the output layer. After training, the coefficient of determination is 0.985 calculated from the test set. Compared to the actual outputs in the test set, the predictions are shown in Fig. 13, in which the diagonal line represents the ideal prediction.

5. Conclusions

This paper briefly introduced ANNs and some variations, such as PINNs, CNNs, and RNNs. We also reviewed the applications of those neural networks on some advanced material science research topics, including multiscale modeling, inverse problems, material image processing, and history-dependent constitutive modeling. In addition, we conducted a case study in which metal-ceramic composite materials were modeled and simulated at the microscale to generate the dataset. Then, we developed and trained neural networks to predict material failure at the macroscale. The dataset and codes of this case study are provided on the github (https://github.com/jwli0728/ANNs-in-Material-Science). We believe that, with computer and computer technology development, the further development of AI/ML/DL algorithms will enhance scientific understanding in different disciplines, including materials discovery and design.

Funding and acknowledgment

Xiao gratefully acknowledges the support from the National Science Foundation (#2104383) and the US Department of Education (ED#P116S210005).

Author contributions

Xiao drafted the paper. Li did the case study. Bordas and Kim reviewed and revised the paper.

Conflict of interest

The authors declare that the research was conducted in the absence of any commercial or financial relationships that could be construed as a potential conflict of interest.

References

Akbari, Z., Mirzadeh, H., & Cabrera, J.-M. (2015). A simple constitutive model for predicting flow stress of medium carbon microalloyed steel during hot deformation. *Materials & Design, 77*, 126–131. https://doi.org/10.1016/J.MATDES.2015.04.005.

Al-Haik, M. S., Garmestani, H., & Savran, A. (2004). Explicit and implicit viscoplastic models for polymeric composite. *International Journal of Plasticity*. https://doi.org/10.1016/j.ijplas.2003.11.017 [Preprint].

Alber, M., et al. (2019). Integrating machine learning and multiscale modeling—Perspectives, challenges, and opportunities in the biological, biomedical, and behavioral sciences. *npj Digital Medicine, 2*(1), 115. https://doi.org/10.1038/s41746-019-0193-y.

American Society for Metals., W. (1979). *Properties and selection—Nonferrous alloys and pure metals* (9th ed., Vol. 1). Metals Park Ohio: American Society for Metals. Available at: ⟨https://www.worldcat.org/title/metals-handbook-2-properties-and-selection-nonferrons-alloys-and-pure-metals/oclc/634910433?referer=di&ht=edition⟩ (Accessed May 29, 2019).

Arroyo, M., & Belytschko, T. (2003). A finite deformation membrane based on inter-atomic potentials for the transverse mechanics of nanotubes. *Mechanics of Materials, 35*(3–6), 193–215. https://doi.org/10.1016/S0167-6636(02)00270-3.

Attarian, S., & Xiao, S. (2022). Investigating the strength of Ti/TiB interfaces at multiple scales using density functional theory, molecular dynamics, and cohesive zone modeling. *Ceramics International*. https://doi.org/10.1016/J.CERAMINT.2022.07.259 [Preprint].

Balokas, G., Czichon, S., & Rolfes, R. (2018). Neural network assisted multiscale analysis for the elastic properties prediction of 3D braided composites under uncertainty. *Composite Structures, 183*(1), 550–562. https://doi.org/10.1016/j.compstruct.2017.06.037.

Bar, H. N., Bhat, M. R., & Murthy, C. R. L. (2004). Identification of failure modes in GFRP using PVDF sensors: ANN approach. *Composite Structures, 65*(2), 231–237. https://doi.org/10.1016/J.COMPSTRUCT.2003.10.019.

Belytschko, T., Liu, W. K., & Moran, B. (2000). *Nonlinear finite elements for continua and structures*. Wiley. ⟨https://books.google.com/books/about/Nonlinear_finite_elements_for_continua_a.html?id=C6goAQAAMAAJ⟩.

Bhaduri, A., Gupta, A., & Graham-Brady, L. (2022). Stress field prediction in fiber-reinforced composite materials using a deep learning approach. *Composites Part B: Engineering, 238*, 109879. https://doi.org/10.1016/J.COMPOSITESB.2022.109879.

Birman, V., et al. (2008). Response of spatially tailored structures to thermal loading. *Journal of Engineering Mathematics, 61*(2–4), 201–217. https://doi.org/10.1007/s10665-007-9151-9.

Bobaru, F., & Zhang, G. (2015). Why do cracks branch? A peridynamic investigation of dynamic brittle fracture. *International Journal of Fracture, 196*(1–2), 59–98. https://doi.org/10.1007/s10704-015-0056-8.

Boettinger, W. J., et al. (2003). *Phase-Field Simulation of Solidification, 32*, 163–194. https://doi.org/10.1146/ANNUREV.MATSCI.32.101901.155803.

Budarapu, P. R., & Rabczuk, T. (2017). Multiscale methods for fracture: A review. *Journal of the Indian Institute of Science 2017, 97*(3), 339–376. https://doi.org/10.1007/S41745-017-0041-5.

Cai, M., Hasanbeig, M., et al. (2021). Modular deep reinforcement learning for continuous motion planning with temporal logic. *IEEE Robotics and Automation Letters, 6*(4), 7973–7980 Available at: http://arxiv.org/abs/2102.12855 (Accessed April 8, 2021).

Cai, M., Xiao, S., et al. (2021). Optimal probabilistic motion planning with potential infeasible LTL constraints. *IEEE Transactions on Automatic Control*. https://doi.org/10.1109/TAC.2021.3138704 [Preprint].

Cecen, A., et al. (2018). Material structure-property linkages using three-dimensional convolutional neural networks. *Acta Materialia, 146*, 76–84. https://doi.org/10.1016/J.ACTAMAT.2017.11.053.

Chandrashekhara, K., Okafor, A. C., & Jiang, Y. P. (1998). Estimation of contact force on composite plates using impact-induced strain and neural networks. *Composites Part B: Engineering, 29*(4), 363–370. https://doi.org/10.1016/S1359-8368(98)00003-1.

Cho, K., et al. (2014). Learning phrase representations using RNN encoder-decoder for statistical machine translation. In *EMNLP 2014–2014 conference on empirical methods in natural language processing, proceedings of the conference* (pp. 1724–1734). Association for Computational Linguistics (ACL). Available at: ⟨https://doi.org/10.3115/v1/d14-1179⟩.

Choudhary, K., et al. (2022). Recent advances and applications of deep learning methods in materials science. *npj Computational Materials 2022, 8*(1), 1–26. https://doi.org/10.1038/s41524-022-00734-6.

Dreiseitl, S., & Ohno-Machado, L. (2002). Logistic regression and artificial neural network classification models: A methodology review. *Journal of Biomedical Informatics, 35*(5–6), 352–359. https://doi.org/10.1016/S1532-0464(03)00034-0.

Ericksen, J. L. (1984). *The cauchy and born hypotheses for crystals. Phase transformations and material instabilities in solids*. Elsevier61–77. https://doi.org/10.1016/B978-0-12-309770-5.50008-4.

Freitag, S., Graf, W., & Kaliske, M. (2013). A material description based on recurrent neural networks for fuzzy data and its application within the finite element method. *Computers & Structures, 124*, 29–37. https://doi.org/10.1016/J.COMPSTRUC.2012.11.011.

Furukawa, T., & Hoffman, M. (2004). Accurate cyclic plastic analysis using a neural network material model. *Engineering Analysis with Boundary Elements, 28*(3), 195–204. https://doi.org/10.1016/S0955-7997(03)00050-X.

Ghaffari, M. A., et al. (2019). Peridynamics with corrected boundary conditions and its implementation in multiscale modeling of rolling contact fatigue. *Journal of Multiscale Modelling, 10*(01), 1841003. https://doi.org/10.1142/s1756973718410032.

Ghaffari, M. A., Zhang, Y., & Xiao, S. (2018). Multiscale modeling and simulation of rolling contact fatigue. *International Journal of Fatigue, 108*, 9–17. https://doi.org/10.1016/J.IJFATIGUE.2017.11.005.

Ghaffari, M. A., Zhang, Y., & Xiao, S. P. (2017). Molecular dynamics modeling and simulation of lubricant between sliding solids. *Journal of Micromechanics and Molecular Physics, 2*(2), 1750009.

Gooneie, A., Schuschnigg, S., & Holzer, C. (2017). A review of multiscale computational methods in polymeric materials. *Polymers, 9*(1), https://doi.org/10.3390/POLYM9010016.

Grabowski, K., et al. (2017). Multiscale electro-mechanical modeling of carbon nanotube composites. *Computational Materials Science, 135*, 169–180. https://doi.org/10.1016/J.COMMATSCI.2017.04.019.

Griffiths, D. J. (1995). *Introduction to quantum mechanics*. Prentice Hall.

Güne, A., et al. (2018). Automatic differentiation in machine learning: A survey. *Journal of Machine Learning Research, 18*, 1–43. https://doi.org/10.5555/3122009.3242010.

Haghighat, E., et al. (2020). A deep learning framework for solution and discovery in solid mechanics. Available at: ⟨https://doi.org/10.48550/arxiv.2003.02751⟩.

Heider, D., Piovoso, M. J., & Gillespie, J. W. (2003). A neural network model-based open-loop optimization for the automated thermoplastic composite tow-placement system. *Composites Part A: Applied Science and Manufacturing, 34*(8), 791–799. https://doi.org/10.1016/S1359-835X(03)00120-9.

Himanen, L., et al. (2019). Data-driven materials science: Status, challenges, and perspectives. *Advanced Science, 6*(21), 1900808. https://doi.org/10.1002/ADVS.201900808.

Janežič, M., Klemenc, J., & Fajdiga, M. (2010). A neural-network approach to describe the scatter of cyclic stress–strain curves. *Materials & Design, 31*(1), 438–448. https://doi.org/10.1016/J.MATDES.2009.05.044.

Jarrah, M. A., Al-Assaf, Y., & Kadi, H. E. (2002). Neuro-fuzzy modeling of fatigue life prediction of unidirectional glass fiber/epoxy composite laminates. *Journal of Composite Materials, 36*(6), 685–700. https://doi.org/10.1177/0021998302036006176.

Jung, J., & Seok, J. (2016). Fatigue crack growth analysis in layered heterogeneous material systems using peridynamic approach. *Composite Structures, 152*, 403–407. https://doi.org/10.1016/J.COMPSTRUCT.2016.05.077.

El Kadi, H. (2006). Modeling the mechanical behavior of fiber-reinforced polymeric composite materials using artificial neural networks—A review. *Composite Structures, 73*(1), 1–23. https://doi.org/10.1016/J.COMPSTRUCT.2005.01.020.

El Kadi, H., & Al-Assaf, Y. (2002). Prediction of the fatigue life of unidirectional glass fiber/epoxy composite laminae using different neural network paradigms. *Composite Structures, 55*(2), 239–246. https://doi.org/10.1016/S0263-8223(01)00152-0.

Kanouté, P., et al. (2009). Multiscale methods for composites: A review. *Archives of Computational Methods in Engineering, 16*(1), 31–75. https://doi.org/10.1007/S11831-008-9028-8.

Kröner, E. (1977). Bounds for effective elastic moduli of disordered materials. *Journal of the Mechanics and Physics of Solids, 25*(2), 137–155. https://doi.org/10.1016/0022-5096(77)90009-6.

Le, B. A., Yvonnet, J., & He, Q.-C. (2015). Computational homogenization of nonlinear elastic materials using neural networks. *International Journal for Numerical Methods in Engineering, 104*(12), 1061–1084. https://doi.org/10.1002/nme.4953.

Li, S., & Liu, W. K. (2002). Meshfree and particle methods and their applications. *Applied Mechanics Reviews, 55*(1), 1–34. https://doi.org/10.1115/1.1431547.

Liu, Z., & Wu, C. T. (2019). Exploring the 3D architectures of deep material network in data-driven multiscale mechanics. *Journal of the Mechanics and Physics of Solids, 127*, 20–46. https://doi.org/10.1016/j.jmps.2019.03.004.

Liu, Z., Wu, C. T., & Koishi, M. (2019). A deep material network for multiscale topology learning and accelerated nonlinear modeling of heterogeneous materials. *Computer Methods in Applied Mechanics and Engineering, 345*, 1138–1168. https://doi.org/10.1016/J.CMA.2018.09.020.

Lu, C., et al. (2012). Multi-scale modeling of shock interaction with a cloud of particles using an artificial neural network for model representation. *Procedia IUTAM, 3*, 25–52. https://doi.org/10.1016/J.PIUTAM.2012.03.003.

Madenci, E., & Oterkus, S. (2016). Ordinary state-based peridynamics for plastic deformation according to von Mises yield criteria with isotropic hardening. *Journal of the Mechanics and Physics of Solids, 86*, 192–219. https://doi.org/10.1016/J.JMPS.2015.09.016.

Mianroodi, J. R., et al. (2022). Lossless multi-scale constitutive elastic relations with artificial intelligence. *npj Computational Materials 2022, 8*(1), 1–12. https://doi.org/10.1038/s41524-022-00753-3.

Miller, R. E., & Tadmor, E. B. (2009). A unified framework and performance benchmark of fourteen multiscale atomistic/continuum coupling methods. *Modelling and Simulation in Materials Science and Engineering, 17*(5), 053001. https://doi.org/10.1088/0965-0393/17/5/053001.

Mitchell, T. M. (1997). *Machine learning*. New York: McGraw-Hill.

Möller, B., & Beer, M. (2004). *Fuzzy randomness: Uncertainty in civil engineering and computational mechanics*. Springer.

Morgan, D., & Jacobs, R. (2020). Opportunities and challenges for machine learning in materials science. *Annual Review of Materials Research, 50*, 71–103. https://doi.org/10.1146/ANNUREV-MATSCI-070218-010015.

Munro, R. G. (2000). Material properties of titanium diboride. *Journal of research of the National Institute of Standards and Technology, 105*(5), 709–720. https://doi.org/10.6028/jres.105.057.

Nishawala, V. V., & Ostoja-Starzewski, M. (2017). Peristatic solutions for finite one- and two-dimensional systems. *Mathematics and Mechanics of Solids, 22*(8), 1639–1653. https://doi.org/10.1177/1081286516641180.

Oeser, M., & Freitag, S. (2009). Modeling of materials with fading memory using neural networks. *International Journal for Numerical Methods in Engineering, 78*(7), 843–862. https://doi.org/10.1002/nme.2518.

Olivito, R. S. (2003). A neural diagnostic system for measuring strain in FRP composite materials. *Cement and Concrete Composites, 25*(7), 703–709. https://doi.org/10.1016/S0958-9465(02)00103-8.

Patil, A. S., et al. (2019). Effect of TiB2 addition on the microstructure and wear resistance of Ti-6Al-4V alloy fabricated through direct metal laser sintering (DMLS). *Journal of Alloys and Compounds, 777*, 165–173. https://doi.org/10.1016/J.JALLCOM.2018.10.308.

Peng, H. J., et al. (2017). Review on high-loading and high-energy lithium–sulfur batteries. *Advanced Energy Materials*, 7(24), 1700260. https://doi.org/10.1002/AENM.201700260.

Pollice, R., et al. (2021). Data-driven strategies for accelerated materials design. *Accounts of Chemical Research, 54*(4), 849–860. https://doi.org/10.1021/ACS.ACCOUNTS.0C00785.

Rabczuk, T., Belytschko, T., & Xiao, S. P. (2004). Stable particle methods based on Lagrangian kernels. *Computer Methods in Applied Mechanics and Engineering, 193*(12–14), 1035–1063. https://doi.org/10.1016/J.CMA.2003.12.005.

Rahman, M. M., et al. (2017). A fully coupled space–time multiscale modeling framework for predicting tumor growth. *Computer Methods in Applied Mechanics and Engineering, 320*, 261–286. https://doi.org/10.1016/J.CMA.2017.03.021.

Raissi, M., Perdikaris, P., & Karniadakis, G. E. (2019). Physics-informed neural networks: A deep learning framework for solving forward and inverse problems involving nonlinear partial differential equations. *Journal of Computational Physics, 378*, 686–707. https://doi.org/10.1016/J.JCP.2018.10.045.

Rao, C., & Liu, Y. (2020). Three-dimensional convolutional neural network (3D-CNN) for heterogeneous material homogenization. *Computational Materials Science, 184*, 109850. https://doi.org/10.1016/J.COMMATSCI.2020.109850.

Ray, S., & Cooney, R. P. (2018). *Thermal degradation of polymer and polymer compositesHandbook of environmental degradation of materials* (3rd ed.). William Andrew Publishing 185–206. Available at: https://doi.org/10.1016/B978-0-323-52472-8.00009-5.

Rizvi, Z. H., Nikolić, M., & Wuttke, F. (2019). Lattice element method for simulations of failure in bio-cemented sands. *Granular Matter, 21*(2), 1–14. https://doi.org/10.1007/S10035-019-0878-6/FIGURES/13.

Rodrigues, J. F., et al. (2021). Big data and machine learning for materials science. *Discover Materials, 1*(1), 1–27. https://doi.org/10.1007/S43939-021-00012-0.

Rutherford, K. L., et al. (1996). Abrasive wear resistance of TiN/NbN multi-layers: Measurement and neural network modelling. *Surface and Coatings Technology, 86–87*, 472–479. https://doi.org/10.1016/S0257-8972(96)02956-8.

Sabokpa, O., et al. (2012). Artificial neural network modeling to predict the high temperature flow behavior of an AZ81 magnesium alloy. *Materials & Design, 39*, 390–396. https://doi.org/10.1016/J.MATDES.2012.03.002.

Sainath, T. N., et al. (2015). Convolutional, long short-term memory, fully connected deep neural networks. In *ICASSP, IEEE international conference on acoustics, speech and signal processing—Proceedings* (pp. 4580–4584). Institute of Electrical and Electronics Engineers Inc. Available at: https://doi.org/10.1109/ICASSP.2015.7178838.

Samanta, A., et al. (2019). Atomistic simulation of diffusion bonding of dissimilar materials undergoing ultrasonic welding. *International Journal of Advanced Manufacturing Technology, 103*(1–4), 879–890. https://doi.org/10.1007/s00170-019-03582-9.

Schmidhuber, J. (2015). Deep Learning in neural networks: An overview. In *Neural Networks* (pp. 85–117). Elsevier Ltd. Available at: ⟨https://doi.org/10.1016/j.neunet.2014.09.003⟩.

Schulz, H., & Behnke, S. (2012). Deep learning: Layer-wise learning of feature hierarchies. *KI—Kunstliche Intelligenz, 26*(4), 357–363. https://doi.org/10.1007/S13218-012-0198-Z/FIGURES/4.

Shukla, K., et al. (2020). Physics-informed neural network for ultrasound nondestructive quantification of surface breaking cracks. *Journal of Nondestructive Evaluation, 39*(3), 1–20. https://doi.org/10.1007/S10921-020-00705-1/FIGURES/16.

Silling, S. A. (2000). Reformulation of elasticity theory for discontinuities and long-range forces. *Journal of the Mechanics and Physics of Solids, 48*(1), 175–209. https://doi.org/10.1016/S0022-5096(99)00029-0.

Silling, S. A., et al. (2007). Peridynamic states and constitutive modeling. *Journal of Elasticity, 88*(2), 151–184. https://doi.org/10.1007/s10659-007-9125-1.

Silling, S. A. (2010). Linearized theory of peridynamic states. *Journal of Elasticity, 99*(1), 85–111. https://doi.org/10.1007/s10659-009-9234-0.

Silling, S. A., & Askari, A. (2014). *Peridynamic model for fatigue cracking*. Available at: ⟨http://prod.sandia.gov/techlib/access-control.cgi/2014/1418590.pdf⟩ (Accessed June 27, 2018).

Silling, S. A., & Lehoucq, R. B. (2010). *Peridynamic theory of solid mechanics. Advances in Applied Mechanics*. Elsevier. https://doi.org/10.1016/S0065-2156(10)44002-8.

Singh, K., Rajput, S. K., & Mehta, Y. (2016). Modeling of the hot deformation behavior of a high phosphorus steel using artificial neural networks. *Materials Discovery, 6*, 1–8. https://doi.org/10.1016/J.MD.2017.03.001.

Smyser, C. P., & Chandrashekhara, K. (1997). Robust vibration control of composite beams using piezoelectric devices and neural networks. *Smart Materials and Structures, 6*(2), 178–189. https://doi.org/10.1088/0964-1726/6/2/007.

Stöcker, J., et al. (2022). A novel self-adversarial training scheme for enhanced robustness of inelastic constitutive descriptions by neural networks. *Computers & Structures, 265*, 106774. https://doi.org/10.1016/J.COMPSTRUC.2022.106774.

Subramanian, N., Rai, A., & Chattopadhyay, A. (2015). Atomistically informed stochastic multiscale model to predict the behavior of carbon nanotube-enhanced nanocomposites. *Carbon, 94*, 661–672. https://doi.org/10.1016/J.CARBON.2015.07.051.

Sutskever, I., Vinyals, O., & Le, Q. V. (2014). Sequence to sequence learning with neural networks. In *NIPS'14: Proceedings of the 27th* international conference on neural information processing systems (pp. 3104–3112). Montreal, CA. Available at: ⟨https://doi.org/10.5555/2969033.296917⟩.

Tadmor, E., Phillips, R., & Ortiz, M. (2000). Hierarchical modeling in the mechanics of materials. *International Journal of Solids and Structures, 37*(1–2), 379–389. https://doi.org/10.1016/S0020-7683(99)00095-5.

Tadmor, E. B., & Miller, R. E. (2011). *Modeling materials: Continuum, atomistic, and multiscale techniques*. Cambridge University Press.

Tadmor, E. B., & Miller, R. E. (2017). Benchmarking, validation and reproducibility of concurrent multiscale methods are still needed. *Modelling and Simulation in Materials Science and Engineering, 25*(7), 071001. https://doi.org/10.1088/1361-651X/aa834f.

Torquato, S., & Haslach, H. (2002). Random heterogeneous materials: Microstructure and macroscopic properties. *Applied Mechanics Reviews, 55*(4), B62–B63. https://doi.org/10.1115/1.1483342.

Tripathi, M. K., Kumar, R., & Tripathi, R. (2020). Big-data driven approaches in materials science: A survey. *Materials Today: Proceedings, 26*, 1245–1249. https://doi.org/10.1016/J.MATPR.2020.02.249.

Tuhami, A. E., & Xiao, S. (2022). Multiscale modeling of metal-ceramic spatially tailored materials via Gaussian process regression and peridynamics. *International Journal of Computational Methods*. https://doi.org/10.1142/S0219876222500256 [Preprint].

Ulmer II, C. W., et al. (1998). Computational neural networks and the rational design of polymeric materials: The next generation polycarbonates. *Computational and Theoretical Polymer Science, 8*(3–4), 311–321. https://doi.org/10.1016/S1089-3156(98)00035-X.

Unger, J. F., & Könke, C. (2009). Neural networks as material models within a multiscale approach. *Computers & Structures, 87*(19–20), 1177–1186. https://doi.org/10.1016/J.COMPSTRUC.2008.12.003.

Valoor, M. T., & Chandrashekhara, K. (2000). A thick composite-beam model for delamination prediction by the use of neural networks. *Composites Science and Technology, 60*(9), 1773–1779. https://doi.org/10.1016/S0266-3538(00)00063-4.

Versino, D., Tonda, A., & Bronkhorst, C. A. (2017). Data driven modeling of plastic deformation. *Computer Methods in Applied Mechanics and Engineering, 318*, 981–1004. https://doi.org/10.1016/J.CMA.2017.02.016.

Wagner, G. J., & Liu, W. K. (2003). Coupling of atomistic and continuum simulations using a bridging scale decomposition. *Journal of Computational Physics, 190*(1), 249–274. https://doi.org/10.1016/S0021-9991(03)00273-0.

Wang, K., & Sun, W. C. (2018). A multiscale multi-permeability poroplasticity model linked by recursive homogenizations and deep learning. *Computer Methods in Applied Mechanics and Engineering, 334*, 337–380. https://doi.org/10.1016/j.cma.2018.01.036.

Wang, Z., et al. (2022). Data-driven materials innovation and applications. *Advanced materials (Deerfield Beach, FL)*, 2104113. https://doi.org/10.1002/ADMA.202104113.

White, D. A., et al. (2019). Multiscale topology optimization using neural network surrogate models. *Computer Methods in Applied Mechanics and Engineering, 346*, 1118–1135. https://doi.org/10.1016/J.CMA.2018.09.007.
Wiley, D. E., Manning, W. R., & Hunter, O. (1969). Elastic properties of polycrystalline TiB2, ZrB2 and HfB2 from room temperature to 1300 K. *Journal of the Less Common Metals, 18*(2), 149–157. https://doi.org/10.1016/0022-5088(69)90134-9.
Xiao, S., et al. (2008). Reliability analysis of carbon nanotubes using molecular dynamics with the aid of grid computing. *Journal of Computational and Theoretical Nanoscience, 5*(4), 528–534. https://doi.org/10.1166/jctn.2008.2495.
Xiao, S., et al. (2020). A machine-learning-enhanced hierarchical multiscale method for bridging from molecular dynamics to continua. *Neural Computing and Applications, 32*(18), 14359–14373. https://doi.org/10.1007/S00521-019-04480-7.
Xiao, S., et al. (2021). Machine learning in multiscale modeling of spatially tailored materials with microstructure uncertainties. *Computers and Structures, 249*, 106511. https://doi.org/10.1016/j.compstruc.2021.106511.
Xiao, S., Andersen, D. R., & Yang, W. (2008). Design and analysis of nanotube-based memory cells. *Nanoscale Research Letters, 3*, 416–420. https://doi.org/10.1007/s11671-008-9167-8.
Xiao, S., & Hou, W. (2007a). Multiscale modeling and simulation of nanotube-based torsional oscillators. *Nanoscale Research Letters, 2*, 54–59. https://doi.org/10.1007/s11671-006-9030-8.
Xiao, S., & Hou, W. (2007b). Studies of nanotube-based resonant oscillators through multiscale modeling and simulation. *Physical Review B, 75*(12), 125414. https://doi.org/10.1103/PhysRevB.75.125414.
Xiao, S., & Yang, W. (2005). A nanoscale meshfree particle method with the implementation of the quasicontinuum method. *International Journal of Computational Methods, 02*(03), 293–313. https://doi.org/10.1142/S0219876205000533.
Xiao, S., & Yang, W. (2006). Temperature-related Cauchy–Born rule for multiscale modeling of crystalline solids. *Computational Materials Science, 37*(3), 374–379. https://doi.org/10.1016/J.COMMATSCI.2005.09.007.
Xiao, S. P., & Belytschko, T. (2004). A bridging domain method for coupling continua with molecular dynamics. *Computer Methods in Applied Mechanics and Engineering, 193*(17–20), 1645–1669. https://doi.org/10.1016/J.CMA.2003.12.053.
Yaghoobi, A., & Chorzepa, M. G. (2017). Fracture analysis of fiber reinforced concrete structures in the micropolar peridynamic analysis framework. *Engineering Fracture Mechanics, 169*, 238–250. https://doi.org/10.1016/J.ENGFRACMECH.2016.11.004.
Yang, W., & Xiao, S. (2008). Extension of the temperature-related Cauchy–Born rule: Material stability analysis and thermo-mechanical coupling. *Computational Materials Science, 41*(4), 431–439. https://doi.org/10.1016/J.COMMATSCI.2007.04.023.
Yilin, G., Fuh Ying Hsi, J., & Wen Feng, L. (2021). Multiscale topology optimisation with nonparametric microstructures using three-dimensional convolutional neural network (3D-CNN) models. *Virtual and Physical Prototyping, 16*(3), 306–317. https://doi.org/10.1080/17452759.2021.1913783.
Zhang, E., Yin, M., & Karniadakis, G. E. (2020). Physics-informed neural networks for nonhomogeneous material identification in elasticity imaging. Available at: ⟨https://doi.org/10.48550/arxiv.2009.04525⟩.
Zhang, Z., & Friedrich, K. (2003). Artificial neural networks applied to polymer composites: A review. *Composites Science and Technology, 63*(14), 2029–2044. https://doi.org/10.1016/S0266-3538(03)00106-4.
Zhang, Z., & Gu, G. X. (2021). Physics-informed deep learning for digital materials. *Theoretical and Applied Mechanics Letters, 11*(1), 100220. https://doi.org/10.1016/j.taml.2021.100220.

Zhang, Z., Klein, P., & Friedrich, K. (2002). Dynamic mechanical properties of PTFE based short carbon fibre reinforced composites: Experiment and artificial neural network prediction. *Composites Science and Technology, 62*(7–8), 1001–1009. https://doi.org/10.1016/S0266-3538(02)00036-2.
Zheng, B., et al. (2022). Physics-informed machine learning model for computational fracture of quasi-brittle materials without labelled data. *International Journal of Mechanical Sciences, 223*, 107282. https://doi.org/10.1016/J.IJMECSCI.2022.107282.
Zhu, Q., & Ni, T. (2017). Peridynamic formulations enriched with bond rotation effects. *International Journal of Engineering Science, 121*, 118–129. https://doi.org/10.1016/j.ijengsci.2017.09.004.
Zhu, Y., et al. (2022). Intelligent traffic light via policy-based deep reinforcement learning. *International Journal of Intelligent Transportation Systems Research,* 1–11. https://doi.org/10.1007/S13177-022-00321-5.
Zopf, C., & Kaliske, M. (2017). Numerical characterisation of uncured elastomers by a neural network based approach. *Computers & Structures, 182*, 504–525. https://doi.org/10.1016/J.COMPSTRUC.2016.12.012.

CHAPTER TWO

Coupled bulk and lattice waves in crystals seen as micromorphic anisotropic continua, with an application to crystals of the tetragonal group☆

Fabrizio Daví*
DICEA and ICRYS, Ancona, Italy
*Corresponding author. e-mail address: davi@univpm.it

Contents

1. Introduction

1.1 Motivation

In a crystal lattice, the constituent atoms can vibrate relative to their equilibrium positions and these vibrations are known as *lattice waves* or

☆ *to the memory of Gianfranco Capriz (1925–2022)*

Advances in Applied Mechanics, Volume 57
ISSN 0065-2156, https://doi.org/10.1016/bs.aams.2023.09.002

phonons. At the same time, when a crystal is subjected to an external force, it can undergo *bulk* or *acoustic waves*. The problem of coupled bulk and lattice waves in crystals arises when these two types of waves interact with each other and this interaction can have a significant effect on the behavior of the crystal (Hussein et al., 2014). Such a coupling between can occur in various ways: for example, when an acoustic wave travels through a crystal, it can interact with the lattice vibrations, leading to a scattering of the acoustic wave and the creation of new phonons. Similarly, the lattice vibrations can also influence the propagation of acoustic waves, leading to changes in their speed or direction.

The study of coupled bulk and lattice waves is important in many fields, including materials science, solid-state physics, and engineering: the knowledge of the frequency spectrum with the associated dispersion relation, the coupling between acoustic and lattice modes and the amount of mechanical energy associated with these modes becomes mandatory for designing new materials with specific properties or for improving the performance of existing materials.

A recent and promising field of application for this topic concerns the study of coupled bulk and lattice vibrations in scintillating crystals, that is crystals which convert ionizing radiations into photons within the visible range. Massive scintillating crystals were used to detect particle collisions in the CMS calorimeter at CERN, Geneve (Lecoq & Annekov, 2006) and shall be used in the FAIR accelerator at GSI, Darmstadt (Erni et al., 2013) and can also be used into security and medical imaging devices. Amongst many other problems concerning quality control and efficiency, one of the major issues related with the prolongated use of these crystals is the radiation damage which displaces atoms and reduces crystal efficiency and the radiation/photons ratio (*vid. e.g.* Dormenev et al., 2005): amongst various techniques for radiation damage recovery one of the most promising is the laser-induced ultrasound lattice vibrations (Montalto et al., 2023).

It becomes mandatory, therefore, to completely understand the coupled waves propagation problem in order to evaluate the frequency range of bulk and lattice vibrations, the amount of energy which is lost in the coupling between lattice and bulk and how much of the incoming energy makes the lattice to vibrate and around which modes. The problem is remarkably complex since most crystals exhibit strong anisotropy, like the monoclinic Cerium doped $Lu_xY_{2-x}SiO_5$:Ce (LYSO) (Mengucci et al., 2015), the hexagonal $LaBr_3$ (Singh, 2010) and the tetragonal $PbWO_4$ (PWO) (Annenkova & Korzhik, 2002).

In classical lattice vibrations, the material is treated as a collection of discrete particles (atoms or molecules) that are connected by springs. The dynamics of each particle is governed by Newton's laws of motion, and the wave propagation is described in terms of the collective motion of the particles. Such a classical approach (Born & Huang, 1954; Dove, 1993) is particularly useful for describing wave phenomena in crystalline materials: however a continuum mechanics approach looks more suited to describe the interactions between lattice and acoustic waves in massive crystals, in the order of 10^{-2} meters as the scintillators are.

It seems thus a natural choice to model the crystal as a *continuum with affine structure* (Capriz, 1989) or *micromorphic continuum* (Eringen, 1999), since such a model appears as a reasonable compromise between the microscopical aspects related to lattice vibrations and the macroscopic vibrations. We remark that there is a connection between waves in micromorphic continua and classical lattice vibrations: indeed both can be used to describe wave propagation in materials but with the continuum theory we can capture a wider range of wave phenomena.

Micromorphic continua were the object of a long-standing and never-fading attention, from a series of pionieering papers (Mindlin, 1964, 1965; Mindlin & Tiersten, 1963), to the first steps toward sistematization in (Capriz & Podio-Guidugli, 1976) and (Capriz & Podio-Guidugli, 1982), to the fundamental treatises (Capriz, 1989) and (Eringen, 1999). One of the most studied problems related to micromorphic continua was that of wave propagation, first studied in (Mindlin, 1964) for isotropic materials.

In recent years this topic obtained a revamped attention, motivated by the study of metamaterials and complex structures (*see e.g.* Hussein, 2018, 2019) by the means of both the classical approach as in (Berezovsky & Engelbrecht, 2011; Berezovsky et al., 2013) or the relaxed one first proposed into (Neff et al., 2014): it is important to remark that the majority of these results however concern isotropic materials, with some limited exceptions.

Here for classical micromorphic continua, we extend to the general case of anisotropic materials the previously known results of wave propagation into isotropic material, and then specialize them to crystals of the tetragonal group.

The article is organized as follows: in §.2 we write the balance law as proposed into (Capriz, 1989) and then, by using the results of (Capriz & Podio-Guidugli, 1982) we show that they are fully equivalent to those given in (Mindlin, 1964); upon the assumption of linearized kinematics and by using linear constitutive relations as in (Mindlin, 1964), we arrive at the propagation condition for the macroscopic progressive waves coupled with microdistortions lattice waves. Such propagation condition, which depends on the *wavenumber* ξ and on two dimensionless parameters which relates the various length-scales of the problem, is completely described by a 12×12 Hermitian matrix whose blocks represents various kind of generalized acoustic tensors and whose eigenvector represents macroscopic displacements coupled with lattice microdistortions. In the general case of triclinic crystals we show that there ever exist three acoustic and nine optic waves and we also give an insight into the structure of dispersion relations: moreover by a suitable scaling in terms of the dimensionless parameters we show that the problem admits two physically meaningful limit cases, namely the *Long wavelength approximation* which represents the propagation phenomena in a body in which we are "zooming-out" away from the crystal lattice, and the *Microvibration* case, where on the converse we are "zooming-in" into the crystal lattice. These two limit cases first introduced into (Mindlin, 1964), besides representing two physical picture of the phenomena, give an insight into the general propagation problem with the cut-off optical frequencies given by the microvibration frequencies.

In the §.3 the constitutive tensors are specialized to crystals of the Tetragonal point group by using (Authier, 2003) for third- and fourth-order tensors and the results given in (Fieschi, 1953) (*see also* Olive, 2013, 2014) for fifth- and sixth-order tensors. There are two reason for the choice of tetragonal crystals: the first one is that the reduction of independent constitutive parameter for tetragonal micromorphic bodies allow for some explicit solutions of the propagation condition and to a at least qualitative representation of the dispersion relations. The second reason is that we are interested to damage recovery in the tetragonal PWO crystals which are currently used in the FAIR accelerator (Novotny et al., 2011).

In the §.4 we begin to study the two limiting cases and then, separately, the wave propagation in crystals belonging to the low-symmetric tetragonal classes 4, $\bar{4}$, $4/m$ and to the high-symmetric classes $4mm$, 422, $\bar{4}2m$ and $4/mmm$, in the case of waves propagating along the tetragonal c-axis. In all cases we give a complete description of the frequency spectrum and of the vibration modes in terms of the constitutive parameters. Then we look at the waves propagating in directions orthogonal to the c-axis where the problem is in general fully coupled and does not allow for an explicit representation for the eigencouples but in some special cases. The results were also presented in a tabular form.

As far as we know, this is the most complete analysis for the wave propagation problem in classical Tetragonal micromorphic continua: of course, to make such an analyisis a predictive tool for experimental applications we need to know a complete set of parameters (43 for the low-symmetric classes 4, $\bar{4}$ and $4/m$ and 25 for the high-symmetry classes). These parameters, that are very difficult to obtain and evaluated by the means of a set of simple experimen, can be obtained at the present only by two viable approach. One is the numerical homogeneization approach used into (d'Agostino et al., 2020) for relaxed micromorphic tetragonal continua undergoing plane strain, whereas the other is by the means of homogeneization and identification methods from lattice dymanics, as it was done for cubic Diamond crystals and Silicon into (Moosavian, 2020).

It there exists a major criticism to the approach we follow in this article: as it was correctly pointed out into (Barbagallo et al., 2017), such a formal treatment has some limitations since it depends indeed on a large number of parameters whose experimental identification can be both difficult and elusive. With respect to such a correct criticism we can say that in any case we have a general framework which should allow to design correct experiments aimed to parameter identification; moreover, as it is show into (Moosavian, 2020; Shodja, 2020), by homogenization techniques we can still estimate the micromorphic model constitutive parameters by starting from classical lattice dynamics.

1.2 Notation

1.2.1 Vectors, tensors and cartesian components

Let $\mathcal{V}$ be the three-dimensional vector space whose elements we denote $\mathbf{v} \in \mathcal{V}$ and let Lin be the space of the second order tensors $\mathbf{A} \in \text{Lin}$, $\mathbf{A}\colon \mathcal{V} \rightarrow \mathcal{V}$. We denote $\mathbf{A}^T$ the transpose of $\mathbf{A}$ such that

$$\mathbf{A}\mathbf{u}\cdot\mathbf{v} = \mathbf{A}^T\mathbf{v}\cdot\mathbf{u}, \qquad \forall \mathbf{u}, \mathbf{v} \in \mathcal{V}; \tag{1}$$

we shall also denote Sym and Skw the subspaces of Lin of the symmetric ($\mathbf{A} = \mathbf{A}^T$) and skew-symmetric ($\mathbf{A} = -\mathbf{A}^T$) tensors respectively.

Let $\mathfrak{Lin}$ be the space of third-order tensors $\mathsf{P}\colon \text{Lin} \rightarrow \mathcal{V}$ and for all $\mathsf{P} \in \mathfrak{Lin}$ we denote the transpose $\mathsf{P}^T\colon \mathcal{V} \rightarrow \text{Lin}$ as:

$$\mathsf{P}[\mathbf{A}]\cdot\mathbf{v} = \mathsf{P}^T\mathbf{v}\cdot\mathbf{A}, \qquad \forall \mathbf{v} \in \mathcal{V}, \forall \mathbf{A} \in \text{Lin}. \tag{2}$$

We shall also make use of fourth-order tensors $\mathbb{C}\colon \text{Lin} \rightarrow \text{Lin}$, fifth-order tensors $\mathcal{F}\colon \mathfrak{Lin} \rightarrow \text{Lin}$ and sixth-order tensors $\mathfrak{H}\colon \mathfrak{Lin} \rightarrow \mathfrak{Lin}$ whose transpose are defined by the means of

$$
\begin{aligned}
\mathbb{C}[\mathbf{A}]\cdot\mathbf{B} &= \mathbb{C}^T[\mathbf{B}]\cdot\mathbf{A}, \quad \forall\mathbf{A},\mathbf{B}\in\text{Lin},\\
\mathcal{F}[\mathbf{A}]\cdot\mathsf{P} &= \mathcal{F}^T[\mathsf{P}]\cdot\mathbf{A}, \quad \forall\mathbf{A}\in\text{Lin},\ \forall\mathsf{P}\in\mathfrak{Lin},\\
\mathfrak{H}[\mathsf{P}]\cdot\mathsf{Q} &= \mathfrak{H}^T[\mathsf{Q}]\cdot\mathsf{P}, \quad \forall\mathsf{P},\mathsf{Q}\in\mathfrak{Lin}.
\end{aligned} \tag{3}
$$

For $\{\mathbf{e}_k\}$, $k = 1, 2, 3$ an orthonormal basis in $\mathcal{V}$, we define the components of the aforementioned elements by:

$$
\begin{aligned}
v_k &= \mathbf{v}\cdot\mathbf{e}_k,\\
A_{kj} &= \mathbf{A}\mathbf{e}_j\cdot\mathbf{e}_k = \mathbf{A}\cdot\mathbf{e}_k\otimes\mathbf{e}_j,\\
\mathsf{P}_{ihk} &= \mathsf{P}[\mathbf{e}_h\otimes\mathbf{e}_k]\cdot\mathbf{e}_i = \mathsf{P}\cdot\mathbf{e}_i\otimes\mathbf{e}_h\otimes\mathbf{e}_k,\\
\mathbb{C}_{ijhk} &= \mathbb{C}[\mathbf{e}_h\otimes\mathbf{e}_k]\cdot\mathbf{e}_i\otimes\mathbf{e}_j, \quad i,j,h,k,m,p = 1,2,3,\\
\mathcal{F}_{ijhkm} &= \mathcal{F}[\mathbf{e}_h\otimes\mathbf{e}_k\otimes\mathbf{e}_m]\cdot\mathbf{e}_i\otimes\mathbf{e}_j,\\
\mathfrak{H}_{ijkhmp} &= \mathfrak{H}[\mathbf{e}_h\otimes\mathbf{e}_m\otimes\mathbf{e}_p]\cdot[\mathbf{e}_i\otimes\mathbf{e}_j\otimes\mathbf{e}_k].
\end{aligned} \tag{4}
$$

We shall also made use of the orthonormal base $\{\mathbf{W}_k\}$, $k = 1, \ldots, 9$ in Lin:

$$
\begin{aligned}
\mathbf{W}_1 &= \mathbf{e}_1\otimes\mathbf{e}_1 \quad \mathbf{W}_2 = \mathbf{e}_2\otimes\mathbf{e}_2 \quad \mathbf{W}_3 = \mathbf{e}_3\otimes\mathbf{e}_3,\\
\mathbf{W}_4 &= \mathbf{e}_2\otimes\mathbf{e}_3 \quad \mathbf{W}_5 = \mathbf{e}_3\otimes\mathbf{e}_1 \quad \mathbf{W}_6 = \mathbf{e}_1\otimes\mathbf{e}_2,\\
\mathbf{W}_7 &= \mathbf{e}_3\otimes\mathbf{e}_2 \quad \mathbf{W}_8 = \mathbf{e}_1\otimes\mathbf{e}_3 \quad \mathbf{W}_9 = \mathbf{e}_2\otimes\mathbf{e}_1,
\end{aligned} \tag{5}
$$

whereas the orthonormal bases $\{\hat{\mathbf{W}}_k\}$, $k = 1, \ldots, 6$ in Sym and $\{\bar{\mathbf{W}}_k\}$, $k = 4, 5, 6$ in Skw are respectively defined as:

$$
\begin{aligned}
\hat{\mathbf{W}}_k &= \mathbf{W}_k, \quad k = 1,2,3,\\
\hat{\mathbf{W}}_4 &= \frac{1}{\sqrt{2}}(\mathbf{W}_4+\mathbf{W}_7), \quad \hat{\mathbf{W}}_5 = \frac{1}{\sqrt{2}}(\mathbf{W}_5+\mathbf{W}_8),\\
\hat{\mathbf{W}}_6 &= \frac{1}{\sqrt{2}}(\mathbf{W}_6+\mathbf{W}_9),
\end{aligned} \tag{6}
$$

and

$$
\begin{aligned}
\bar{\mathbf{W}}_4 &= \frac{1}{\sqrt{2}}(\mathbf{W}_4-\mathbf{W}_7), \quad \bar{\mathbf{W}}_5 = \frac{1}{\sqrt{2}}(\mathbf{W}_5-\mathbf{W}_8),\\
\bar{\mathbf{W}}_6 &= \frac{1}{\sqrt{2}}(\mathbf{W}_6-\mathbf{W}_9).
\end{aligned} \tag{7}
$$

In terms of the bases (5)–(7) we can also represent the fourth-order tensors components with the so-called Voigt two-index notation, namely *e.g.* for (5):

$$
\mathbb{C}_{ij} = \mathbb{C}[\mathbf{W}_j]\cdot\mathbf{W}_i, \quad i,j = 1, \ldots, 9. \tag{8}
$$

We also find useful to decompose the space $\mathcal{V} \oplus \text{Lin}$ into the following subspaces:

$$\begin{aligned} \mathcal{V}_\perp &\equiv \text{span}\{\mathbf{e}_1, \mathbf{e}_2\}; \quad \mathcal{V}_\| \equiv \text{span}\{\mathbf{e}_3\}; \\ \mathcal{Z}_1 &\equiv \text{span}\{\mathbf{W}_1, \mathbf{W}_2, \mathbf{W}_3\}, \quad \mathcal{Z}_2 \equiv \text{span}\{\mathbf{W}_6, \mathbf{W}_9\}, \\ \mathcal{Z}_3 &\equiv \text{span}\{\mathbf{W}_4, \mathbf{W}_5, \mathbf{W}_7, \mathbf{W}_8\}, \end{aligned} \tag{9}$$

with $\mathcal{V} \oplus \text{Lin} \equiv \mathcal{V}_\perp \oplus \mathcal{V}_\| \oplus \mathcal{Z}_1 \oplus \mathcal{Z}_2 \oplus \mathcal{Z}_3$.

1.2.2 Lattice vectors

Let $\{\mathbf{a}_\Theta\}$, $\Theta = 1, 2, 3$ be the unit vectors associated with the crystallographic directions and such that $\mathbf{a}_1 \times \mathbf{a}_2 \cdot \mathbf{a}_3 > 0$. If we denote with $\mathbf{G}$ the crystal lattice deformation such that the deformed lattice vectors $\{\bar{\mathbf{a}}_\Theta\}$, $\Theta = 1, 2, 3$ are given by Fig. 1.

$$\bar{\mathbf{a}}_\Theta = \mathbf{G}\mathbf{a}_\Theta, \quad \Theta = 1, 2, 3, \quad \det \mathbf{G} > 0, \tag{10}$$

then it admits the following representation in the base $\{\mathbf{a}_\Theta\}$

$$\mathbf{G} = G_{\Theta\Lambda}\mathbf{a}_\Theta \otimes \mathbf{a}_\Lambda, \quad \Theta, \Lambda = 1, 2, 3, \tag{11}$$

in such a way that:

$$\bar{\mathbf{a}}_\Gamma = (\mathbf{a}_\Lambda \cdot \mathbf{a}_\Gamma)\, G_{\Theta\Lambda}\mathbf{a}_\Theta, \quad \Theta, \Lambda, \Gamma = 1, 2, 3. \tag{12}$$

The vectors of the base $\{\mathbf{a}_\Theta\}$ can be represented in terms of the vectors of the base $\{\mathbf{e}_i\}$ (which can be identified with the crystallographic directions of a pseudo-cubic lattice) as

$$\mathbf{a}_\Theta = H_{i\Theta}\mathbf{e}_i, \quad i, \Theta = 1, 2, 3, \quad \det[H_{i\Theta}] > 0, \tag{13}$$

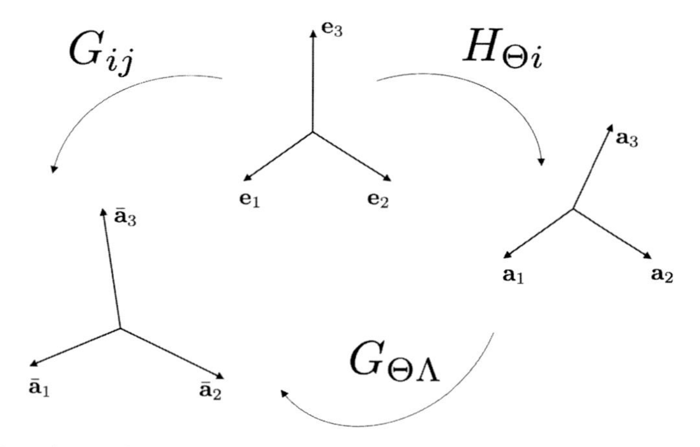

Fig. 1 The relations between $\{\mathbf{e}_k\}$, $\{\mathbf{a}_\Theta\}$ and $\{\bar{\mathbf{a}}_\Theta\}$.

and accordingly by (11) and (13) we have

$$\mathbf{G} = G_{\Theta\lambda} H_{i\Theta} H_{j\Lambda} \mathbf{e}_i \otimes \mathbf{e}_j = G_{ij} \mathbf{e}_i \otimes \mathbf{e}_j, \tag{14}$$

and the components of the lattice deformation with respect of the pseudo-cubic lattice are

$$G_{ij} = G_{\Theta\Lambda} H_{i\Theta} H_{j\Lambda}. \tag{15}$$

1.2.2.1 Example

Let

$$\mathbf{G} = \mathbf{a}_\Theta \otimes \mathbf{a}_\Theta + \gamma \mathbf{a}_1 \otimes \mathbf{a}_2, \tag{16}$$

be the shear deformation of a monoclinic lattice with monoclinic angle β:

$$\begin{aligned} \mathbf{a}_1 &= \mathbf{e}_1, \\ \mathbf{a}_2 &= \mathbf{e}_2, \\ \mathbf{a}_3 &= \cos\beta \mathbf{e}_1 + \sin\beta \mathbf{e}_3. \end{aligned} \tag{17}$$

From (11), (17) and (16) then we get the following non-zero components of $\mathbf{G}$ in the base $\{\mathbf{e}_k\}$:

$$\begin{aligned} G_{11} &= 1 + \cos^2\beta, \quad G_{22} = 1, \quad G_{33} = \sin^2\beta, \\ G_{12} &= \gamma, \quad G_{13} = G_{31} = \sin\beta\cos\beta. \end{aligned} \tag{18}$$

1.2.3 Lattice eigenmodes tensors

Finally, in order to describe the infinitesimal lattice vibrations we shall make use of the following seven modes, represented in the pseudo-cubic lattice Fig. 2:

- Non uniform dilatation:

$$\mathbf{D}_1 = \alpha \mathbf{W}_1 + \beta \mathbf{W}_2 + \gamma \mathbf{W}_3; \tag{19}$$

- Dilatation along $\mathbf{e}_3$ and uniform plane strain in the plane orthogonal to $\mathbf{e}_3$:

$$\mathbf{D}_2 = \alpha (\mathbf{I} - \mathbf{W}_3) + \gamma \mathbf{W}_3; \tag{20}$$

- Traceless plane strain orthogonal to $\mathbf{e}_3$:

$$\mathbf{D}_3 = \mathbf{W}_1 - \mathbf{W}_2; \tag{21}$$

- Shear between $\mathbf{e}_1$ and $\mathbf{e}_2$:

$$\mathbf{S}_{12} = \alpha \hat{\mathbf{W}}_6; \tag{22}$$

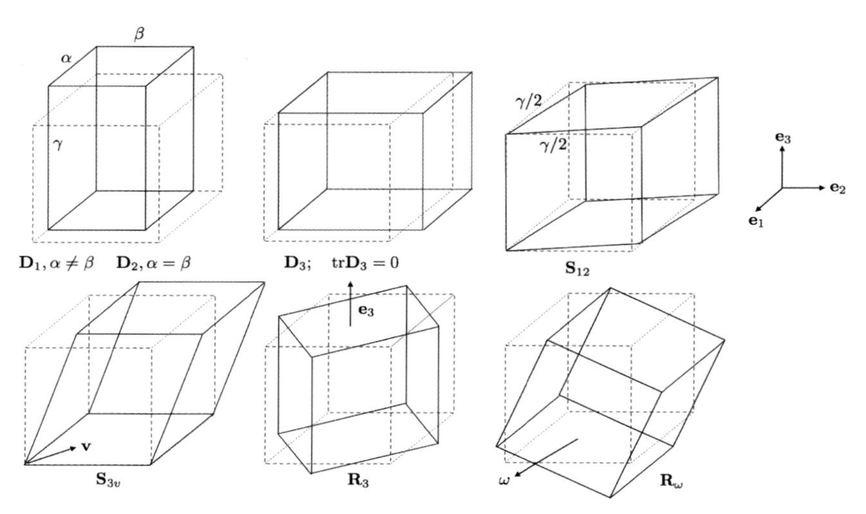

Fig. 2 The seven lattice eigenmodes.

- Shear between $\mathbf{e}_3$ and the direction $\mathbf{v} = \alpha\mathbf{e}_1 + \beta\mathbf{e}_2$:

$$\mathbf{S}_{3v} = -\alpha\hat{\mathbf{W}}_4 + \beta\hat{\mathbf{W}}_5; \tag{23}$$

- Rigid rotation around the direction $\mathbf{e}_3$:

$$\mathbf{R}_3 = \omega_3\bar{\mathbf{W}}_6; \tag{24}$$

- Rigid rotation around the direction $\boldsymbol{\omega} = \omega_1\mathbf{e}_1 + \omega_2\mathbf{e}_2$:

$$\mathbf{R}_\omega = \omega_2\bar{\mathbf{W}}_5 - \omega_1\bar{\mathbf{W}}_4. \tag{25}$$

2. Crystal as a micromorphic continuum

2.1 Balance laws

Let $\mathcal{B}$ a region of the Euclidean three-dimensional space we pointwise identify with the reference configuration of a crystal, and let x be a point of $\mathcal{B}$. We assume that at each point $x \in \mathcal{B}$ is defined a crystal lattice $\{\mathbf{a}_1, \mathbf{a}_2, \mathbf{a}_3\}$ with $\mathbf{a}_1 \times \mathbf{a}_2 \cdot \mathbf{a}_3 \geq 1$.

As in Mindlin (1964) we assume that at each point $x \in \mathcal{B}$ and at each time $t \in [0, \tau)$ it is well-defined the *motion* by the means of the two fields:

$$y = y(x, t), \qquad \mathbf{G} = \mathbf{G}(x, t) \tag{26}$$

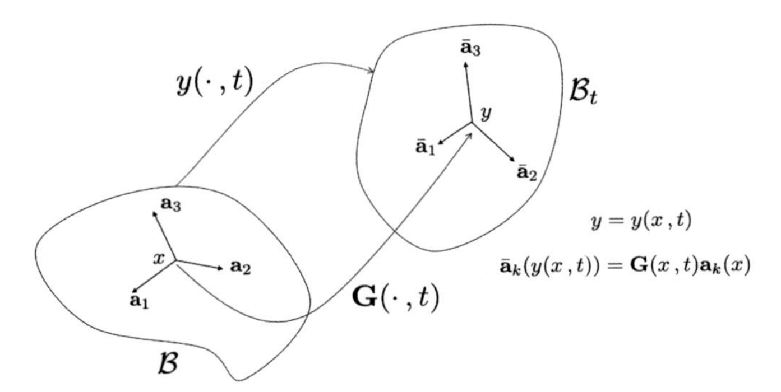

Fig. 3 The reference and deformed continua and lattices.

provided y is locally injective and orientation-preserving and $\mathbf{G}$ is orientation-preserving, namely Fig. 3:

$$\det \mathbf{F} > 0, \quad \mathbf{F}(x, t) = \nabla y(x, t), \quad \det \mathbf{G} > 0, \tag{27}$$

in such a way that at each point $y \in \mathcal{B}_t \equiv y(\mathcal{B}, t)$ the deformed crystal lattice $\{\bar{\mathbf{a}}_1, \bar{\mathbf{a}}_2, \bar{\mathbf{a}}_3\}$ is given by:

$$\bar{\mathbf{a}}_k = \mathbf{G}\mathbf{a}_k, \quad k = 1, 2, 3. \tag{28}$$

We identify $\mathcal{B}$, endowed with the motion (26), with a *continuum with affine structure* (Capriz, 1989) or *micromorphic* (Eringen, 1968), whose underlying manifold is $\mathcal{M} = \mathrm{Lin}^+$ and whose balance laws are given by:

- the balance of macroforces:

$$\mathrm{div}\mathbf{T}^T + \mathbf{b} = \rho\dot{\mathbf{v}}, \tag{29}$$

  where $\mathbf{T}$ is the (non-symmetric) Cauchy stress, $\mathbf{b}$ is the volume force density, ρ is the mass density and $\mathbf{v}$ the material velocity;
- the balance of microforces:

$$\mathrm{div}\mathsf{T} - \mathbf{K} + \mathbf{B} = \rho\ddot{\mathbf{G}}\mathbf{J}, \tag{30}$$

  where the third-order tensor T represents the microstress, $\mathbf{K}$ the interactive microforce, $\mathbf{B}$ the density of volume micro forces and $\mathbf{J}$ the microinertia tensor
- the balance of couples:

$$\mathrm{skw}(-\mathbf{T} + \mathbf{G}\mathbf{K}^T + (\mathrm{grad}\mathbf{G})\mathsf{T}^T) = \mathbf{0}. \tag{31}$$

A different set of balance laws was provided in Mindlin (1964): in order to recover these balance laws from (29) to (31) first of all we notice that $\text{skw}\mathbf{T} = -\text{skw}\mathbf{T}^T$ and set:

$$\bar{\mathbf{T}} = \mathbf{T}^T + \mathbf{G}\mathbf{K}^T + (\text{grad}\mathbf{G})\mathsf{T}^T \in \text{Sym}; \tag{32}$$

then, by following (Mindlin, 1964), we define the *relative stress* as:

$$\mathbf{S} = -(\mathbf{G}\mathbf{K}^T + (\text{grad}\mathbf{G})\mathsf{T}^T), \tag{33}$$

in such a way that

$$\mathbf{T}^T = \bar{\mathbf{T}} + \mathbf{S}. \tag{34}$$

As second step we multiply the transpose of (30) for $\mathbf{G}$ to obtain:

$$\mathbf{G}(\text{div}\mathsf{T})^T - \mathbf{G}\mathbf{K}^T + \mathbf{G}\mathbf{B}^T = \rho\mathbf{G}(\ddot{\mathbf{G}}\mathbf{J})^T, \tag{35}$$

and since

$$\mathbf{G}(\text{div}\mathsf{T})^T = \text{div}(\mathbf{G}\mathsf{T}^T) - (\text{grad}\mathbf{G})\mathsf{T}^T, \tag{36}$$

then with the aid of (34) from (29), (35) and (36) we recover equations (4.1) of (Mindlin, 1964):

$$\begin{aligned} \text{div}(\bar{\mathbf{T}} + \mathbf{S}) + \mathbf{b} &= \rho\dot{\mathbf{v}}, \\ \text{div}\mathsf{H} + \mathbf{S} + \bar{\mathbf{B}} &= \rho\mathbf{G}\mathbf{J}\ddot{\mathbf{G}}^T, \end{aligned} \tag{37}$$

where

$$\mathsf{H} = \mathbf{G}\mathsf{T}^T, \quad \bar{\mathbf{B}} = \mathbf{G}\mathbf{B}^T. \tag{38}$$

2.2 Linearized kinematics and constitutive relations

2.2.1 Linearized kinematical measures

As it is shown in Capriz (1989), Eringen (1968, 1999), there are many appropriate kinematical measures for a constitutive theory of micromorphic continua; here we choose those proposed into (Eringen, 1999), eqn. (1.5.11):

$$\begin{aligned} &\mathbf{E} = \frac{1}{2}(\mathbf{F}^T\mathbf{F} - \mathbf{I}), \quad \mathbf{M} = \mathbf{F}^T\mathbf{G}^{-1} - \mathbf{I}, \\ &\mathsf{G} = \mathbf{G}^{-1}\,\text{grad}\mathbf{G}, \end{aligned} \tag{39}$$

where $\mathbf{E}$ is the Green-Lagrange deformation measure.

We assume that both the deformation gradient and the lattice deformation can be decomposed additively into:

$$\mathbf{F} = \mathbf{I} + \nabla\mathbf{u}, \quad \mathbf{G} = \mathbf{I} + \mathbf{L}, \tag{40}$$

where $\mathbf{u}(x) = y(x) - x$ is the *displacement vector* and $\mathbf{L}$ is the *microdisplacement* or *microdistortion* (Barbagallo et al., 2017). If we assume that

$$\varepsilon = \sup\{\|\nabla\mathbf{u}\|, \|\mathbf{L}\|\}, \tag{41}$$

then the kinematical measures (39) can be rewritten, to within higher-order terms into ε, as:

$$\mathbf{E} = \mathrm{sym}\nabla\mathbf{u}, \quad \mathbf{M} = \nabla\mathbf{u}^T - \mathbf{L}, \quad \mathsf{G} = \nabla\mathbf{L}; \tag{42}$$

the tensor $\mathbf{M}$ is called the *relative strain* (Mindlin, 1964) or the *relative distortion* (Barbagallo et al., 2017).

2.2.2 Equivalent measures of interaction

By using (40) and (41) into (33) and (38) we have:

$$\mathsf{H} = \mathsf{T} + O(\varepsilon), \quad \mathbf{S} = -\mathbf{K}^T + O(\varepsilon),$$
$$\bar{\mathbf{B}} = \mathbf{B}^T + O(\varepsilon), \tag{43}$$

and:

$$\mathbf{G}\mathbf{J}\ddot{\mathbf{G}}^T = \mathbf{J}\ddot{\mathbf{L}}^T + O(\varepsilon^2). \tag{44}$$

Accordingly, from the balance law $(37)_2$ we obtain, to within higher-order terms, the balance of microforces (30) with $\ddot{\mathbf{L}}$ in place of $\ddot{\mathbf{G}}$:

$$\mathrm{div}\mathsf{T} - \mathbf{K} + \mathbf{B} = \rho\mathbb{J}[\ddot{\mathbf{L}}], \tag{45}$$

where the fourth-order microinertia tensor $\mathbb{J}$ is defined as:

$$\mathbb{J}[\ddot{\mathbf{L}}] = \mathbf{J}\ddot{\mathbf{L}}^T, \quad \mathbb{J}_{ijhk} = \delta_{ih}J_{jk}; \tag{46}$$

the macroforces balance (29) can instead be rewritten as

$$\mathrm{div}(\bar{\mathbf{T}} - \mathbf{K}^T) + \mathbf{b} = \rho\ddot{\mathbf{u}}. \tag{47}$$

In the sequel we shall use H, $\mathbf{S}$ and $\bar{\mathbf{T}}$ as the measures of interaction associated to a linearized kinematics, since in this case they are equivalent, to within higher-order terms, to T, $\mathbf{K}$ and $\mathbf{T}$.

2.2.3 Constitutive relations

We assume, as a constitutive hypothesis, a linear dependence of the measures of interaction $\bar{\mathbf{T}}$, $\mathbf{S}$ and H on the linearized kinematical variables (39) and write (*cf.* eqn. (5.3) of (Mindlin, 1964)):

$$\begin{aligned} \bar{\mathbf{T}} &= \mathbb{C}[\mathbf{E}] + \mathbb{D}[\mathbf{M}] + L_c\mathcal{F}[\mathsf{G}], \\ \mathbf{S} &= \mathbb{D}^T[\mathbf{E}] + \mathbb{B}[\mathbf{M}] + L_c\mathcal{G}[\mathsf{G}], \\ \mathsf{H} &= L_c\mathcal{F}^T[\mathbf{E}] + L_c\mathcal{G}^T[\mathbf{M}] + L_c^2\mathfrak{H}[\mathsf{G}], \end{aligned} \tag{48}$$

where:

- $\mathbb{C}$: Sym $\to$ Sym, $\mathbb{C} = \mathbb{C}^T$ is the fourth-order elasticity tensor, whose components obey:

$$\mathbb{C}_{ijhk} = \mathbb{C}_{jihk} = \mathbb{C}_{ijkh} = \mathbb{C}_{hkij}, \tag{49}$$

  and there are at most 21 independent components.
- The fourth-order tensor $\mathbb{B}$: Lin $\to$ Lin, $\mathbb{B} = \mathbb{B}^T$, whose independent components are at most 45:

$$\mathbb{B}_{ijhk} = \mathbb{B}_{hkij}. \tag{50}$$

- The fourth-order tensor $\mathbb{D}$: Lin $\to$ Sym has 54 independent components:

$$\mathbb{D}_{ijhk} = \mathbb{D}_{jihk}, \quad (\mathbb{D}^T)_{ijhk} = \mathbb{D}_{hkij}. \tag{51}$$

- The fifth-order tensor $\mathcal{F}$: $\mathfrak{Lin} \to$ Sym withy 162 components and which obey:

$$\mathcal{F}_{ijhkm} = \mathcal{F}_{jihkm}, \quad (\mathcal{F}^T)_{ijhkm} = \mathcal{F}_{hkmij}. \tag{52}$$

- The fifth-order tensor $\mathcal{G}$: $\mathfrak{Lin} \to$ Lin whose 243 components obey:

$$(\mathcal{G}^T)_{ijhkm} = \mathcal{G}_{hkmij}. \tag{53}$$

- The sixth-order tensor $\mathfrak{H} = \mathfrak{H}^T$ has at most 378 independent components $\mathfrak{H}_{ijkhmn}$ which obey:

$$\mathfrak{H}_{ijkhmn} = \mathfrak{H}_{hmnijk}. \tag{54}$$

In the constitutive relation (48), in order to make all the measures of interaction of the dimension of a stress (Force/Area), we introduced the micromorphic *correlation length* $L_c > 0$: this length scale is the first we need to introduce into the model and is related to the non-local effects associated with the gradient of the microdistorsion tensor.

Indeed for $L_c \to 0$ we are considering large samples of crystals (Neff et al., 2017) whereas the limit $L_c \to \infty$ acts as a zoom into the microstructure (*cf.* Barbagallo et al., 2017): we shall made these statements more rigorous in the next subsection.

We define the elastic energy density $\mathcal{E} = \mathcal{E}(\mathbf{E}, \mathbf{M}, \mathsf{G})$

$$\begin{aligned}\mathcal{E}(\mathbf{E}, \mathbf{M}, \mathsf{G}) &= \frac{1}{2}(\bar{\mathbf{T}}\cdot\mathbf{E} + \mathbf{S}\cdot\mathbf{M} + \mathsf{H}\cdot\mathsf{G}) \\ &= \frac{1}{2}(\mathbb{C}[\mathbf{E}]\cdot\mathbf{E} + \mathbb{B}[\mathbf{M}]\cdot\mathbf{M}) + \mathbb{D}[\mathbf{M}]\cdot\mathbf{E} \\ &\quad + L_c(\mathcal{F}[\mathsf{G}]\cdot\mathbf{E} + \mathcal{G}[\mathsf{G}]\cdot\mathbf{M}) + \frac{1}{2}L_c^2\mathfrak{H}[\mathsf{G}]\cdot\mathsf{G};\end{aligned} \tag{55}$$

as in (Mindlin, 1964), the requirement that the energy density is positive implies that $\mathbb{C}$, $\mathbb{B}$ and $\mathfrak{H}$ are positive definite:

$$\begin{aligned}&\mathbb{C}[\mathbf{A}]\cdot\mathbf{A} > 0, \quad \mathbb{B}[\mathbf{A}]\cdot\mathbf{A} > 0, \quad \forall\mathbf{A} \in \mathrm{Lin}/\{\mathbf{0}\}, \\ &\mathfrak{H}[\mathsf{A}]\cdot\mathsf{A} > 0, \quad \forall\mathsf{A} \in \mathfrak{Lin}/\{\mathbf{0}\},\end{aligned} \tag{56}$$

as well as

$$\begin{aligned}&\det(\mathbb{B} - \mathbb{D}^T\mathbb{C}^{-1}\mathbb{D}) > 0, \quad \det(\mathbb{C} - \mathbb{D}\mathbb{B}^{-1}\mathbb{D}^T) > 0, \\ &\det(\mathfrak{H} - \mathcal{F}^T\mathbb{C}^{-1}\mathcal{F}) > 0, \quad \det(\mathfrak{H} - \mathcal{G}^T\mathbb{B}^{-1}\mathcal{G}) > 0,\end{aligned} \tag{57}$$

and

$$\det\begin{bmatrix}\mathbb{C} & \mathbb{D} & L_c\,\mathcal{F} \\ \mathbb{D}^T & \mathbb{B} & L_c\,\mathcal{G} \\ L_c\,\mathcal{F}^T & L_c\,\mathcal{G}^T & L_c^2\,\mathfrak{H}\end{bmatrix} > 0. \tag{58}$$

In the most general case, that of crystals of the Triclinic group, these constitutive relations require the knowledge of 903 material constants, subject to the restrictions (56)–(58), whereas in the simplest case of Isotropic materials these constants reduce to 18 independent at most (Mindlin, 1964).

In the next subsection we shall give a general and formal treatment of waves propagation in a crystal without any of the restrictions given by crystal symmetries.

2.3 Wave propagation

2.3.1 Coupled bulk and lattice waves

The starting point for the description of the microscopic crystal lattice vibrations coupled with the macroscopic bulk vibrations are the balance laws (37), written in terms of the linearized kinematics (42) by the means of the constitutive relations (48) and with zero volume macro- and microforces:

$$\begin{aligned} \mathrm{div}(\bar{\mathbf{T}} + \mathbf{S}) &= \rho \ddot{\mathbf{u}}, \\ \mathrm{div}\mathsf{H} + \mathbf{S} &= \rho \mathbb{J}[\ddot{\mathbf{L}}]. \end{aligned} \tag{59}$$

We seek for (59) progressive plane wave solutions of the form:

$$\begin{aligned} \mathbf{u}(x, t) &= \mathbf{a} e^{i\sigma}, \quad \mathbf{a} \in \mathcal{V}, \\ \sigma &= \xi \mathbf{x} \cdot \mathbf{m} - \omega t, \\ \mathbf{L}(x, t) &= \mathbf{C} e^{i\sigma}, \quad \mathbf{C} \in \mathrm{Lin}, \end{aligned} \tag{60}$$

where ω is the frequency, $\mathbf{m}$ the direction of propagation, $\xi = \lambda^{-1} > 0$ the wavenumber with λ the wavelength and where $\mathbf{a}$ and $\mathbf{C}$ denote respectively the displacement and microdistortion amplitudes which, in the general case, are complex-valued.

We find at this point necessary to introduce, besides the characteristic lenght scale L_c, two further length scales: the *macroscopic length* $L_m > 0$ associated to the displacement amplitude:

$$\mathbf{a} = L_m \mathbf{a}_o, \tag{61}$$

with $\mathbf{a}_o$ a dimensionless vector, and the *lattice length* $L_l > 0$ which allows to write the micro-inertia tensor as

$$\mathbb{J} = L_l^2 \mathbb{J}_o, \tag{62}$$

with $\mathbb{J}_o$ a dimensionless fourth-order tensor (Barbagallo et al., 2017).

We then introduce two dimensionless parameters

$$\zeta_1 = \frac{L_c}{L_m}, \quad \zeta_2 = \frac{L_l}{L_m}, \tag{63}$$

whose limiting values describe the following situations:

- in the limit $\zeta_1 \to 0$ we are considering large samples of crystals in which the magnitude of lattice deformation is negligible with respect to the that of the macroscopic deformations;
- in the limit $\zeta_1 \to \infty$ we are instead zooming into the microstructure and conversely are the of macroscopic deformations that can be neglected;

- in the case $\zeta_2 \to 0$ we are instead neglecting the contribution of micro-inertia with respect to the macroscopic inertia.

  Since

$$
\begin{aligned}
\nabla \mathbf{u} &= i\xi L_m \mathbf{a}_o \otimes \mathbf{m} e^{i\sigma}, \quad \nabla \mathbf{L} = i\xi \mathbf{C} \otimes \mathbf{m} e^{i\sigma},\\
\ddot{\mathbf{u}} &= -\omega^2 L_m \mathbf{a}_o e^{i\sigma}, \quad \ddot{\mathbf{L}} = -\omega^2 \mathbf{C} e^{i\sigma},
\end{aligned} \tag{64}
$$

then we are led, by (42), (48), (59), (63) and (64), to the propagation conditions written in terms of the two lengths $\lambda = \xi^{-1}$, L_m and of the two dimensionless parameters ζ_1, ζ_2:

$$
\begin{aligned}
&\xi^2 \mathbf{A}(\mathbf{m})\mathbf{a}_o + (\xi^2\zeta_1 \mathsf{P}(\mathbf{m}) + i\xi L_m^{-1} \mathsf{Q}(\mathbf{m}))[\mathbf{C}] = \omega^2 \mathbf{a}_o,\\
&(\xi^2\zeta_1 \mathsf{P}^T(\mathbf{m}) - i\xi L_m^{-1} \mathsf{Q}^T(\mathbf{m}))\mathbf{a} + (\xi^2\zeta_1^2 \mathbb{A}(\mathbf{m}) + L_m^{-2}\bar{\mathbb{B}})[\mathbf{C}]\\
&= \omega^2 \zeta_2^2 \mathbb{J}_o[\mathbf{C}],
\end{aligned} \tag{65}
$$

where we defined the following acousto-optic tensors

$$
\begin{aligned}
\mathbf{A}(\mathbf{m})\mathbf{a} &= \rho^{-1}(\mathbb{C}[\mathbf{a} \otimes \mathbf{m}] + \mathbb{D}[\mathbf{m} \otimes \mathbf{a}]\\
&\quad + \mathbb{D}^T[\mathbf{a} \otimes \mathbf{m}] + \mathbb{B}[\mathbf{m} \otimes \mathbf{a}]),\\
\mathsf{P}(\mathbf{m})[\mathbf{C}] &= \rho^{-1}(\mathcal{F} + \mathcal{G})[\mathbf{C} \otimes \mathbf{m}]\mathbf{m},\\
\mathsf{Q}(\mathbf{m})[\mathbf{C}] &= \rho^{-1}(\mathbb{D} + \mathbb{B})[\mathbf{C}]\mathbf{m},\\
\mathbb{A}(\mathbf{m})[\mathbf{C}] &= \rho^{-1}\mathfrak{H}[\mathbf{C} \otimes \mathbf{m}]\mathbf{m},\\
\bar{\mathbb{B}} &= \rho^{-1}\mathbb{B},
\end{aligned} \tag{66}
$$

whose representation in components is

$$
\begin{aligned}
A_{ij} &= \rho^{-1}(\mathbb{C}_{iljk} m_k m_l + \mathbb{D}_{ilhj} m_h m_l + \mathbb{D}_{iljk} m_k m_l + \mathbb{B}_{ilhj} m_h m_l),\\
\mathsf{P}_{ihk} &= \rho^{-1}(\mathcal{F}_{ijhkp} + \mathcal{G}_{ijhkp}) m_p m_j,\\
\mathsf{Q}_{ijh} &= \rho^{-1}(\mathbb{D}_{ijhk} + \mathbb{B}_{ijhk}) m_k,\\
\mathbb{A}_{ijhk} &= \rho^{-1}\mathfrak{H}_{ijlhkp} m_l m_p,\\
\bar{\mathbb{B}}_{ijhk} &= \rho^{-1}\mathbb{B}_{ijhk}.
\end{aligned} \tag{67}
$$

We call $\mathbb{A}(\mathbf{m})$ the microacousto-optic tensor and $\mathbf{A}(\mathbf{m})$ the generalized acoustic tensor: the latter, in the absence of the microstructure, reduces to the acoustic tensor for linearly elastic bodies, *vid. e.g.* (Gurtin, 1972), §.70; we notice that, since $\mathbb{C} = \mathbb{C}^T$, $\mathbb{B} = \mathbb{B}^T$ and $\mathfrak{H} = \mathfrak{H}^T$, then from (67) we have that:

$$
\mathbf{A} \in \mathrm{Sym}, \quad \mathbb{A} = \mathbb{A}^T. \tag{68}
$$

We define the two 12 × 12 hermitian block matrix $A(\xi) = A^*(\xi)$, which we call the *Acousto-optic matrix*, the 12 × 12 symmetric block matrix $J = J^T$ and the 12 dimension eigenvector w:

$$\begin{bmatrix} \xi^2\mathbf{A} & \xi^2\zeta_1\mathsf{P} + i\xi L_m^{-1}\mathsf{Q} \\ \xi^2\zeta_1\mathsf{P}^T - i\xi L_m^{-1}\mathsf{Q}^T & \xi^2\zeta_1^2\mathbb{A} + L_m^{-2}\bar{\mathbb{B}} \end{bmatrix} J \equiv \begin{bmatrix} \mathbf{I} & \mathbf{0} \\ \mathbf{0} & \zeta_2^2\mathbb{J}_o \end{bmatrix} w \equiv \begin{bmatrix} \mathbf{a}_o \\ \mathbf{C} \end{bmatrix}, \tag{69}$$

in such a way that we can rewrite the propagation condition (65) as

$$(A - \omega^2 J)\, w = 0. \tag{70}$$

We require, that A be positive-definite for all $\xi > 0$ (the semidefiniteness as in (Neff et al., 2017) being required when $\xi \to 0$) and then, since $\mathbb{A}$ and $\mathbb{B}$ are positive-definite by (56) and the definition $(66)_4$, the positive-definiteness of A implies that also $\mathbf{A}(\mathbf{m})$ be positive definite; accordingly the eigenvalue problem (70) admits the twelve eigencouples with real eigenvalues

$$(\omega_k,\, w_k), \qquad < J w_h,\, w_k > = \delta_{hk}, \qquad h,\, k = 1,\, \ldots, 12, \tag{71}$$

where with the notation

$$<a,\, b> = \sum_{h=1}^{m} a_h\, b_h^*, \tag{72}$$

we denote the euclidean inner product on a m-dimensional complex space C^m.

The characteristic equation associated with the propagation condition (70) is

$$\det(A(\xi) - \omega^2 J) = 0, \tag{73}$$

and since the components of A are functions of the wavenumber ξ, then also the eigencouples are:

$$\xi \mapsto (\omega_k(\xi),\, w_k(\xi)), \qquad k = 1,\, \ldots, 12. \tag{74}$$

The functional dependences between ω and ξ are called the *dispersion relations* and in terms of these relations we can define the *phase* v_k^p and *group velocity* v_k^g as:

$$v_k^p(\xi) = \frac{\omega_k(\xi)}{\xi}, \qquad v_k^g(\xi) = \frac{d\omega_k(\xi)}{d\xi}, \qquad k = 1,\, \ldots, 12. \tag{75}$$

As pointed out into (Madeo et al., 2015), waves in micromorphic continua can be classified into:

- *Acoustic waves*, whose frequencies $\omega_k(\xi)$ goes to zero for $\xi \to 0$;
- *Optic waves* for which the limit for $\xi \to 0$ is different from zero: the limit $\omega_k(0)$ is called the *cut-off frequency* with group velocities $v_k^g(0) = 0$;

- *Standing waves* those associated to immaginary values $\xi = \pm ik$, $k > 0$ for some frequencies:

$$\mathbf{u}(\mathbf{x}, t) = \mathbf{a}\, e^{\mp k\mathbf{x}\cdot\mathbf{m}} e^{-i\omega t}, \quad \mathbf{L}(\mathbf{x}, t) = \mathbf{C}\, e^{\mp k\mathbf{x}\cdot\mathbf{m}} e^{-i\omega t}; \tag{76}$$

these waves do not propagate, but keep oscillating increasing or decreasing in a given, limited, region within the crystal.

In view of this classification, we begin to study the behavior of the eigencouples of (70) as $\xi \to 0$ by using some results obtained into (Lancaster, 1964).

First of all, provided the components of A are analytic functions of $\xi \geq 0$, there exists a neighborhood of $\xi = \xi_o$ where the eigenvalues $\omega_k(\xi)$ are regular and where their derivatives are well-defined. We use this result, in the case of simple eigenvalues, to give an explicit formula for the group velocities (Lancaster, 1964, Thm. 5):

$$v_k^g(\xi_o) = \frac{1}{2\omega_k(\xi_o)} < \left.\frac{\mathrm{d}A}{\mathrm{d}\xi}\right|_{\xi=\xi_o} w_k(\xi_o),\, w_k(\xi_o) >,$$

$$(k \text{ not summed}); \tag{77}$$

moreover the eigenvectors $w_k(\xi)$ are differentiable functions of the wave-number.

Now, since for $\xi \to 0$ the matrix A is real with three multiple eigenvalues $\omega = 0$, nine non-zero (simple) eigenvalues $\hat{\omega}_j$ and twelve real eigenvectors $\hat{w}_k$, then we can use Theorem 2 of (Lancaster, 1964) to show that (73) admits three and only three zero eigenvalues as ξ approaches zero: moreover by the differentiability of eigenvalues and eigenvectors we have that:

$$\hat{\omega}_k = \lim_{\xi\to 0} \omega_k(\xi), \quad \hat{w}_k = \lim_{\xi\to 0} w_k(\xi), \tag{78}$$

and for the non zero eigenvalues we have from (77)

$$v_j^g(0) = \frac{1}{2\hat{\omega}_j} < \left.\frac{\mathrm{d}A}{\mathrm{d}\xi}\right|_{\xi=0} \hat{w}_j,\, \hat{w}_j > = 0, \quad (j \text{ not summed}). \tag{79}$$

Therefore, by the application of the results obtained in (Lancaster, 1964) to our case, we obtain that in any anisotropic micromorphic continuum:

- There exist three Acoustic waves and nine Optic waves;
- The cut-off frequencies for the Optic waves are the limit as $\xi \to 0$ of the eigenvalues of (70);

- The frequencies for the Acoustic waves are the eigenvalues of (70) which goes to zero in the limit for $\xi \to 0$;
- The eigenvectors for $\xi \to 0$ are real.

As far as the standing waves are concerned, we notice that by (76) it is easy to show that A would be not positive-definite and therefore we can conclude that no standing waves are possible within this anisotropic model. Indeed as it was observed into (Neff et al., 2014; Madeo, 2017; Madeo et al., 2016, 2015) for the isotropic case, standing waves are associated with band-gap material and are not possible within the classical micromorphic model: they appears instead in the relaxed micromorphic model proposed into (Neff et al., 2014).

The solutions of the eigenvalues problem (70) depend, besides the parameter ξ, also on the three parameters L_m, ζ_1 and ζ_2 whose limiting values, as we already remarked, corresponds to different physical scales: therefore, besides the complete condition given by (65), we shall study into some details these two limit cases.

2.3.2 The long wavelength approximation: macroscopic waves

If we let $\zeta_1 \to 0$ and $\zeta_2 \to 0$ we are at the same time "zooming-out" from the crystal and disregarding the microinertia; such an approximation is called the *long wavelength approximation* (Barbagallo et al., 2017).

The propagation condition (65) reduces in this case, to within higher-order terms in ζ_1 and ζ_2, to:

$$\begin{aligned} \xi^2 \mathbf{A}(\mathbf{m})\mathbf{a}_o + i\xi L_m^{-1}\mathbf{Q}(\mathbf{m}))[\mathbf{C}] &= \omega^2 \mathbf{a}_o, \\ - i\xi \mathbf{Q}^T(\mathbf{m})\mathbf{a}_o + L_m^{-1}\bar{\mathbb{B}}[\mathbf{C}] &= \mathbf{0}. \end{aligned} \tag{80}$$

By the positive-definiteness of $\mathbb{B}$ then from $(80)_2$ we are led to the relation between the macroscopic and the lattice deformation amplitudes:

$$\mathbf{C} = iL_m \xi \mathbb{B}^{-1}\mathbf{Q}^T(\mathbf{m})\mathbf{a}_o, \tag{81}$$

and therefore from $(80)_1$ we obtain the classical continuum propagation condition

$$\hat{\mathbf{A}}(\mathbf{m})\mathbf{a}_o = c^2\mathbf{a}_o, \quad \omega = c\xi, \quad c = v^p = v^g, \tag{82}$$

where this time the acoustic tensor $\hat{\mathbf{A}}(\mathbf{m})$ (which is independent on the macroscopic length L_m) is defined as:

$$\begin{aligned} \hat{\mathbf{A}}(\mathbf{m})\mathbf{a}_o &= (\mathbf{A}(\mathbf{m}) - \mathbf{Q}(\mathbf{m})\mathbb{B}^{-1}\mathbf{Q}^T(\mathbf{m}))\mathbf{a}_o \\ &= \hat{\mathbb{C}}[\mathbf{a}_o \otimes \ \mathbf{m}]\mathbf{m}, \end{aligned} \tag{83}$$

where by (57) the fourth-order tensor $\hat{\mathbb{C}}$

$$\hat{\mathbb{C}} = \hat{\mathbb{C}}^T = \mathbb{C} - \mathbb{D}\mathbb{B}^{-1}\mathbb{D}^T; \tag{84}$$

is positive-definite and hence the acoustic tensor $\hat{\mathbf{A}}(\mathbf{m})$ is positive-definite too. It is easy to show by its definition that $\hat{\mathbb{C}}$ admits the same representation as $\mathbb{C}$ according to the crystal symmetry: therefore, for the solution of (82) for anisotropic crystal of various symmetry groups we can refer to the classical acoustic propagation solutions provided either in (Sadaki, 1941) or in (Achenbach, 1973).

The three eigencouples of (82) represents acoustic waves; however in this approximation $\hat{\mathbf{A}}(\mathbf{m})$ is not the acoustic tensor of the linear elasticity and the presence of the microstructure makes the propagation velocities smaller than in linearly elastic bodies: moreover we have also three microdistortions associated to the eigenvectors of (82) by the means of (81); these microdistorsions are purely immaginary and depend on the ratio $L_m/\lambda = L_m\xi$ between the macroscopic scale and the wavelength.

2.3.3 The limit $L_c \to \infty$: microvibrations

If we let $L_c \to \infty$, for fixed $L_m \approx L_l$, then we are zooming into the crystal; into the limit the constitutive relations $(48)_{1,2}$ remain finite for any choice of material only if $\nabla\mathbf{L} = \mathbf{0}$. In this case, from $(60)_2$ we have that $\xi = 0$ and the propagation condition $(65)_1$ leads to a solution $\omega = 0$ with multiplicity three. Since by $(60)_1$ $\mathbf{u}$ reduces to a rigid motion, then without loss of generality we can set:

$$\mathbf{u}(\mathbf{x}, t) = \mathbf{0}, \quad \mathbf{L}(t) = \mathbf{C}\, e^{i\omega t}, \tag{85}$$

and from the propagation condition $(65)_2$ then we are led to the characteristic equation

$$\det(\mathbb{B} - \rho\omega^2\mathbb{J}) = \mathbf{0}. \tag{86}$$

The microvibration solutions (85) of (65) described by the propagation condition (86) extend to the general anisotropic case the results first obtained (Mindlin, 1964) for isotropic materials (*vid. also* (Madeo et al., 2015)). Propagation condition (86) admits nine eigencouples

$$(\omega_k, \mathbf{C}_k), \quad k = 1, \ldots, 9, \tag{87}$$

with real eigentensors $\mathbf{C}_k$ and whose eigenvalues represent the cut-off frequencies for the propagation condition (70), as previously described into §.2.3.

3. The acousto-optic tensors for Tetragonal crystals

As we already remarked an explicit solution for the propagation condition (70) and an explicit determination for the dispersion relations (74) is not possible in the general anisotropic case, nor it would be particularly useful, since the associated kinematics would be fully coupled.

However many of the components of both the acousto-optic tensors $\mathbf{A}$ and $\mathbb{A}$, as well as of P and Q may vanish according to both the crystal symmetry group and the propagation direction $\mathbf{m}$ and it make sense to obtain explicit solutions for special cases of symmetry and directions of propagation. The simplest case of isotropic material, that depends on 18 constitutive parameters only, was studied in full-length into (Mindlin, 1964) and further analized and extended to a relaxed micromorphic model into (Neff et al., 2014,2017).

Here we give a complete description of the components of the acousto-optic tensors (66) provided we assume that the direction spanned by the base vector $\mathbf{e}_3$ is directed as the tetragonal c-axis.

3.1 The high-order tensors for the Tetragonal group

3.1.1 The fourth-order tensors $\mathbb{C}$, $\mathbb{D}$, $\mathbb{B}$ and $\mathbb{J}$

The non-zero components for all the classes of the Tetragonal point group are given in tabular form into (Authier, 2003) for the tensors $\mathbb{C}$ and $\mathbb{B}$ whereas those of $\mathbb{D}$ can be obtained with the additional conditions induced by the symmetries of the first two components.

We list the tabular form of these tensors in the Voigt's notation $1 = 11$, $2 = 22$, $3 = 33$, $4 = 23$, $5 = 31$, $6 = 12$, $7 = 32$, $8 = 13$, $9 = 21$, which for tensors mapping from and into Sym reduces to $4 = 23 = 32$, $5 = 13 = 31$ and $6 = 12 = 21$.

3.1.1.1 The elasticity tensor $\mathbb{C}$

Classes 4, $\bar{4}$ and $4/m$ (7 independent components):

$$
[\mathbb{C}] \equiv \begin{bmatrix} \mathbb{C}_{11} & \mathbb{C}_{12} & \mathbb{C}_{13} & 0 & 0 & \mathbb{C}_{16} \\ \cdot & \mathbb{C}_{11} & \mathbb{C}_{13} & 0 & 0 & -\mathbb{C}_{16} \\ \cdot & \cdot & \mathbb{C}_{33} & 0 & 0 & 0 \\ \cdot & \cdot & \cdot & \mathbb{C}_{44} & 0 & 0 \\ \cdot & \cdot & \cdot & \cdot & \mathbb{C}_{44} & 0 \\ \cdot & \cdot & \cdot & \cdot & \cdot & \mathbb{C}_{66} \end{bmatrix}; \tag{88}
$$

Classes $4mm$, 422, $4/mmm$ and $\bar{4}2m$ (6 independent components):

$$[\mathbb{C}] \equiv \begin{bmatrix} \mathbb{C}_{11} & \mathbb{C}_{12} & \mathbb{C}_{13} & 0 & 0 & 0 \\ \cdot & \mathbb{C}_{11} & \mathbb{C}_{13} & 0 & 0 & 0 \\ \cdot & \cdot & \mathbb{C}_{33} & 0 & 0 & 0 \\ \cdot & \cdot & \cdot & \mathbb{C}_{44} & 0 & 0 \\ \cdot & \cdot & \cdot & \cdot & \mathbb{C}_{44} & 0 \\ \cdot & \cdot & \cdot & \cdot & \cdot & \mathbb{C}_{66} \end{bmatrix}; \quad (89)$$

the tensor $\mathbb{C}_{micro}$ admits the same non-null components as (88) and (89).

3.1.1.2 The tensor $\mathbb{D}$

Classes 4, $\bar{4}$ and $4/m$ (14 independent components):

$$[\mathbb{D}] \equiv \begin{bmatrix} \mathbb{D}_{11} & \mathbb{D}_{12} & \mathbb{D}_{13} & 0 & 0 & \mathbb{D}_{16} & 0 & 0 & -\mathbb{D}_{26} \\ \mathbb{D}_{12} & \mathbb{D}_{11} & \mathbb{D}_{13} & 0 & 0 & \mathbb{D}_{26} & 0 & 0 & -\mathbb{D}_{16} \\ \mathbb{D}_{31} & \mathbb{D}_{31} & \mathbb{D}_{33} & 0 & 0 & \mathbb{D}_{36} & 0 & 0 & -\mathbb{D}_{36} \\ 0 & 0 & 0 & \mathbb{D}_{44} & \mathbb{D}_{45} & 0 & \mathbb{D}_{55} & -\mathbb{D}_{54} & 0 \\ 0 & 0 & 0 & \mathbb{D}_{54} & \mathbb{D}_{55} & 0 & -\mathbb{D}_{45} & \mathbb{D}_{44} & 0 \\ \mathbb{D}_{61} & -\mathbb{D}_{61} & 0 & 0 & 0 & \mathbb{D}_{66} & 0 & 0 & \mathbb{D}_{66} \\ 0 & 0 & 0 & \mathbb{D}_{44} & \mathbb{D}_{45} & 0 & \mathbb{D}_{55} & -\mathbb{D}_{54} & 0 \\ 0 & 0 & 0 & \mathbb{D}_{54} & \mathbb{D}_{55} & 0 & -\mathbb{D}_{45} & \mathbb{D}_{44} & 0 \\ \mathbb{D}_{61} & -\mathbb{D}_{61} & 0 & 0 & 0 & \mathbb{D}_{66} & 0 & 0 & \mathbb{D}_{66} \end{bmatrix}; \quad (90)$$

Classes $4mm$, 422, $4/mmm$ and $\bar{4}2m$ (8 independent components):

$$[\mathbb{D}] \equiv \begin{bmatrix} \mathbb{D}_{11} & \mathbb{D}_{12} & \mathbb{D}_{13} & 0 & 0 & 0 & 0 & 0 & 0 \\ \mathbb{D}_{12} & \mathbb{D}_{11} & \mathbb{D}_{13} & 0 & 0 & 0 & 0 & 0 & 0 \\ \mathbb{D}_{31} & \mathbb{D}_{31} & \mathbb{D}_{33} & 0 & 0 & 0 & 0 & 0 & 0 \\ 0 & 0 & 0 & \mathbb{D}_{44} & 0 & 0 & \mathbb{D}_{55} & 0 & 0 \\ 0 & 0 & 0 & 0 & \mathbb{D}_{55} & 0 & 0 & \mathbb{D}_{44} & 0 \\ 0 & 0 & 0 & 0 & 0 & \mathbb{D}_{66} & 0 & 0 & \mathbb{D}_{66} \\ 0 & 0 & 0 & \mathbb{D}_{44} & 0 & 0 & \mathbb{D}_{55} & 0 & 0 \\ 0 & 0 & 0 & 0 & \mathbb{D}_{55} & 0 & 0 & \mathbb{D}_{44} & 0 \\ 0 & 0 & 0 & 0 & 0 & \mathbb{D}_{66} & 0 & 0 & \mathbb{D}_{66} \end{bmatrix}; \quad (91)$$

3.1.1.3 The tensor $\mathbb{B}$

Classes 4, $\bar{4}$ and $4/m$ (13 independent components):

$$[\mathbb{B}] \equiv \begin{bmatrix} \mathbb{B}_{11} & \mathbb{B}_{12} & \mathbb{B}_{13} & 0 & 0 & \mathbb{B}_{16} & 0 & 0 & -\mathbb{B}_{26} \\ \cdot & \mathbb{B}_{11} & \mathbb{B}_{13} & 0 & 0 & \mathbb{B}_{26} & 0 & 0 & -\mathbb{B}_{16} \\ \cdot & \cdot & \mathbb{B}_{33} & 0 & 0 & \mathbb{B}_{36} & 0 & 0 & -\mathbb{B}_{36} \\ \cdot & \cdot & \cdot & \mathbb{B}_{44} & \mathbb{B}_{45} & 0 & \mathbb{B}_{47} & 0 & 0 \\ \cdot & \cdot & \cdot & \cdot & \mathbb{B}_{55} & 0 & 0 & \mathbb{B}_{47} & 0 \\ \cdot & \cdot & \cdot & \cdot & \cdot & \mathbb{B}_{66} & 0 & 0 & \mathbb{B}_{69} \\ \cdot & \cdot & \cdot & \cdot & \cdot & \cdot & \mathbb{B}_{55} & -\mathbb{B}_{45} & 0 \\ \cdot & \cdot & \cdot & \cdot & \cdot & \cdot & \cdot & \mathbb{B}_{44} & 0 \\ \cdot & \cdot & \cdot & \cdot & \cdot & \cdot & \cdot & \cdot & \mathbb{B}_{66} \end{bmatrix}; \tag{92}$$

Classes $4mm$, 422, $4/mmm$ and $\bar{4}2m$ (9 independent components):

$$[\mathbb{B}] \equiv \begin{bmatrix} \mathbb{B}_{11} & \mathbb{B}_{12} & \mathbb{B}_{13} & 0 & 0 & 0 & 0 & 0 & 0 \\ \cdot & \mathbb{B}_{11} & \mathbb{B}_{13} & 0 & 0 & 0 & 0 & 0 & 0 \\ \cdot & \cdot & \mathbb{B}_{33} & 0 & 0 & 0 & 0 & 0 & 0 \\ \cdot & \cdot & \cdot & \mathbb{B}_{44} & 0 & 0 & \mathbb{B}_{47} & 0 & 0 \\ \cdot & \cdot & \cdot & \cdot & \mathbb{B}_{55} & 0 & 0 & \mathbb{B}_{47} & 0 \\ \cdot & \cdot & \cdot & \cdot & \cdot & \mathbb{B}_{66} & 0 & 0 & \mathbb{B}_{69} \\ \cdot & \cdot & \cdot & \cdot & \cdot & \cdot & \mathbb{B}_{55} & 0 & 0 \\ \cdot & \cdot & \cdot & \cdot & \cdot & \cdot & \cdot & \mathbb{B}_{44} & 0 \\ \cdot & \cdot & \cdot & \cdot & \cdot & \cdot & \cdot & \cdot & \mathbb{B}_{66} \end{bmatrix}; \tag{93}$$

3.1.1.4 The fourth-order inertia tensor $\mathbb{J}$

Let $\mathbf{J}$ be the micro-inertia tensor whose components are $J_{ij} = J_{ji}$, $i, j = 1, 2, 3$: then, by $(46)_2$, the matrix $\mathbb{J} = \mathbb{J}^T$ is

$$[\mathbb{J}] \equiv \begin{bmatrix} J_{11} & 0 & 0 & 0 & 0 & J_{12} & 0 & J_{13} & 0 \\ \cdot & J_{22} & 0 & J_{23} & 0 & 0 & 0 & 0 & J_{12} \\ \cdot & \cdot & J_{33} & 0 & J_{13} & 0 & J_{23} & 0 & 0 \\ \cdot & \cdot & \cdot & J_{33} & 0 & 0 & 0 & 0 & J_{13} \\ \cdot & \cdot & \cdot & \cdot & J_{11} & 0 & J_{12} & 0 & 0 \\ \cdot & \cdot & \cdot & \cdot & \cdot & J_{22} & 0 & J_{23} & 0 \\ \cdot & \cdot & \cdot & \cdot & \cdot & \cdot & J_{22} & 0 & 0 \\ \cdot & \cdot & \cdot & \cdot & \cdot & \cdot & \cdot & J_{33} & 0 \\ \cdot & \cdot & \cdot & \cdot & \cdot & \cdot & \cdot & \cdot & J_{11} \end{bmatrix}; \tag{94}$$

for Tetragonal crystals, provided we identify the c-axis with the direction $\mathbf{e}_3$, we have $J_{11} = J_{22}$ and $J_{ij} = 0,\ i \neq j$: accordingly (94) reduces to:

$$[\mathbb{J}] \equiv \begin{bmatrix} J_{11} & 0 & 0 & 0 & 0 & 0 & 0 & 0 & 0 \\ \cdot & J_{11} & 0 & 0 & 0 & 0 & 0 & 0 & 0 \\ \cdot & \cdot & J_{33} & 0 & 0 & 0 & 0 & 0 & 0 \\ \cdot & \cdot & \cdot & J_{33} & 0 & 0 & 0 & 0 & 0 \\ \cdot & \cdot & \cdot & \cdot & J_{11} & 0 & 0 & 0 & 0 \\ \cdot & \cdot & \cdot & \cdot & \cdot & J_{11} & 0 & 0 & 0 \\ \cdot & \cdot & \cdot & \cdot & \cdot & \cdot & J_{11} & 0 & 0 \\ \cdot & \cdot & \cdot & \cdot & \cdot & \cdot & \cdot & J_{33} & 0 \\ \cdot & \cdot & \cdot & \cdot & \cdot & \cdot & \cdot & \cdot & J_{11} \end{bmatrix}. \tag{95}$$

3.1.2 The fifth-order tensors $\mathcal{G}$ and $\mathcal{F}$

A detailed study of the symmetries for fifth- and sixth-order tensor was done into (Fieschi, 1953): for the Tetragonal group the symmetries are different between classes, and accordingly we study them in detail beginning with the fifth-order tensor $\mathcal{G}$; we also follow (Fieschi, 1953) into the use of the notation

$$\mathcal{G}_{11112}\,[5], \tag{96}$$

to denote all the 5 possible combinations of the index, namely: $\mathcal{G}_{11112}$, $\mathcal{G}_{11121}$, $\mathcal{G}_{11211}$, $\mathcal{G}_{12111}$ and $\mathcal{G}_{21111}$.

Since $\mathcal{G}$ is an odd tensors, then for the two centrosymmetric classes $4/m$ and $4mm$ their components vanishes altogether: for the other classes we have the following restrictions.

- Class 4: for this class we have 61 independent components:

$$\begin{gathered} \mathcal{G}_{33333}\,[1], \quad \mathcal{G}_{11113} = \mathcal{G}_{22223}\,[5], \quad \mathcal{G}_{11333} = \mathcal{G}_{22333}\,[10], \\ \mathcal{G}_{11123} = -\mathcal{G}_{22213}\,[20], \quad \mathcal{G}_{33312} = -\mathcal{G}_{33321}\,[10], \\ \mathcal{G}_{11223} = \mathcal{G}_{22113}\,[15]. \end{gathered} \tag{97}$$

- Class $\bar{4}$: for this class we have the following restrictions into (97)

$$\mathcal{G}_{33333} = 0, \quad \mathcal{G}_{33312} = \mathcal{G}_{33321} = 0, \tag{98}$$

which means that $(97)_{1,5}$ must be zero and the independent components reduce to 50.

- Classes $4/mmm$, $\bar{4}2m$, 422: for these classes the tensors $\mathcal{G}$ splits into polar and axial ones.
- The polar tensors have has 61 non-zero components for all classes:

$$\begin{aligned} &\mathcal{G}_{33333}\,[1] \qquad \mathcal{G}_{22223} = \mathcal{G}_{11113}\,[5], \\ &\mathcal{G}_{33322} = \mathcal{G}_{33311}\,[10], \quad \mathcal{G}_{22113} = \mathcal{G}_{11223}\,[15], \end{aligned} \tag{99}$$

  31 being the independent ones.
- The axial tensors have 60 components with only 30 independent for the class $4/mmm$:

$$\mathcal{G}_{22213} = -\mathcal{G}_{11123}\,[20], \quad \mathcal{G}_{33312} = -\mathcal{G}_{33321}\,[10]; \tag{100}$$

  the 60 components (30 independent) for the class $\bar{4}2m$ are

$$\mathcal{G}_{22213} = \mathcal{G}_{11123}\,[20], \quad \mathcal{G}_{33312} = \mathcal{G}_{33321}\,[10]; \tag{101}$$

  whereas for the classe 422 we have $\mathcal{G} = 0$.

To obtain the number of the independent components for the tensor $\mathcal{F}$, we recall that it obeys $\mathcal{F}_{ijhkm} = \mathcal{F}_{jihkm}$: hence the number of independent components reduces to 41 components for the class 4, 33 components for class $\bar{4}$. For the polar classes $4/mmm$, $\bar{4}2m$ and 422 the components of $\mathcal{F}$ reduce to 42 (21 independent), whereas for the axial classes $4/mmm$, $\bar{4}2m$ they reduce to 30 with 15 independent components.

3.1.3 The sixth-order tensor $\mathfrak{H}$

Also for sixth-order tensors the symmetries changes for different classes. Following (Fieschi, 1953) we have, by using the same convention we used for the fifth-order tensor, that the for the classes 4, $\bar{4}$ and $4/m$ the non-null components are:

$$\begin{aligned} &\mathfrak{H}_{111111} = \mathfrak{H}_{222222}\,[1] \quad \mathfrak{H}_{333333}\,[1] \quad \mathfrak{H}_{111112} = -\mathfrak{H}_{222221}\,[6], \\ &\mathfrak{H}_{111122} = \mathfrak{H}_{222211}\,[15], \quad \mathfrak{H}_{111133} = \mathfrak{H}_{222233}\,[15], \\ &\mathfrak{H}_{333311} = \mathfrak{H}_{333322}\,[15], \quad \mathfrak{H}_{333312} - \mathfrak{H}_{333321}\,[15], \\ &\mathfrak{H}_{111222}\,[20], \quad \mathfrak{H}_{222333} = -\mathfrak{H}_{333222}\,[10], \\ &\mathfrak{H}_{111332} = \mathfrak{H}_{222331}\,[60], \quad \mathfrak{H}_{221133} = \mathfrak{H}_{112233}\,[45]; \end{aligned} \tag{102}$$

however, the number of non-zero and independent component obtained into (Fieschi, 1953) refers to a tensor with no major symmetries, whereas in our case $\mathfrak{H} = \mathfrak{H}^T$ and hence there are only 108 independent components into (102).

For the classes 422, $4mm$, $4/mmm$ and for the polar tensors of the class $\bar{4}2m$ we have the further restrictions into (102):

$$\begin{aligned}\mathfrak{H}_{122222} &= \mathfrak{H}_{211111} = \mathfrak{H}_{333312} = \mathfrak{H}_{333321} = 0,\\ \mathfrak{H}_{222331} &= \mathfrak{H}_{111332} = \mathfrak{H}_{222333} = \mathfrak{H}_{333222} = 0,\end{aligned} \tag{103}$$

which further reduce the number of independent components.

3.2 The acousto-optic tensors

In this subsection we shall list the components of the various acousto-optic tensors $\mathbf{A}(\mathbf{m})$, $\mathrm{P}(\mathbf{m})$, $\mathrm{Q}(\mathbf{m})$ and $\mathbb{A}(\mathbf{m})$ defined by (66) and which appear into the matrix A (69).

3.2.1 The generalized acoustic tensor A(m)

From relation $(67)_1$ we have the components of the second-order acoustic tensor $\mathbf{A}(\mathbf{m})$ for the classes 4, $\bar{4}$ and $4/m$ of the Tetragonal group:

$$\begin{aligned}
A_{11} &= \rho^{-1}((\mathbb{C}_{11} + \mathbb{B}_{11} + 2\mathbb{D}_{11})m_1^2\\
&\quad + (\mathbb{C}_{66} + \mathbb{B}_{69} + 2\mathbb{D}_{66})m_2^2\\
&\quad + (\mathbb{C}_{44} + \mathbb{B}_{47} + \mathbb{D}_{55} + \mathbb{D}_{44})m_3^2\\
&\quad + (\mathbb{C}_{16} + \mathbb{B}_{61} - \mathbb{B}_{26} + \mathbb{D}_{16} + 2\mathbb{D}_{61} - \mathbb{D}_{26})m_1m_2)\\
A_{22} &= \rho^{-1}((\mathbb{C}_{66} + \mathbb{B}_{69} + 2\mathbb{D}_{66})m_1^2\\
&\quad + (\mathbb{C}_{11} + \mathbb{B}_{11} + 2\mathbb{D}_{11})m_2^2\\
&\quad + (\mathbb{C}_{44} + \mathbb{B}_{47} + \mathbb{D}_{55} + \mathbb{D}_{44})m_3^2\\
&\quad - (\mathbb{C}_{16} + \mathbb{B}_{61} - \mathbb{B}_{26} + \mathbb{D}_{16} + 2\mathbb{D}_{61} - \mathbb{D}_{26})m_1m_2)\\
A_{33} &= \rho^{-1}((\mathbb{C}_{44} + \mathbb{D}_{44} + \mathbb{D}_{55} + \mathbb{B}_{47})(m_1^2 + m_2^2)\\
&\quad + (\mathbb{C}_{33} + 2\mathbb{D}_{33} + \mathbb{B}_{33})m_3^2)\\
A_{23} &= \rho^{-1}((-\mathbb{B}_{36} + \mathbb{D}_{45} - \mathbb{D}_{54})m_1m_3\\
&\quad + (\mathbb{C}_{44} + \mathbb{B}_{13} + \mathbb{B}_{44} + 2\mathbb{D}_{13} + \mathbb{D}_{44} + \mathbb{D}_{55})m_2m_3)\\
A_{13} &= \rho^{-1}((\mathbb{C}_{44} + \mathbb{B}_{13} + \mathbb{B}_{55} + 2\mathbb{D}_{13} + \mathbb{D}_{44} + \mathbb{D}_{55})m_1m_3\\
&\quad + (\mathbb{B}_{36} + \mathbb{D}_{54} - \mathbb{D}_{45})m_2m_3)\\
A_{12} &= \rho^{-1}((\mathbb{C}_{16} + \mathbb{B}_{16} + \mathbb{D}_{16} - \mathbb{D}_{26})m_1^2\\
&\quad + (-\mathbb{C}_{16} + \mathbb{B}_{62} - 2\mathbb{D}_{61})m_2^2\\
&\quad + (-\mathbb{B}_{45} + \mathbb{D}_{54} - \mathbb{D}_{45})m_3^2\\
&\quad + (\mathbb{C}_{66} + 2\mathbb{C}_{12} + \mathbb{B}_{12} + \mathbb{B}_{66}\\
&\quad + 2(\mathbb{D}_{12} + \mathbb{D}_{66}))m_1m_2).
\end{aligned} \tag{104}$$

For the classes $4mm$, 422, $4/mmm$ and $\bar{4}2m$ relations (104) simplify into:

$$
\begin{aligned}
A_{11} &= \rho^{-1}((\mathbb{C}_{11} + \mathbb{B}_{11} + 2\mathbb{D}_{11})m_1^2 + (\mathbb{C}_{66} + \mathbb{B}_{69} + 2\mathbb{D}_{66})m_2^2 \\
&\quad + (\mathbb{C}_{44} + \mathbb{B}_{44} + \mathbb{D}_{55} + \mathbb{D}_{44})m_3^2) \\
A_{22} &= \rho^{-1}((\mathbb{C}_{66} + \mathbb{B}_{69} + 2\mathbb{D}_{66})m_1^2 + (\mathbb{C}_{11} + \mathbb{B}_{11} + 2\mathbb{D}_{11})m_2^2 \\
&\quad + (\mathbb{C}_{44} + \mathbb{B}_{44} + \mathbb{D}_{55} + \mathbb{D}_{44})m_3^2) \\
A_{33} &= \rho^{-1}((\mathbb{C}_{44} + \mathbb{D}_{44} + \mathbb{D}_{55} + \mathbb{B}_{44})(m_1^2 + m_2^2) \\
&\quad + (\mathbb{C}_{33} + 2\mathbb{D}_{33} + \mathbb{B}_{33})m_3^2) \\
A_{23} &= \rho^{-1}((\mathbb{C}_{44} + 2\mathbb{D}_{13} + \mathbb{D}_{44} + \mathbb{D}_{55})m_2 m_3 + (\mathbb{B}_{13} + \mathbb{B}_{44})m_2 m_3), \\
A_{13} &= \rho^{-1}((\mathbb{C}_{44} + 2\mathbb{D}_{13} + \mathbb{D}_{44} + \mathbb{D}_{55})m_1 m_3 + (\mathbb{B}_{13} + \mathbb{B}_{55})m_1 m_3), \\
A_{12} &= \rho^{-1}(\mathbb{C}_{66} + 2\mathbb{C}_{12} + 2\mathbb{D}_{12} + 2\mathbb{D}_{66} + \mathbb{B}_{12} + \mathbb{B}_{66})m_1 m_2. \qquad (105)
\end{aligned}
$$

We remark that in the special cases when the direction of propagation $\mathbf{m}$ is either parallel ($\mathbf{m} = \mathbf{e}_3$) or orthogonal ($\mathbf{m}_\perp \cdot \mathbf{e}_3 = 0$) to the tetragonal c-axis the matrix of $\mathbf{A}(\mathbf{m})$ has one of the following simpler tabular representation:

$$
[\mathbf{A}(\mathbf{m}_\perp)] \equiv \begin{bmatrix} \bullet & \bullet & 0 \\ \cdot & \bullet & 0 \\ \cdot & \cdot & \bullet \end{bmatrix}, \qquad \text{all classes};
$$

$$
[\mathbf{A}(\mathbf{e}_3)] \equiv \begin{bmatrix} \bullet & \bullet & 0 \\ \cdot & \bullet & 0 \\ \cdot & \cdot & \bullet \end{bmatrix}, \qquad \text{classes } 4,\ \bar{4},\ 4/m;
$$

$$
[\mathbf{A}(\mathbf{e}_3)] \equiv \begin{bmatrix} \bullet & 0 & 0 \\ \cdot & \bullet & 0 \\ \cdot & \cdot & \bullet \end{bmatrix}, \qquad \text{classes } 4mmm,\ 4/mm,\ 422,\ \bar{4}2m. \qquad (106)
$$

3.2.1.1 The long-wavelength approximation tensor $\hat{\mathbf{A}}(\mathbf{m})$

The definition of the acoustic tensor $\hat{\mathbf{A}}(\mathbf{m})$, which rules the macroscopic waves propagation in the long-wavelength approximation, leads to the same result as in the classical linearly elastic case but with the components of $\hat{\mathbb{C}}$ in place of those of $\mathbb{C}$; accordingly, since $A_{ij} = \rho^{-1}\mathbb{C}_{iljk} m_k m_l$, then for the classes 4, $\bar{4}$ and $4/m$ we have the explicit representation:

$$
\begin{aligned}
\hat{A}_{11} &= \rho^{-1}(\hat{\mathbb{C}}_{11}m_1^2 + \hat{\mathbb{C}}_{66}m_2^2 + \hat{\mathbb{C}}_{16}m_1 m_2 + \hat{\mathbb{C}}_{44}m_3^2), \\
\hat{A}_{22} &= \rho^{-1}(\hat{\mathbb{C}}_{11}m_2^2 + \hat{\mathbb{C}}_{66}m_1^2 - \hat{\mathbb{C}}_{16}m_1 m_2 + \hat{\mathbb{C}}_{44}m_3^2), \\
\hat{A}_{33} &= \rho^{-1}(m_3^2\hat{\mathbb{C}}_{33} + \hat{\mathbb{C}}_{44}(m_1^2 + m_2^2)), \\
\hat{A}_{23} &= \rho^{-1}\hat{\mathbb{C}}_{44}m_2 m_3, \\
\hat{A}_{13} &= \rho^{-1}\hat{\mathbb{C}}_{44}m_1 m_3, \\
\hat{A}_{12} &= \rho^{-1}(\hat{\mathbb{C}}_{16}(m_1^2 - m_2^2) + (\hat{\mathbb{C}}_{66} + 2\hat{\mathbb{C}}_{12})m_1 m_2); \qquad (107)
\end{aligned}
$$

the components for the classes $4mm$, 422, $4/mmm$ and $\bar{4}2m$ can be obtained by taking $\hat{\mathbb{C}}_{16} = 0$ into (107). The same consideration which led to the representation (106) still holds.

3.2.2 The third-order tensor Q(m)

3.2.2.1 Classes 4, $\bar{4}$ and 4/m

For these classes the third-order tensor $\mathsf{Q}(\mathbf{m})$ has 21 independent components, that when calculated by the means of $(67)_3$ read:

$$
\begin{aligned}
\mathsf{Q}_{111} &= \rho^{-1}((\mathbb{D}_{11} + \mathbb{B}_{11})m_1 + (\mathbb{D}_{16} + \mathbb{B}_{16})m_2),\\
\mathsf{Q}_{122} &= \rho^{-1}((\mathbb{D}_{66} + \mathbb{B}_{69})m_1 - (\mathbb{D}_{61} - \mathbb{B}_{26})m_2),\\
\mathsf{Q}_{133} &= \rho^{-1}((\mathbb{D}_{55} + \mathbb{B}_{47})m_1 - (\mathbb{D}_{45} + \mathbb{B}_{45})m_2),\\
\mathsf{Q}_{112} &= \rho^{-1}((\mathbb{D}_{12} + \mathbb{B}_{12})m_1 - (\mathbb{D}_{26} + \mathbb{B}_{26})m_2),\\
\mathsf{Q}_{121} &= \rho^{-1}((\mathbb{D}_{16} + \mathbb{B}_{16})m_1 + (\mathbb{D}_{66} + \mathbb{B}_{66})m_2),\\
\mathsf{Q}_{123} &= \rho^{-1}\mathbb{B}_{36}m_3,\\
\mathsf{Q}_{131} &= \rho^{-1}(\mathbb{D}_{44} + \mathbb{B}_{44})m_3,\\
\mathsf{Q}_{132} &= \rho^{-1}\mathbb{D}_{54}m_3,\\
\mathsf{Q}_{113} &= \rho^{-1}(\mathbb{D}_{13} + \mathbb{B}_{13})m_3,\\
\mathsf{Q}_{211} &= \rho^{-1}((\mathbb{D}_{26} - \mathbb{B}_{26})m_1 + (\mathbb{D}_{66} + \mathbb{B}_{69})m_2),\\
\mathsf{Q}_{222} &= \rho^{-1}(-(\mathbb{D}_{16} + \mathbb{B}_{16})m_1 + (\mathbb{D}_{11} + \mathbb{B}_{11})m_2),\\
\mathsf{Q}_{233} &= \rho^{-1}((\mathbb{D}_{45} + \mathbb{B}_{45})m_1 + (\mathbb{D}_{55} + \mathbb{B}_{47})m_2),\\
\mathsf{Q}_{212} &= \rho^{-1}((\mathbb{D}_{66} + \mathbb{B}_{66})m_1 - (\mathbb{D}_{61} + \mathbb{B}_{16})m_2),\\
\mathsf{Q}_{221} &= \rho^{-1}((\mathbb{D}_{12} + \mathbb{B}_{12})m_1 + (\mathbb{D}_{26} + \mathbb{B}_{26})m_2),\\
\mathsf{Q}_{311} &= \rho^{-1}(\mathbb{D}_{44} + \mathbb{B}_{47})m_3,\\
\mathsf{Q}_{333} &= \rho^{-1}(\mathbb{D}_{33} + \mathbb{B}_{33})m_3,\\
\mathsf{Q}_{312} &= \rho^{-1}(\mathbb{D}_{54} + \mathbb{B}_{45})m_3,\\
\mathsf{Q}_{323} &= \rho^{-1}(\mathbb{D}_{45}m_1 + (\mathbb{D}_{55} + \mathbb{B}_{55})m_2),\\
\mathsf{Q}_{331} &= \rho^{-1}((\mathbb{D}_{31} + \mathbb{B}_{13})m_1 + (\mathbb{D}_{36} + \mathbb{B}_{36})m_2),\\
\mathsf{Q}_{332} &= \rho^{-1}(-(\mathbb{D}_{36} + \mathbb{B}_{36})m_1 + (\mathbb{D}_{31} + \mathbb{B}_{13})m_2),\\
\mathsf{Q}_{313} &= \rho^{-1}((\mathbb{D}_{55} + \mathbb{B}_{55})m_1 - \mathbb{D}_{45}m_2),
\end{aligned}
\tag{108}
$$

with the restrictions

$$\begin{aligned} \mathsf{Q}_{232} &= \mathsf{Q}_{131}, \quad \mathsf{Q}_{223} = \mathsf{Q}_{113}, \quad \mathsf{Q}_{322} = \mathsf{Q}_{311}, \\ \mathsf{Q}_{213} &= -\mathsf{Q}_{123}, \quad \mathsf{Q}_{231} = -\mathsf{Q}_{132}, \quad \mathsf{Q}_{321} = -\mathsf{Q}_{312}. \end{aligned} \tag{109}$$

and whose tabular representation is

$$[\mathsf{Q}] = \begin{bmatrix} (\mathsf{Q}_{111}) & (\mathsf{Q}_{122}) & (\mathsf{Q}_{133}) & \mathsf{Q}_{123} & \mathsf{Q}_{131} & (\mathsf{Q}_{112}) & \mathsf{Q}_{132} & \mathsf{Q}_{113} & (\mathsf{Q}_{121}) \\ (\mathsf{Q}_{211}) & (\mathsf{Q}_{222}) & (\mathsf{Q}_{233}) & \mathsf{Q}_{113} & -\mathsf{Q}_{132} & (\mathsf{Q}_{212}) & \mathsf{Q}_{131} & -\mathsf{Q}_{123} & (\mathsf{Q}_{221}) \\ \mathsf{Q}_{311} & \mathsf{Q}_{311} & \mathsf{Q}_{333} & (\mathsf{Q}_{323}) & (\mathsf{Q}_{331}) & \mathsf{Q}_{312} & (\mathsf{Q}_{332}) & (\mathsf{Q}_{313}) & -\mathsf{Q}_{312} \end{bmatrix}; \tag{110}$$

in (110) the 14 components which depend solely on m_1, m_2 are represented within brackets $(\cdot)$, with the other 13 components depending only on m_3.

3.2.2.2 Classes $4mm$, 422, $4/mmm$ and $\bar{4}2m$

For these classes we have

$$\begin{aligned} \mathsf{Q}_{123} &= \mathsf{Q}_{132} = \mathsf{Q}_{231} = \mathsf{Q}_{213} = \mathsf{Q}_{312} = \mathsf{Q}_{321} = 0, \\ \mathsf{Q}_{113} &= \mathsf{Q}_{223}, \quad \mathsf{Q}_{311} = \mathsf{Q}_{322}, \quad \mathsf{Q}_{331} = \mathsf{Q}_{332}, \end{aligned} \tag{111}$$

and we are left with 18 independent components:

$$\begin{aligned} \mathsf{Q}_{111} &= \rho^{-1}(\mathbb{D}_{11} + \mathbb{B}_{11})\, m_1 \\ \mathsf{Q}_{122} &= \rho^{-1}(\mathbb{D}_{66} + \mathbb{B}_{69})\, m_1 \\ \mathsf{Q}_{133} &= \rho^{-1}(\mathbb{D}_{55} + \mathbb{B}_{47})\, m_1 \\ \mathsf{Q}_{131} &= \rho^{-1}(\mathbb{D}_{44} + \mathbb{B}_{44})\, m_3 \\ \mathsf{Q}_{112} &= \rho^{-1}(\mathbb{D}_{12} + \mathbb{B}_{12})\, m_2 \\ \mathsf{Q}_{113} &= \rho^{-1}(\mathbb{D}_{13} + \mathbb{B}_{13})\, m_3 \\ \mathsf{Q}_{121} &= \rho^{-1}(\mathbb{D}_{66} + \mathbb{B}_{66})\, m_2 \\ \mathsf{Q}_{211} &= \rho^{-1}(\mathbb{D}_{66} + \mathbb{B}_{69})\, m_2 \\ \mathsf{Q}_{222} &= \rho^{-1}(\mathbb{D}_{11} + \mathbb{B}_{11})\, m_2 \\ \mathsf{Q}_{233} &= \rho^{-1}(\mathbb{D}_{55} + \mathbb{B}_{47})\, m_2 \\ \mathsf{Q}_{212} &= \rho^{-1}(\mathbb{D}_{66} + \mathbb{B}_{66})\, m_1 \\ \mathsf{Q}_{232} &= \rho^{-1}(\mathbb{D}_{44} + \mathbb{B}_{44})\, m_3 \\ \mathsf{Q}_{221} &= \rho^{-1}(\mathbb{D}_{21} + \mathbb{B}_{12})\, m_1 \end{aligned}$$

$$
\begin{aligned}
\mathsf{Q}_{311} &= \rho^{-1}(\mathbb{D}_{44} + \mathbb{B}_{47})\, m_3 \\
\mathsf{Q}_{333} &= \rho^{-1}(\mathbb{D}_{33} + \mathbb{B}_{33})\, m_3 \\
\mathsf{Q}_{323} &= \rho^{-1}(\mathbb{D}_{55} + \mathbb{B}_{45})\, m_2 \\
\mathsf{Q}_{331} &= \rho^{-1}(\mathbb{D}_{31} + \mathbb{B}_{13})\, m_1 \\
\mathsf{Q}_{313} &= \rho^{-1}(\mathbb{D}_{55} + \mathbb{B}_{55})\, m_1
\end{aligned}
\tag{112}
$$

with the following tabular representation, where we put in square brackets $[\cdot]$ those depending on m_1 and in round brackets $(\cdot)$ those which depends only on m_2, the remaining depending on m_3:

$$
[\mathsf{Q}] = \begin{bmatrix}
[\mathsf{Q}_{111}] & [\mathsf{Q}_{122}] & [\mathsf{Q}_{133}] & 0 & \mathsf{Q}_{131} & (\mathsf{Q}_{112}) & \mathsf{Q}_{131} & \mathsf{Q}_{113} & (\mathsf{Q}_{121}) \\
(\mathsf{Q}_{211}) & (\mathsf{Q}_{222}) & (\mathsf{Q}_{233}) & \mathsf{Q}_{113} & 0 & [\mathsf{Q}_{212}] & 0 & 0 & [\mathsf{Q}_{221}] \\
\mathsf{Q}_{311} & \mathsf{Q}_{311} & \mathsf{Q}_{333} & (\mathsf{Q}_{323}) & [\mathsf{Q}_{331}] & 0 & [\mathsf{Q}_{313}] & [\mathsf{Q}_{313}] & 0
\end{bmatrix}.
\tag{113}
$$

3.2.3 The third-order tensor P(m)

First of all we recall that for the centrosymmetric classes $4/m$ and $4mm$ and for the axial class 422, since $\mathcal{F} = \mathcal{G} = \mathbf{0}$ then $\mathrm{P} = \mathbf{0}$.

3.2.3.1 Classes 4 and $\bar{4}$

The components for these classes are evaluated by the means of $(67)_2$ and (97): here we denote here $\mathcal{A}_{ijhkm} = \mathcal{F}_{ijhkm} + \mathcal{G}_{ijhkm}$ and with boldface $\boldsymbol{\mathcal{A}}_{ijhkm}$ those components which vanish for the class $\bar{4}$:

$$
\begin{aligned}
\mathsf{P}_{111} &= \rho^{-1}((\mathcal{A}_{11113} + \mathcal{A}_{13111})\, m_1 m_3 + (\mathcal{A}_{12113} + \mathcal{A}_{13112})\, m_2 m_3), \\
\mathsf{P}_{122} &= \rho^{-1}((\mathcal{A}_{11223} + \mathcal{A}_{13221})\, m_1 m_3 - (\mathcal{A}_{21113} + \mathcal{A}_{23111})\, m_2 m_3), \\
\mathsf{P}_{111} &= \rho^{-1}((\mathcal{A}_{11113} + \mathcal{A}_{13111})\, m_1 m_3 + (\mathcal{A}_{12113} + \mathcal{A}_{13112})\, m_2 m_3), \\
\mathsf{P}_{122} &= \rho^{-1}((\mathcal{A}_{11223} + \mathcal{A}_{13221})\, m_1 m_3 - (\mathcal{A}_{21113} + \mathcal{A}_{23111})\, m_2 m_3), \\
\mathsf{P}_{133} &= \rho^{-1}((\mathcal{A}_{11333} + \mathcal{A}_{13331})\, m_1 m_3 + (\boldsymbol{\mathcal{A}}_{\mathbf{12333}} + \boldsymbol{\mathcal{A}}_{\mathbf{13332}})\, m_2 m_3), \\
\mathsf{P}_{123} &= \rho^{-1}(\mathcal{A}_{11231} m_1^2 - \mathcal{A}_{21131} m_2^2 + \mathcal{A}_{13233} m_3^2 \\
&\qquad + (\mathcal{A}_{11232} + \mathcal{A}_{12231})\, m_1 m_2), \\
\mathsf{P}_{131} &= \rho^{-1}(\mathcal{A}_{11311} m_1^2 + \mathcal{A}_{12312} m_2^2 + \mathcal{A}_{13313} m_3^2 \\
&\qquad + (\mathcal{A}_{11312} + \mathcal{A}_{12311})\, m_1 m_2), \\
\mathsf{P}_{112} &= \rho^{-1}((\mathcal{A}_{11123} + \mathcal{A}_{13121})\, m_1 m_3 + (\mathcal{A}_{12123} + \mathcal{A}_{13122})\, m_2 m_3), \\
\mathsf{P}_{132} &= \rho^{-1}(\mathcal{A}_{11321} m_1^2 - \mathcal{A}_{21311} m_2^2 + \boldsymbol{\mathcal{A}}_{\mathbf{13323}} m_3^2 + (\mathcal{A}_{11322} + \mathcal{A}_{12321})\, m_1 m_2 \\
&\qquad + (\mathcal{A}_{12323} + \mathcal{A}_{13322})\, m_2 m_3), \\
\mathsf{P}_{113} &= \rho^{-1}(\mathcal{A}_{11131} m_1^2 + \mathcal{A}_{12132} m_2^2 + \mathcal{A}_{13133} m_3^2 \\
&\qquad + (\mathcal{A}_{11132} + \mathcal{A}_{12131})\, m_1 m_2), \\
\mathsf{P}_{121} &= \rho^{-1}((\mathcal{A}_{11213} + \mathcal{A}_{13211})\, m_1 m_3 + (\mathcal{A}_{12213} + \mathcal{A}_{13212})\, m_2 m_3),
\end{aligned}
$$

$$
\begin{aligned}
\mathsf{P}_{211} &= \rho^{-1}((\mathcal{A}_{21113}+\mathcal{A}_{23111})m_1m_3+(\mathcal{A}_{11223}+\mathcal{A}_{13221})m_2m_3),\\
\mathsf{P}_{222} &= \rho^{-1}(-(\mathcal{A}_{12113}+\mathcal{A}_{13112})m_1m_3+(\mathcal{A}_{11113}+\mathcal{A}_{13111})m_2m_3),\\
\mathsf{P}_{233} &= \rho^{-1}(-(\mathcal{A}_{\mathbf{12333}}+\mathcal{A}_{\mathbf{13332}})m_1m_3+(\mathcal{A}_{11333}+\mathcal{A}_{13331})m_2m_3),\\
\mathsf{P}_{223} &= \rho^{-1}(\mathcal{A}_{12132}m_1^2+\mathcal{A}_{11131}m_2^2+\mathcal{A}_{13133}m_3^2\\
&\qquad +(\mathcal{A}_{21232}+\mathcal{A}_{22231})m_1m_2),\\
\mathsf{P}_{231} &= \rho^{-1}(\mathcal{A}_{21311}m_1^2-\mathcal{A}_{11321}m_2^2-\mathcal{A}_{\mathbf{13323}}m_3^2+(\mathcal{A}_{12321}+\mathcal{A}_{11322})m_1m_2\\
&\quad +(\mathcal{A}_{21313}+\mathcal{A}_{23311})m_1m_3),\\
\mathsf{P}_{212} &= \rho^{-1}((\mathcal{A}_{12213}+\mathcal{A}_{13212})m_1m_3-(\mathcal{A}_{11213}+(\mathcal{A}_{13211})m_2m_3),\\
\mathsf{P}_{232} &= \rho^{-1}(\mathcal{A}_{12312}m_1^2+\mathcal{A}_{11311}m_2^2+\mathcal{A}_{13313}m_3^2\\
&\qquad -(\mathcal{A}_{12311}+\mathcal{A}_{11312})m_1m_2),\\
\mathsf{P}_{213} &= \rho^{-1}(\mathcal{A}_{21131}m_1^2-\mathcal{A}_{11231}m_2^2-\mathcal{A}_{\mathbf{13233}}m_3^2+(\mathcal{A}_{12231}+\mathcal{A}_{11232})m_1m_2\\
&\quad +(\mathcal{A}_{22133}+\mathcal{A}_{23132})m_2m_3),\\
\mathsf{P}_{221} &= \rho^{-1}((\mathcal{A}_{12123}+\mathcal{A}_{13122})m_1m_3-(\mathcal{A}_{11123}+\mathcal{A}_{13121})m_2m_3),\\
\mathsf{P}_{311} &= \rho^{-1}(\mathcal{A}_{31111}m_1^2+\mathcal{A}_{32112}m_2^2+\mathcal{A}_{33113}m_3^2\\
&\quad +(\mathcal{A}_{31112}+\mathcal{A}_{32111})m_1m_2+(\mathcal{A}_{31113}+\mathcal{A}_{33111})m_1m_3),\\
\mathsf{P}_{322} &= \rho^{-1}(\mathcal{A}_{31221}m_1^2+\mathcal{A}_{31111}m_2^2+\mathcal{A}_{33113}m_3^2\\
&\qquad -(\mathcal{A}_{32111}+\mathcal{A}_{31112})m_1m_2),\\
\mathsf{P}_{333} &= \rho^{-1}(\mathcal{A}_{31331}m_1^2+\mathcal{A}_{31331}m_2^2+\mathcal{A}_{\mathbf{33333}}m_3^2\\
&\qquad +(\mathcal{A}_{\mathbf{31332}}+\mathcal{A}_{\mathbf{32331}})m_1m_2),\\
\mathsf{P}_{323} &= \rho^{-1}((\mathcal{A}_{\mathbf{31233}}+\mathcal{A}_{\mathbf{33231}})m_1m_3+(\mathcal{A}_{31133}+\mathcal{A}_{33131})m_2m_3),\\
\mathsf{P}_{331} &= \rho^{-1}((\mathcal{A}_{31313}+\mathcal{A}_{33311})m_1m_3-(\mathcal{A}_{\mathbf{31323}}+\mathcal{A}_{\mathbf{33321}})m_2m_3),\\
\mathsf{P}_{312} &= \rho^{-1}(\mathcal{A}_{31121}m_1^2-\mathcal{A}_{31211}m_2^2+\mathcal{A}_{33123}m_3^2\\
&\quad +(\mathcal{A}_{31122}+\mathcal{A}_{32121})m_1m_2-(\mathcal{A}_{31213}+\mathcal{A}_{33212})m_2m_3),\\
\mathsf{P}_{332} &= \rho^{-1}((\mathcal{A}_{\mathbf{31323}}+\mathcal{A}_{\mathbf{33321}})m_1m_3+(\mathcal{A}_{32323}+\mathcal{A}_{33322})m_2m_3),\\
\mathsf{P}_{313} &= \rho^{-1}((\mathcal{A}_{31133}+\mathcal{A}_{33131})m_1m_3-(\mathcal{A}_{\mathbf{31233}}+\mathcal{A}_{\mathbf{33231}})m_2m_3),\\
\mathsf{P}_{321} &= \rho^{-1}(\mathcal{A}_{31211}m_1^2+\mathcal{A}_{32212}m_2^2-\mathcal{A}_{\mathbf{33123}}m_3^2\\
&\qquad +(\mathcal{A}_{32121}+\mathcal{A}_{31122})m_1m_2).
\end{aligned}
\tag{114}
$$

In the tabular form of the tensor $\mathsf{P}(\mathbf{m})$ we represent in brackets $(\cdot)$ the components that vanish when $\mathbf{m}\cdot\mathbf{e}_3 = 0$:

$$
[\mathsf{P}] = \begin{bmatrix}
(\mathsf{P}_{111}) & (\mathsf{P}_{122}) & (\mathsf{P}_{133}) & \mathsf{P}_{123} & \mathsf{P}_{131} & (\mathsf{P}_{112}) & \mathsf{P}_{132} & \mathsf{P}_{113} & (\mathsf{P}_{121})\\
(\mathsf{P}_{211}) & (\mathsf{P}_{222}) & (\mathsf{P}_{233}) & \mathsf{P}_{223} & \mathsf{P}_{231} & (\mathsf{P}_{212}) & \mathsf{P}_{232} & \mathsf{P}_{213} & (\mathsf{P}_{221})\\
\mathsf{P}_{311} & \mathsf{P}_{322} & \mathsf{P}_{333} & (\mathsf{P}_{323}) & (\mathsf{P}_{331}) & \mathsf{P}_{312} & (\mathsf{P}_{332}) & (\mathsf{P}_{313}) & \mathsf{P}_{321}
\end{bmatrix}.
\tag{115}
$$

For the class 4 in the case $\mathbf{m} = \mathbf{e}_3$ we also have that

$$
\begin{aligned}
\mathsf{P}_{223} &= \mathsf{P}_{113}, \quad \mathsf{P}_{231} = -\mathsf{P}_{132}, \quad \mathsf{P}_{232} = -\mathsf{P}_{131},\\
\mathsf{P}_{213} &= -\mathsf{P}_{123}, \quad \mathsf{P}_{322} = \mathsf{P}_{311}, \quad \mathsf{P}_{321} = -\mathsf{P}_{312}.
\end{aligned}
\tag{116}
$$

3.2.3.2 Classes $\bar{4}2m$, 422, $4/mmm$ (Polar tensors)

In this case we have

$$
\begin{aligned}
\mathsf{P}_{111} &= \rho^{-1}(\mathcal{A}_{11113} + \mathcal{A}_{13111})\, m_1 m_3,\\
\mathsf{P}_{122} &= \rho^{-1}(\mathcal{A}_{11223} + \mathcal{A}_{13221})\, m_1 m_3,\\
\mathsf{P}_{133} &= \rho^{-1}(\mathcal{A}_{11333} + \mathcal{A}_{13331})\, m_1 m_3,\\
\mathsf{P}_{123} &= \rho^{-1}(\mathcal{A}_{11232} + \mathcal{A}_{12231})\, m_1 m_2,\\
\mathsf{P}_{131} &= \rho^{-1}(\mathcal{A}_{11311} m_1^2 + \mathcal{A}_{12312} m_2^2 + \mathcal{A}_{13313} m_3^2),\\
\mathsf{P}_{112} &= \rho^{-1}(\mathcal{A}_{12123} + \mathcal{A}_{13122})\, m_2 m_3,\\
\mathsf{P}_{132} &= \rho^{-1}((\mathcal{A}_{11322} + \mathcal{A}_{12321})\, m_1 m_2 + (\mathcal{A}_{12323} + \mathcal{A}_{13322})\, m_2 m_3),\\
\mathsf{P}_{113} &= \rho^{-1}(\mathcal{A}_{11131} m_1^2 + \mathcal{A}_{12132} m_2^2 + \mathcal{A}_{13133} m_3^2),\\
\mathsf{P}_{121} &= \rho^{-1}(\mathcal{A}_{12213} + \mathcal{A}_{13212})\, m_2 m_3,\\
\mathsf{P}_{211} &= \rho^{-1}(\mathcal{A}_{11223} + \mathcal{A}_{13221})\, m_2 m_3,\\
\mathsf{P}_{222} &= \rho^{-1}(\mathcal{A}_{11113} + \mathcal{A}_{13111})\, m_2 m_3,\\
\mathsf{P}_{233} &= \rho^{-1}(\mathcal{A}_{11333} + \mathcal{A}_{13331})\, m_2 m_3,\\
\mathsf{P}_{223} &= \rho^{-1}(\mathcal{A}_{12132} m_1^2 + \mathcal{A}_{11131} m_2^2 + \mathcal{A}_{13133} m_3^2\\
&\quad + (\mathcal{A}_{21232} + \mathcal{A}_{22231})\, m_1 m_2),\\
\mathsf{P}_{231} &= \rho^{-1}((\mathcal{A}_{12321} + \mathcal{A}_{11322})\, m_1 m_2 + (\mathcal{A}_{21313} + \mathcal{A}_{23311})\, m_1 m_3),\\
\mathsf{P}_{212} &= \rho^{-1}(\mathcal{A}_{12213} + \mathcal{A}_{13212})\, m_1 m_3,\\
\mathsf{P}_{232} &= \rho^{-1}(\mathcal{A}_{12312} m_1^2 + \mathcal{A}_{11311} m_2^2 + \mathcal{A}_{13313} m_3^2),\\
\mathsf{P}_{213} &= \rho^{-1}((\mathcal{A}_{12231} + \mathcal{A}_{11232})\, m_1 m_2 + (\mathcal{A}_{22133} + \mathcal{A}_{23132})\, m_2 m_3),\\
\mathsf{P}_{221} &= \rho^{-1}(\mathcal{A}_{12123} + \mathcal{A}_{13122})\, m_1 m_3,\\
\mathsf{P}_{311} &= \rho^{-1}(\mathcal{A}_{31111} m_1^2 + \mathcal{A}_{32112} m_2^2 + \mathcal{A}_{33113} m_3^2\\
&\quad + (\mathcal{A}_{31113} + \mathcal{A}_{33111})\, m_1 m_3),\\
\mathsf{P}_{322} &= \rho^{-1}(\mathcal{A}_{31221} m_1^2 + \mathcal{A}_{31111} m_2^2 + \mathcal{A}_{33113} m_3^2),\\
\mathsf{P}_{333} &= \rho^{-1}(\mathcal{A}_{31331}(m_1^2 - m_2^2) + \mathcal{A}_{33333} m_3^2),\\
\mathsf{P}_{323} &= \rho^{-1}(\mathcal{A}_{31133} + \mathcal{A}_{33131})\, m_2 m_3,\\
\mathsf{P}_{331} &= \rho^{-1}(\mathcal{A}_{31313} + \mathcal{A}_{33311})\, m_1 m_3,\\
\mathsf{P}_{312} &= \rho^{-1}((\mathcal{A}_{31122} + \mathcal{A}_{32121})\, m_1 m_2 - (\mathcal{A}_{31213} + \mathcal{A}_{33212})\, m_2 m_3),\\
\mathsf{P}_{332} &= \rho^{-1}(\mathcal{A}_{32323} + \mathcal{A}_{33322})\, m_2 m_3,\\
\mathsf{P}_{313} &= \rho^{-1}(\mathcal{A}_{31133} + \mathcal{A}_{33131})\, m_1 m_3,\\
\mathsf{P}_{321} &= \rho^{-1}(\mathcal{A}_{32121} + \mathcal{A}_{31122})\, m_1 m_2.
\end{aligned}
\tag{117}
$$

We show in the tabular form the components which are different from zero when $\mathbf{m} = \mathbf{e}_3$

$$[\mathsf{P}] = \begin{bmatrix} 0 & 0 & 0 & 0 & \mathsf{P}_{131} & 0 & 0 & \mathsf{P}_{113} & 0 \\ 0 & 0 & 0 & \mathsf{P}_{113} & 0 & 0 & \mathsf{P}_{131} & 0 & 0 \\ \mathsf{P}_{311} & \mathsf{P}_{311} & \mathsf{P}_{333} & 0 & 0 & 0 & 0 & 0 & 0 \end{bmatrix}; \tag{118}$$

whereas when $\mathbf{m}\cdot\mathbf{e}_3 = 0$ we have

$$[\mathsf{P}] = \begin{bmatrix} 0 & 0 & 0 & \mathsf{P}_{123} & \mathsf{P}_{131} & 0 & \mathsf{P}_{132} & \mathsf{P}_{113} & 0 \\ 0 & 0 & 0 & \mathsf{P}_{223} & \mathsf{P}_{231} & 0 & \mathsf{P}_{232} & \mathsf{P}_{213} & 0 \\ \mathsf{P}_{311} & \mathsf{P}_{322} & \mathsf{P}_{333} & 0 & 0 & \mathsf{P}_{312} & 0 & 0 & \mathsf{P}_{321} \end{bmatrix}. \tag{119}$$

3.2.3.3 Classes $4/mmm$ and $\bar{4}2m$ (Axial tensors)

In this case the components of $\mathrm{P}(\mathbf{m})$ for the class $4/mmm$ are

$$\begin{aligned}
\mathsf{P}_{111} &= \rho^{-1}(\mathcal{A}_{12113} + \mathcal{A}_{13112})m_2m_3, \\
\mathsf{P}_{122} &= \rho^{-1}(\mathcal{A}_{12223} + \mathcal{A}_{13222})m_2m_3, \\
\mathsf{P}_{133} &= \rho^{-1}(\mathcal{A}_{12333} + \mathcal{A}_{13332})m_2m_3, \\
\mathsf{P}_{123} &= \rho^{-1}(\mathcal{A}_{11231}(m_1^2 - m_2^2) + \mathcal{A}_{13233}m_3^2), \\
\mathsf{P}_{131} &= \rho^{-1}(\mathcal{A}_{11312} + \mathcal{A}_{12311})m_1m_2, \\
\mathsf{P}_{112} &= \rho^{-1}(\mathcal{A}_{11123} + \mathcal{A}_{13121})m_1m_3, \\
\mathsf{P}_{132} &= \rho^{-1}(\mathcal{A}_{21311}(m_1^2 - m_2^2) + \mathcal{A}_{13323}m_3^2), \\
\mathsf{P}_{113} &= \rho^{-1}(\mathcal{A}_{11132} + \mathcal{A}_{12131})m_1m_2, \\
\mathsf{P}_{121} &= \rho^{-1}(\mathcal{A}_{11123} + \mathcal{A}_{13121})m_2m_3, \\
\mathsf{P}_{211} &= \rho^{-1}(\mathcal{A}_{23111} + \mathcal{A}_{21113})m_1m_3, \\
\mathsf{P}_{222} &= \rho^{-1}(\mathcal{A}_{21223} + \mathcal{A}_{23221})m_1m_3, \\
\mathsf{P}_{233} &= \rho^{-1}(\mathcal{A}_{21333} + \mathcal{A}_{23331})m_1m_2 \\
\mathsf{P}_{223} &= \rho^{-1}(\mathcal{A}_{22231} + \mathcal{A}_{21232})m_1m_2, \\
\mathsf{P}_{231} &= \rho^{-1}(\mathcal{A}_{21311}(m_1^2 - m_2^2) + \mathcal{A}_{23313}m_3^2) \\
\mathsf{P}_{212} &= \rho^{-1}(\mathcal{A}_{22123} + \mathcal{A}_{23122})m_2m_3, \\
\mathsf{P}_{232} &= \rho^{-1}(\mathcal{A}_{22321} + \mathcal{A}_{21322})m_1m_2, \\
\mathsf{P}_{213} &= \rho^{-1}(\mathcal{A}_{12232}(m_1^2 - m_2^2) + \mathcal{A}_{23133}m_3^2) \\
\mathsf{P}_{221} &= \rho^{-1}(\mathcal{A}_{22213} + \mathcal{A}_{23212})m_2m_3, \\
\mathsf{P}_{311} &= \rho^{-1}(\mathcal{A}_{31112} + \mathcal{A}_{32111})m_1m_2 \\
\mathsf{P}_{322} &= \rho^{-1}(\mathcal{A}_{31222} + \mathcal{A}_{32221})m_1m_2, \\
\mathsf{P}_{333} &= \rho^{-1}(\mathcal{A}_{31332} + \mathcal{A}_{32331})m_1m_2,
\end{aligned}$$

$$
\begin{aligned}
\mathsf{P}_{323} &= \rho^{-1}(\mathcal{A}_{33231} + \mathcal{A}_{31233})\, m_1 m_3,\\
\mathsf{P}_{331} &= \rho^{-1}(\mathcal{A}_{33312} + \mathcal{A}_{32331})\, m_2 m_3,\\
\mathsf{P}_{312} &= \rho^{-1}(\mathcal{A}_{31121}(m_1^2 - m_2^2) + \mathcal{A}_{33123} m_3^2)\\
\mathsf{P}_{332} &= \rho^{-1}(\mathcal{A}_{33321} + \mathcal{A}_{31323})\, m_1 m_3,\\
\mathsf{P}_{313} &= \rho^{-1}(\mathcal{A}_{33132} + \mathcal{A}_{32133})\, m_2 m_3,\\
\mathsf{P}_{321} &= \rho^{-1}(\mathcal{A}_{31211}(m_1^2 - m_2^2) + \mathcal{A}_{33213} m_3^2);
\end{aligned}
\tag{120}
$$

those for the class $\bar{4}2m$ are obtained by replacing the term $(m_1^2 - m_2^2)$ into P_{123}, P_{132}, P_{231}, P_{213}, P_{312} and P_{321} with the term $(m_1^2 + m_2^2)$.

The tabular representation of $\mathrm{P}(\mathbf{m})$ now is:

$$
[\mathsf{P}] = \begin{bmatrix}
[\mathsf{P}_{111}] & [\mathsf{P}_{122}] & [\mathsf{P}_{133}] & \mathsf{P}_{123} & (\mathsf{P}_{131}) & [\mathsf{P}_{112}] & \mathsf{P}_{132} & \mathsf{P}_{113} & [\mathsf{P}_{121}]\\
[\mathsf{P}_{211}] & [\mathsf{P}_{222}] & (\mathsf{P}_{233}) & (\mathsf{P}_{223}) & \mathsf{P}_{231} & [\mathsf{P}_{212}] & (\mathsf{P}_{232}) & \mathsf{P}_{213} & [\mathsf{P}_{221}]\\
(\mathsf{P}_{311}) & (\mathsf{P}_{322}) & \mathsf{P}_{333} & [\mathsf{P}_{323}] & [\mathsf{P}_{331}] & \mathsf{P}_{312} & [\mathsf{P}_{332}] & [\mathsf{P}_{313}] & \mathsf{P}_{321}
\end{bmatrix},
\tag{121}
$$

where the terms in round brackets $(\cdot)$ vanish for $\mathbf{m} = \mathbf{e}_3$ whereas those in square brackets $[\cdot]$ vanish in both cases $\mathbf{m} = \mathbf{e}_3$ and $\mathbf{m}\cdot\mathbf{e}_3 = 0$.

3.2.4 The fourth-order tensor $\mathbb{A}(\mathbf{m})$

3.2.4.1 Classes 4, and $\bar{4}$ 4/*m*

The components follows by $(67)_4$ and (102) and read:

$$
\begin{aligned}
\mathbb{A}_{11} &= \rho^{-1}(\mathfrak{H}_{111111} m_1^2 + \mathfrak{H}_{112112} m_2^2 + \mathfrak{H}_{113113} m_3^2 + 2\mathfrak{H}_{111112} m_1 m_2),\\
\mathbb{A}_{22} &= \rho^{-1}(\mathfrak{H}_{112112} m_1^2 + \mathfrak{H}_{111111} m_2^2 + \mathfrak{H}_{113113} m_3^2 - 2\mathfrak{H}_{111112} m_1 m_2),\\
\mathbb{A}_{33} &= \rho^{-1}(\mathfrak{H}_{331331}(m_1^2 + m_2^2) + \mathfrak{H}_{333333} m_3^2 + 2\mathfrak{H}_{331332} m_1 m_2),\\
\mathbb{A}_{44} &= \rho^{-1}(\mathfrak{H}_{231231} m_1^2 + \mathfrak{H}_{131131} m_2^2 + \mathfrak{H}_{133133} m_3^2 - 2\mathfrak{H}_{131132} m_1 m_2),\\
\mathbb{A}_{55} &= \rho^{-1}(\mathfrak{H}_{311311} m_1^2 + \mathfrak{H}_{312312} m_2^2 + \mathfrak{H}_{313313} m_3^2 + 2\mathfrak{H}_{311312} m_1 m_2),\\
\mathbb{A}_{66} &= \rho^{-1}(\mathfrak{H}_{121121} m_1^2 + \mathfrak{H}_{211211} m_2^2 + \mathfrak{H}_{123123} m_3^2 + 2\mathfrak{H}_{121122} m_1 m_2),\\
\mathbb{A}_{77} &= \rho^{-1}(\mathfrak{H}_{312312} m_1^2 + \mathfrak{H}_{311311} m_2^2 + \mathfrak{H}_{313313} m_3^2 - 2\mathfrak{H}_{311312} m_1 m_2),\\
\mathbb{A}_{88} &= \rho^{-1}(\mathfrak{H}_{131131} m_1^2 + \mathfrak{H}_{231231} m_2^2 + \mathfrak{H}_{133133} m_3^2 + 2\mathfrak{H}_{131132} m_1 m_2),\\
\mathbb{A}_{99} &= \rho^{-1}(\mathfrak{H}_{211211} m_1^2 + \mathfrak{H}_{212212} m_2^2 + \mathfrak{H}_{123123} m_3^2 + 2\mathfrak{H}_{211212} m_1 m_2),\\
\mathbb{A}_{12} &= \rho^{-1}(\mathfrak{H}_{111221} m_1^2 + \mathfrak{H}_{221111} m_2^2 + \mathfrak{H}_{113223} m_3^2 + 2\mathfrak{H}_{111222} m_1 m_2),\\
\mathbb{A}_{13} &= \rho^{-1}(\mathfrak{H}_{111331} m_1^2 + \mathfrak{H}_{112332} m_2^2 + \mathfrak{H}_{113333} m_3^2 + 2\mathfrak{H}_{111332} m_1 m_2),\\
\mathbb{A}_{14} &= \rho^{-1}(2\mathfrak{H}_{111233} m_1 m_3 + 2\mathfrak{H}_{112233} m_2 m_3),\\
\mathbb{A}_{15} &= \rho^{-1}(2\mathfrak{H}_{111313} m_1 m_3 + 2\mathfrak{H}_{112313} m_2 m_3),\\
\mathbb{A}_{16} &= \rho^{-1}(\mathfrak{H}_{111121} m_1^2 + \mathfrak{H}_{112122} m_2^2 + \mathfrak{H}_{113123} m_3^2 + 2\mathfrak{H}_{111122} m_1 m_2),\\
\mathbb{A}_{17} &= \rho^{-1}(2\mathfrak{H}_{111323} m_1 m_3 + 2\mathfrak{H}_{112323} m_2 m_3),\\
\mathbb{A}_{18} &= \rho^{-1}(2\mathfrak{H}_{111133} m_1 m_3 + 2\mathfrak{H}_{112133} m_2 m_3),
\end{aligned}
$$

$$
\begin{aligned}
\mathbb{A}_{19} &= \rho^{-1}(\mathfrak{H}_{111211}m_1^2 + \mathfrak{H}_{112212}m_2^2 + \mathfrak{H}_{113213}m_3^2 + 2\mathfrak{H}_{111212}m_1m_2),\\
\mathbb{A}_{23} &= \rho^{-1}(\mathfrak{H}_{112332}m_1^2 + \mathfrak{H}_{111331}m_2^2 + \mathfrak{H}_{113333}m_3^2 - 2\mathfrak{H}_{112331}m_1m_2),\\
\mathbb{A}_{24} &= -\rho^{-1}(2\mathfrak{H}_{112133}m_1m_3 + 2\mathfrak{H}_{111133}m_2m_3),\\
\mathbb{A}_{25} &= \rho^{-1}(2\mathfrak{H}_{112323}m_1m_3 - 2\mathfrak{H}_{111323}m_2m_3),\\
\mathbb{A}_{26} &= \rho^{-1}(\mathfrak{H}_{221121}m_1^2 - \mathfrak{H}_{111211}m_2^2 - \mathfrak{H}_{113213}m_3^2 + 2\mathfrak{H}_{112211}m_1m_2),\\
\mathbb{A}_{27} &= -\rho^{-1}(2\mathfrak{H}_{112313}m_1m_3 + 2\mathfrak{H}_{111313}m_2m_3),\\
\mathbb{A}_{28} &= \rho^{-1}(2\mathfrak{H}_{112233}m_1m_3 - 2\mathfrak{H}_{111233}m_2m_3),\\
\mathbb{A}_{29} &= \rho^{-1}(\mathfrak{H}_{221211}m_1^2 - \mathfrak{H}_{111121}m_2^2 - \mathfrak{H}_{113123}m_3^2 + 2\mathfrak{H}_{112121}m_1m_2),\\
\mathbb{A}_{34} &= \rho^{-1}(2\mathfrak{H}_{331233}m_1m_3 + 2\mathfrak{H}_{331133}m_2m_3),\\
\mathbb{A}_{35} &= \rho^{-1}(2\mathfrak{H}_{331313}m_1m_3 - 2\mathfrak{H}_{331323}m_2m_3),\\
\mathbb{A}_{36} &= \rho^{-1}(\mathfrak{H}_{331121}m_1^2 - \mathfrak{H}_{331211}m_2^2 + \mathfrak{H}_{333123}m_3^2 + 2\mathfrak{H}_{331122}m_1m_2),\\
\mathbb{A}_{37} &= \rho^{-1}(2\mathfrak{H}_{331323}m_1m_3 + 2\mathfrak{H}_{331313}m_2m_3),\\
\mathbb{A}_{38} &= \rho^{-1}(2\mathfrak{H}_{331133}m_1m_3 - 2\mathfrak{H}_{331233}m_2m_3),\\
\mathbb{A}_{39} &= \rho^{-1}(\mathfrak{H}_{331211}m_1^2 - \mathfrak{H}_{331121}m_2^2 - \mathfrak{H}_{333123}m_3^2 + 2\mathfrak{H}_{331212}m_1m_2),\\
\mathbb{A}_{45} &= \rho^{-1}(\mathfrak{H}_{231311}m_1^2 - \mathfrak{H}_{131321}m_2^2 - \mathfrak{H}_{133323}m_3^2 + 2\mathfrak{H}_{231312}m_1m_2),\\
\mathbb{A}_{46} &= \rho^{-1}(2\mathfrak{H}_{132213}m_1m_3 - 2\mathfrak{H}_{131213}m_2m_3),\\
\mathbb{A}_{47} &= \rho^{-1}(\mathfrak{H}_{132312}m_1^2 + \mathfrak{H}_{131311}m_2^2 + \mathfrak{H}_{133313}m_3^2 - 2\mathfrak{H}_{132311}m_1m_2),\\
\mathbb{A}_{48} &= \rho^{-1}(\mathfrak{H}_{231131}m_1^2 - \mathfrak{H}_{131231}m_2^2 - 2\mathfrak{H}_{132231}m_1m_2),\\
\mathbb{A}_{49} &= \rho^{-1}(2\mathfrak{H}_{132123}m_1m_3 - 2\mathfrak{H}_{131123}m_2m_3),\\
\mathbb{A}_{56} &= \rho^{-1}(2\mathfrak{H}_{311123}m_1m_3 + 2\mathfrak{H}_{312123}m_2m_3),\\
\mathbb{A}_{57} &= \rho^{-1}(\mathfrak{H}_{311321}m_1^2 - \mathfrak{H}_{321311}m_2^2 + 2\mathfrak{H}_{311322}m_1m_2),\\
\mathbb{A}_{58} &= \rho^{-1}(\mathfrak{H}_{311131}m_1^2 + \mathfrak{H}_{132312}m_2^2 + \mathfrak{H}_{313133}m_3^2 + 2\mathfrak{H}_{311132}m_1m_2),\\
\mathbb{A}_{59} &= \rho^{-1}(2\mathfrak{H}_{311213}m_1m_3 + 2\mathfrak{H}_{312213}m_2m_3),\\
\mathbb{A}_{67} &= \rho^{-1}(2\mathfrak{H}_{121323}m_1m_3 + 2\mathfrak{H}_{122323}m_2m_3),\\
\mathbb{A}_{68} &= \rho^{-1}(2\mathfrak{H}_{121133}m_1m_3 + 2\mathfrak{H}_{122133}m_2m_3),\\
\mathbb{A}_{69} &= \rho^{-1}(\mathfrak{H}_{121211}m_1^2 + \mathfrak{H}_{211121}m_2^2 + \mathfrak{H}_{123213}m_3^2 + 2\mathfrak{H}_{121212}m_1m_2),\\
\mathbb{A}_{78} &= \rho^{-1}(\mathfrak{H}_{311131}m_1^2 + \mathfrak{H}_{312132}m_2^2 + \mathfrak{H}_{313133}m_3^2 + 2\mathfrak{H}_{311132}m_1m_2),\\
\mathbb{A}_{79} &= \rho^{-1}(2\mathfrak{H}_{311213}m_1m_3 + 2\mathfrak{H}_{123321}m_2m_3),\\
\mathbb{A}_{89} &= \rho^{-1}(2\mathfrak{H}_{131213}m_1m_3 + 2\mathfrak{H}_{132213}m_2m_3).
\end{aligned}
\tag{122}
$$

From (122) we have that when $\mathbf{m} = \mathbf{e}_3\mathbb{A}$ has the same 13 independent components as $\mathbb{B}$; for $\mathbf{m}\cdot\mathbf{e}_3 = 0$ we have instead that there are 26 independent components.

3.2.4.2 Classes 422, *4mm*, *4/mmm* and polar tensors of the class $\bar{4}2m$

By using (103) into (122) we get the following components:

$$
\begin{aligned}
\mathbb{A}_{11} &= \rho^{-1}(\mathfrak{H}_{111111}m_1^2 + \mathfrak{H}_{112112}m_2^2 + \mathfrak{H}_{113113}m_3^2), \quad \mathbb{A}_{14} = \rho^{-1}2\mathfrak{H}_{112233}m_2m_3,\\
\mathbb{A}_{22} &= \rho^{-1}(\mathfrak{H}_{112112}m_1^2 + \mathfrak{H}_{111111}m_2^2 + \mathfrak{H}_{113113}m_3^2), \quad \mathbb{A}_{15} = \rho^{-1}2\mathfrak{H}_{111313}m_1m_3,\\
\mathbb{A}_{33} &= \rho^{-1}(\mathfrak{H}_{331331}(m_1^2 + m_2^2) + \mathfrak{H}_{333333}m_3^2), \quad \mathbb{A}_{16} = \rho^{-1}2\mathfrak{H}_{111122}m_1m_2,\\
\mathbb{A}_{44} &= \rho^{-1}(\mathfrak{H}_{231231}m_1^2 + \mathfrak{H}_{131131}m_2^2 + \mathfrak{H}_{133133}m_3^2), \quad \mathbb{A}_{17} = \rho^{-1}2\mathfrak{H}_{112323}m_2m_3,\\
\mathbb{A}_{55} &= \rho^{-1}(\mathfrak{H}_{311311}m_1^2 + \mathfrak{H}_{312312}m_2^2 + \mathfrak{H}_{313313}m_3^2), \quad \mathbb{A}_{18} = \rho^{-1}2\mathfrak{H}_{111133}m_1m_3,
\end{aligned}
$$

$$
\begin{aligned}
\mathbb{A}_{66} &= \rho^{-1}(\mathfrak{H}_{121121}m_1^2 + \mathfrak{H}_{211211}m_2^2 + \mathfrak{H}_{123123}m_3^2), \quad \mathbb{A}_{19} = \rho^{-1}2\mathfrak{H}_{111212}m_1m_2,\\
\mathbb{A}_{77} &= \rho^{-1}(\mathfrak{H}_{312312}m_1^2 + \mathfrak{H}_{311311}m_2^2 + \mathfrak{H}_{313313}m_3^2), \quad \mathbb{A}_{24} = -\rho^{-1}2\mathfrak{H}_{111133}m_2m_3,\\
\mathbb{A}_{88} &= \rho^{-1}(\mathfrak{H}_{131131}m_1^2 + \mathfrak{H}_{231231}m_2^2 + \mathfrak{H}_{133133}m_3^2), \quad \mathbb{A}_{25} = \rho^{-1}2\mathfrak{H}_{112323}m_1m_3,\\
\mathbb{A}_{99} &= \rho^{-1}(\mathfrak{H}_{211211}m_1^2 + \mathfrak{H}_{212212}m_2^2 + \mathfrak{H}_{123123}m_3^2), \quad \mathbb{A}_{26} = \rho^{-1}2\mathfrak{H}_{112211}m_1m_2,\\
\mathbb{A}_{12} &= \rho^{-1}(\mathfrak{H}_{111221}m_1^2 + \mathfrak{H}_{221111}m_2^2 + \mathfrak{H}_{113223}m_3^2), \quad \mathbb{A}_{27} = -\rho^{-1}2\mathfrak{H}_{111313}m_2m_3,\\
\mathbb{A}_{13} &= \rho^{-1}(\mathfrak{H}_{111331}m_1^2 + \mathfrak{H}_{112332}m_2^2 + \mathfrak{H}_{113333}m_3^2), \quad \mathbb{A}_{28} = \rho^{-1}2\mathfrak{H}_{112233}m_1m_3,\\
\mathbb{A}_{23} &= \rho^{-1}(\mathfrak{H}_{112332}m_1^2 + \mathfrak{H}_{111331}m_2^2 + \mathfrak{H}_{113333}m_3^2), \quad \mathbb{A}_{29} = \rho^{-1}2\mathfrak{H}_{112121}m_1m_2,\\
\mathbb{A}_{47} &= \rho^{-1}(\mathfrak{H}_{132312}m_1^2 + \mathfrak{H}_{131311}m_2^2 + \mathfrak{H}_{133313}m_3^2), \quad \mathbb{A}_{34} = \rho^{-1}2\mathfrak{H}_{331133}m_2m_3,\\
\mathbb{A}_{56} &= \rho^{-1}(2\mathfrak{H}_{311123}m_1m_3 + 2\mathfrak{H}_{312123}m_2m_3),\\
\mathbb{A}_{58} &= \rho^{-1}(\mathfrak{H}_{311131}m_1^2 + \mathfrak{H}_{132312}m_2^2 + \mathfrak{H}_{313133}m_3^2), \quad \mathbb{A}_{36} = \rho^{-1}2\mathfrak{H}_{331122}m_1m_2,\\
\mathbb{A}_{69} &= \rho^{-1}(\mathfrak{H}_{121211}m_1^2 + \mathfrak{H}_{211121}m_2^2 + \mathfrak{H}_{123213}m_3^2), \quad \mathbb{A}_{37} = \rho^{-1}2\mathfrak{H}_{331313}m_2m_3,\\
\mathbb{A}_{78} &= \rho^{-1}(\mathfrak{H}_{311131}m_1^2 + \mathfrak{H}_{312132}m_2^2 + \mathfrak{H}_{313133}m_3^2), \quad \mathbb{A}_{38} = \rho^{-1}2\mathfrak{H}_{331133}m_1m_3,\\
\mathbb{A}_{39} &= \rho^{-1}2\mathfrak{H}_{331212}m_1m_2, \quad \mathbb{A}_{45} = \rho^{-1}2\mathfrak{H}_{231312}m_1m_2,\\
\mathbb{A}_{46} &= \rho^{-1}2\mathfrak{H}_{132213}m_1m_3, \quad \mathbb{A}_{48} = -\rho^{-1}2\mathfrak{H}_{132231}m_1m_2,\\
\mathbb{A}_{49} &= \rho^{-1}2\mathfrak{H}_{132123}m_1m_3, \quad \mathbb{A}_{57} = \rho^{-1}2\mathfrak{H}_{311322}m_1m_2,\\
\mathbb{A}_{59} &= \rho^{-1}2\mathfrak{H}_{312213}m_2m_3, \quad \mathbb{A}_{67} = \rho^{-1}2\mathfrak{H}_{121323}m_1m_3,\\
\mathbb{A}_{68} &= \rho^{-1}2\mathfrak{H}_{122133}m_2m_3, \quad \mathbb{A}_{79} = \rho^{-1}2\mathfrak{H}_{123321}m_2m_3,\\
\mathbb{A}_{89} &= \rho^{-1}2\mathfrak{H}_{132213}m_2m_3.
\end{aligned} \tag{123}
$$

For the direction of propagation $\mathbf{m} = \mathbf{e}_3$ the tensor $\mathbb{A}$ has the same non-null independent components as$\mathbb{B}$, whereas in the case $\mathbf{m}\cdot\mathbf{e}_3 = 0$ there are again 26 independent components.

At a glance, by looking at (104), (105), (108), (112), (114), (117), (120), (122) and (123) for a generic propagation direction $\mathbf{m}$, the matrix A does not simplify enough and the problem maintains the same complexity as for a crystal of the Triclinic group, that is the eigenvectors w are a linear combination of the elements of the space $\mathcal{V} \oplus$ Lin and it is impossible to obtain an explicit representation of the eigenvalues in terms of the compoents of the acoustic tensors.

However, for the two relevant cases of propagation direction either parallel ($\mathbf{m} = \mathbf{e}_3$) or orthogonal ($\mathbf{m}\cdot\mathbf{e}_3 = 0$) to the tetragonal c-axis, many of these acoustic tensors components vanish and also the number of the independent components reduces too: accordingly in many cases the 12th degree algebraic equation which follows from the propagation condition (70) can be factorized into some lower-degree algebraic equations for which an explicit solution can be given and whose associated kinematics can be understood more easily.

We shall study in detail these two propagation direction and we begin with the two limit propagation problems we obtained into §.2.3.1 (long-wavelength approximation) and §. 2.3.2 (microvibrations). First of all we shall deal with the microvibrations problem which is independent on the direction of propagation.

4. Wave propagation in Tetragonal crystals

In this section we shall study the wave propagation condition (70) for crystals belonging to the Tetragonal symmetry group upon the assumption that the direction spanned by the base vector $\mathbf{e}_3$ is directed as the tetragonal c-axis.

As far as we know the propagation problem in Tetragonal materials was previuosly studied only into (d'Agostino et al., 2020): however their analysis, which concerns a relaxed micromorphic model rather then the classical one, was limited to two-dimensional plane strain; their main focus was indeed the parameter identification by means of numerical homogeneization.

4.1 Microvibrations

4.1.1 Classes 4, $\bar{4}$ and 4/m

We begin with the lower-symmetry tetragonal classes: when we consider the propagation condition (86) with $\mathbb{B}$ and $\mathbb{J}$ given respectively by (92) and (95), we notice that both tensors are reduced by the two subspaces of Lin (Halmos, 1987):

$$\mathcal{U}_1 = \mathcal{Z}_1 \oplus \mathcal{Z}_2, \quad \mathcal{U}_2 = \mathcal{Z}_3, \tag{124}$$

where the subspaces $\mathcal{Z}_k$, $k = 1, 2, 3$ are defined by (9). This implies that:

$$\mathbb{B}[\mathbf{C}_\alpha] \in \mathcal{U}_\alpha, \quad \mathbb{J}[\mathbf{C}_\alpha] \in \mathcal{U}_\alpha, \quad \forall\, \mathbf{C}_\alpha \in \mathcal{U}_\alpha, \quad \alpha = 1, 2. \tag{125}$$

The eigencouples split accordingly into two group: a first one $(\omega_k, \mathbf{C}_k)$, $k = 1, \ldots 5$ with $\mathbf{C}_k \in \mathcal{U}_1$ and whose eigentensors are a combination of the modes $\mathbf{D}_1$, $\mathbf{D}_2$, $\mathbf{D}_3$, $\mathbf{S}_{12}$ and $\mathbf{R}_3$; the second one $(\omega_j, \mathbf{C}_j)$, $j = 6, \ldots 9$ with $\mathbf{C}_j \in \mathcal{U}_2$ and whose eigentensors are a combination of the modes $\mathbf{S}_{3\nu}$ and $\mathbf{R}_\omega$.

If we define the normalized components of $\mathbb{B}$ as follow:

$$\begin{aligned}
a &= \frac{\mathbb{B}_{11}}{\rho J_{11}}, \quad b = \frac{\mathbb{B}_{33}}{\rho J_{33}}, \quad d = \frac{\mathbb{B}_{12}}{\rho\sqrt{J_{11}J_{33}}}, \quad e = \frac{\mathbb{B}_{13}}{\rho\sqrt{J_{11}J_{33}}}, \\
c &= \frac{\mathbb{B}_{66}}{\rho J_{11}}, \quad f = \frac{\mathbb{B}_{69}}{\rho J_{11}}, \quad g = \frac{\mathbb{B}_{16}}{\rho J_{11}}, \quad h = \frac{\mathbb{B}_{26}}{\rho J_{11}}, \quad l = \frac{\mathbb{B}_{36}}{\rho\sqrt{J_{11}J_{33}}}, \\
m &= \frac{\mathbb{B}_{44}}{\rho J_{33}}, \quad n = \frac{\mathbb{B}_{55}}{\rho J_{11}}, \quad p = \frac{\mathbb{B}_{45}}{\rho\sqrt{J_{11}J_{33}}}, \quad q = \frac{\mathbb{B}_{47}}{\rho\sqrt{J_{11}J_{33}}},
\end{aligned} \tag{126}$$

and then we begin our analysis with the subspace $\mathcal{U}_1$.

We notice that the algebraic fifth-degree characteristic equation in ω^2 associated to the eigenvalue problem in $\mathcal{U}_1$ can be factorized into

$$(\omega^4 - B\omega^2 + C)(\omega^6 + D\omega^4 + E\omega^2 + F) = 0, \tag{127}$$

where

$$\begin{aligned}
B &= a + c + f - d,\\
C &= (a-d)(c+f) - (g-h)^2,\\
D &= a + b + c + d - f,\\
E &= 2a(b+c) + bc - f(a+b+d) - 2(e^2 + l^2) - (g+h)^2,\\
F &= (c-b)^2 - (g+h)^2 - 2d^2 - 9e^2 + f^2 + 18l^2,\\
&\quad + 3d(b+c) - 2f(2b+c+d) - 3gh - 4bc.
\end{aligned} \tag{128}$$

The eigenvalues are thus given by

$$\omega_{1,2}^2 = \frac{1}{2}(B \mp \sqrt{B^2 - 4C}), \tag{129}$$

with $\omega_1 < \omega_2$ and, by the means of Cardano's formulae (Zwillinger, 2003), by:

$$\begin{aligned}
\omega_3^2 &= \frac{1}{3}\left(D + 2\sqrt{P}\cos\frac{\theta}{3}\right),\\
\omega_4^2 &= \frac{1}{3}\left(D + 2\sqrt{P}\cos\frac{\theta + 2\pi}{3}\right),\\
\omega_5^2 &= \frac{1}{3}\left(D + 2\sqrt{P}\cos\frac{\theta - 2\pi}{3}\right),
\end{aligned} \tag{130}$$

with either $\omega_4 < \omega_3 < \omega_5$ or $\omega_4 > \omega_3 > \omega_5$ and where:

$$\begin{aligned}
&P = D^2 - 3E, \quad Q = D^3 - 9DE - 27F,\\
&\theta = \cos^{-1}\frac{Q}{2\sqrt[3]{P}}.
\end{aligned} \tag{131}$$

The associated eigentensors combine the modes $\mathbf{D}_1$, $\mathbf{S}_{12}$, $\mathbf{R}_3$ in all but one case, when it combines $\mathbf{D}_2$ and $\mathbf{R}_3$:

$$\begin{aligned}
\mathbf{C}_1 &= \alpha_1\mathbf{W}_1 + \beta_1\mathbf{W}_2 + \gamma_1\mathbf{W}_3 + \delta_1\hat{\mathbf{W}}_6 + \epsilon_1\bar{\mathbf{W}}_6,\\
\mathbf{C}_2 &= \beta_1\mathbf{W}_1 + \alpha_1\mathbf{W}_2 + \gamma_1\mathbf{W}_3 - \delta_1\hat{\mathbf{W}}_6 + \epsilon_1\bar{\mathbf{W}}_6,\\
\mathbf{C}_3 &= \alpha_3(\mathbf{I} - \mathbf{W}_3) + \gamma_3\mathbf{W}_3 + 2\delta_3\bar{\mathbf{W}}_6,\\
\mathbf{C}_4 &= \alpha_4\mathbf{W}_1 + \beta_4\mathbf{W}_2 + \gamma_4\mathbf{W}_3 + \delta_4\hat{\mathbf{W}}_6 + \epsilon_4\bar{\mathbf{W}}_6,\\
\mathbf{C}_5 &= \beta_4\mathbf{W}_1 + \alpha_4\mathbf{W}_2 + \gamma_4\mathbf{W}_3 - \delta_4\hat{\mathbf{W}}_6 + \epsilon_4\bar{\mathbf{W}}_6;
\end{aligned} \tag{132}$$

here the real coefficients α_k, β_k, γ_k, δ_k and ϵ_k, which depend on the components of $\mathbb{B}$, are given explicitly into the Appendix, equations (205).

When we turn our attention to the problem in the subspace $\mathcal{U}_2$, then we obtain two eigenvalues of multiplicity 2:

$$\begin{aligned}\omega_6^2 &= \omega_7^2 = \frac{1}{2}(m+n) - \sqrt{\left(\frac{m-n}{2}\right)^2 + p^2 + q^2},\\ \omega_8^2 &= \omega_9^2 = \frac{1}{2}(m+n) + \sqrt{\left(\frac{m-n}{2}\right)^2 + p^2 + q^2},\end{aligned} \tag{133}$$

whose (non-normalized) eigentensors are

$$\begin{aligned}\mathbf{C}_6 &= p(\hat{\mathbf{W}}_4 - \bar{\mathbf{W}}_4) + (n - q - \omega_6^2)\hat{\mathbf{W}}_5\\ &\quad - (n + q - \omega_6^2)\bar{\mathbf{W}}_5,\\ \mathbf{C}_7 &= (m - q - \omega_6^2)\hat{\mathbf{W}}_4 - (m + q - \omega_6^2)\bar{\mathbf{W}}_4\\ &\quad + p(\hat{\mathbf{W}}_5 - \bar{\mathbf{W}}_5),\\ \mathbf{C}_8 &= -p\mathbf{W}_4 + (m - q - \omega_8^2)\hat{\mathbf{W}}_5 + (m + q - \omega_8^2)\bar{\mathbf{W}}_5,\\ \mathbf{C}_9 &= (n - q - \omega_8^2)\hat{\mathbf{W}}_4 + (n + q - \omega_8^2)\bar{\mathbf{W}}_4 - p\mathbf{W}_5,\end{aligned} \tag{134}$$

each one representing a combination of the modes $\mathbf{S}_{3\nu}$ and $\mathbf{R}_\omega$; from (133) we have

$$\omega_6 = \omega_7 < \omega_8 = \omega_9. \tag{135}$$

All together we have

$$\begin{aligned}&\omega_1 < \omega_2, \quad \omega_4 < \omega_3 < \omega_5, \quad (\text{or } \omega_4 > \omega_3 > \omega_5),\\ &\omega_6 = \omega_7 < \omega_8 = \omega_9,\end{aligned} \tag{136}$$

and besides this we cannot give a complete ordering between these frequencies without the knowledge of the numerical values of components of $\mathbb{B}$.

4.1.2 Classes 4mm, 422, $\bar{4}2m$ and 4/mmm

For these classes the matrix $\mathbb{B}$ is reduced by the three subspaces $\mathcal{Z}_1$, $\mathcal{Z}_2$ and $\mathcal{Z}_3$; in order to find the solution in $\mathcal{Z}_1$ we notice that since $\omega^2 = a - d$ is a root for the cubic characteristic equation, then we can easily obtain:

$$\omega_1^2 = a - d,$$
$$\omega_2^2 = \frac{a+b+d}{2} - \sqrt{\left(\frac{a+d-b}{2}\right)^2 + 2e^2},$$
$$\omega_3^2 = \frac{a+b+d}{2} + \sqrt{\left(\frac{a+d-b}{2}\right)^2 + 2e^2}, \tag{137}$$

with

$$\omega_2 < \omega_3, \quad \omega_1 < \omega_3. \tag{138}$$

The corresponding (non-normalized) eigentensors are:

$$\mathbf{C}_1 = \mathbf{W}_1 - \mathbf{W}_2,$$
$$\mathbf{C}_2 = \alpha_2 \mathbf{W}_1 + \beta_2 \mathbf{W}_2 + \gamma_2 \mathbf{W}_3,$$
$$\mathbf{C}_3 = \beta_2 \mathbf{W}_1 + \alpha_2 \mathbf{W}_2 + \gamma_2 \mathbf{W}_3, \tag{139}$$

with:

$$\alpha_2 = (b - \omega_2^2)d - e^2, \quad \beta_2 = e^2 - b(a - \omega_2^2),$$
$$\gamma_2 = e(a - d - \omega_2^2). \tag{140}$$

In the subspace $\mathcal{Z}_2$ we have two eigencouples associated a shear in the plane orthogonal to the c-axis ($\mathbf{C}_4$) and a rotation about the propagation direction ($\mathbf{C}_5$):

$$(\omega_4^2 = c - f, \ \mathbf{C}_4 = \bar{\mathbf{W}}_6),$$
$$(\omega_5^2 = c + f, \ \mathbf{C}_5 = \hat{\mathbf{W}}_6), \tag{141}$$

whereas the solutions on $\mathcal{Z}_3$ are given by (133) and (134) with $p = 0$; again we have

$$\omega_6 = \omega_7 < \omega_8 = \omega_9. \tag{142}$$

and for $\mathbb{B}_{69} > 0$ also

$$\omega_4 < \omega_5, \tag{143}$$

the inequality being reversed when $\mathbb{B}_{69} < 0$.

The kinematics of microdistortions for these classes is represents in $\mathcal{Z}_1$ by either $\mathbf{D}_1$, or the traceless plane strain $\mathbf{D}_3$; in $\mathcal{Z}_2$ we have instead a shear in the plane orthogonal to the c-axis and a rigid microrotation about the same direction whereas in $\mathcal{Z}_3$ we have the same kinematics as in $\mathcal{U}_2$.

Also for these classes it is not possible to give a complete ordering between all the nine eigenvalues (138), (142) and (143), without the knowledge of the numerical values for the components of $\mathbb{B}$.

4.2 Long-wavelength approximation

4.2.1 Propagation along the c-axis

For the propagation direction $\mathbf{m} = \mathbf{e}_3$ (*i.e.* along the tetragonal *c*-axis: henceforth we shall use *c* and $\mathbf{e}_3$ as synonimus when we describe the material symmetry) the tensor (107) reduces for all classes to the isotropic-like representation:

$$\hat{\mathbf{A}}(\mathbf{e}_3) = \frac{1}{\rho}(\hat{\mathbb{C}}_{44}(\mathbf{I} - \mathbf{W}_3) + \hat{\mathbb{C}}_{33}\mathbf{W}_3), \tag{144}$$

and we have have a longitudinal and two transverse acoustic waves with frequencies:

$$\begin{aligned} \omega_1(\xi) = \omega_2(\xi) = \xi\sqrt{\frac{\hat{\mathbb{C}}_{44}}{\rho}}, \quad & \mathbf{a}_1 = \cos\beta\mathbf{e}_1 + \sin\beta\mathbf{e}_2, \\ & \mathbf{a}_2 = -\sin\beta\mathbf{e}_1 + \cos\beta\mathbf{e}_2, \\ \omega_3(\xi) = \xi\sqrt{\frac{\hat{\mathbb{C}}_{33}}{\rho}}, \quad & \mathbf{a}_3 = \mathbf{e}_3, \end{aligned} \tag{145}$$

By (81), these macroscopic displacements are accompained by a microdistortion associated with the longitudinal wave of frequency $(145)_1$

$$\mathbf{C}_3 = iL_m\xi\mathbb{B}^{-1}\mathbf{Q}^T(\mathbf{e}_3)\mathbf{e}_3. \tag{146}$$

For the classes 4, $\bar{4}$ and $4/m$, the tensor $\mathrm{Q}(\mathbf{e}_3)$ has the tabular representation (110) with the non-null components given by (108) and since $\mathbb{B}^{-1}$ has the same non-null components of (92), then the tensor $\mathbf{C}_3$ is represented by

$$\mathbf{C}_3 = \alpha_3(\mathbf{I} - \mathbf{W}_3) + \beta_3\mathbf{W}_3 + \gamma_3\bar{\mathbf{W}}_6, \tag{147}$$

where:

$$\begin{aligned} \alpha_3 &= \mathrm{Q}_{311}(\mathbb{B}_{11}^{-1} + \mathbb{B}_{12}^{-1}) + \mathrm{Q}_{333}\mathbb{B}_{13}^{-1} + \mathrm{Q}_{312}(\mathbb{B}_{16}^{-1} - \mathbb{B}_{26}^{-1}), \\ \beta_3 &= \mathrm{Q}_{333}\mathbb{B}_{33}^{-1} + 2\mathrm{Q}_{311}\mathbb{B}_{13}^{-1} + 2\mathrm{Q}_{312}\mathbb{B}_{36}^{-1}, \\ \gamma_3 &= \mathrm{Q}_{311}(\mathbb{B}_{16}^{-1} + \mathbb{B}_{26}^{-1}) + \mathrm{Q}_{333}\mathbb{B}_{36}^{-1} + \mathrm{Q}_{312}(\mathbb{B}_{66}^{-1} - \mathbb{B}_{69}^{-1}), \end{aligned} \tag{148}$$

and the components Q_{311}, Q_{333} and Q_{312} are obtained from (108) evaluted for $m_1 = m_2 = 0$ and $m_3 = 1$.

The microdistortions associated to the longitudinal waves are therefore a combination of the modes $\mathbf{D}_2$ and $\mathbf{R}_3$.

The microdistortions accompanied to the transverse waves which propagates along $\mathbf{m} = \mathbf{e}_3$ are instead given by

$$\mathbf{C}_k = iL_m\xi\mathbb{B}^{-1}\mathbf{Q}^T(\mathbf{e}_3)\,\mathbf{a}_k, \quad k = 1, 2, \tag{149}$$

which can be represented as

$$\mathbf{C}_k = \alpha_k\hat{\mathbf{W}}_4 + \beta_k\hat{\mathbf{W}}_5 + \gamma_k\bar{\mathbf{W}}_4 + \delta_k\bar{\mathbf{W}}_5, \quad k = 1, 2, \tag{150}$$

with, for instance, in the case $k = 1$

$$\begin{aligned}
2\alpha_1 &= (\mathrm{Q}_{113}\cos\beta - \mathrm{Q}_{123}\sin\beta)(\mathbb{B}^{-1}_{44} + \mathbb{B}^{-1}_{47}) \\
&\quad + (\mathrm{Q}_{132}\cos\beta - \mathrm{Q}_{131}\sin\beta)(\mathbb{B}^{-1}_{55} + \mathbb{B}^{-1}_{47}) \\
&\quad + \mathbb{B}^{-1}_{45}((\mathrm{Q}_{131} - \mathrm{Q}_{123})\cos\beta - (\mathrm{Q}_{132} + \mathrm{Q}_{113})\sin\beta), \\
2\beta_1 &= \cos\beta(\mathrm{Q}_{131} - \mathrm{Q}_{132})(\mathbb{B}^{-1}_{55} + \mathbb{B}^{-1}_{47}) \\
&\quad + (\mathrm{Q}_{123}\cos\beta + \mathrm{Q}_{113}\sin\beta)(\mathbb{B}^{-1}_{44} + \mathbb{B}^{-1}_{47}) \\
&\quad + \mathbb{B}^{-1}_{45}((\mathrm{Q}_{113} + \mathrm{Q}_{131})\cos\beta - (\mathrm{Q}_{123} + \mathrm{Q}_{132})\sin\beta), \\
2\gamma_1 &= \cos\beta(\mathrm{Q}_{131} - \mathrm{Q}_{132})(\mathbb{B}^{-1}_{55} - \mathbb{B}^{-1}_{47}) \\
&\quad + (\mathrm{Q}_{123}\cos\beta + \mathrm{Q}_{113}\sin\beta)(\mathbb{B}^{-1}_{44} - \mathbb{B}^{-1}_{47}) \\
&\quad + \mathbb{B}^{-1}_{45}((\mathrm{Q}_{131} - \mathrm{Q}_{113})\cos\beta + (\mathrm{Q}_{123} - \mathrm{Q}_{132})\sin\beta), \\
2\delta_1 &= (\mathrm{Q}_{113}\cos\beta - \mathrm{Q}_{123}\sin\beta)(\mathbb{B}^{-1}_{44} - \mathbb{B}^{-1}_{47}) \\
&\quad + (\mathrm{Q}_{132}\cos\beta - \mathrm{Q}_{131}\sin\beta)(\mathbb{B}^{-1}_{55} - \mathbb{B}^{-1}_{47}) \\
&\quad + \mathbb{B}^{-1}_{45}((\mathrm{Q}_{131} + \mathrm{Q}_{123})\cos\beta - (\mathrm{Q}_{132} - \mathrm{Q}_{113})\sin\beta),
\end{aligned}$$

where Q_{123}, Q_{131}, Q_{132} and Q_{113} are given by (108) with $m_3 = 1$ and $m_1 = m_2 = 0$.

Therefore each transverse wave $\mathbf{a}_k$ which propagates along $\mathbf{m} = \mathbf{e}_3$ generates a combination of the modes $\mathbf{S}_{3v}$ and $\mathbf{R}_\omega$.

When we deal with the classes $4mm$, 422, $4/mmm$, $\bar{4}2m$ then $\mathbb{B}^{-1}$ has the same non-null components of (92) and we also have

$$\mathrm{Q}_{312} = \mathrm{Q}_{123} = \mathrm{Q}_{132} = 0; \tag{151}$$

the only difference with the results for the lower-symmetry classes is that $\gamma_3 = 0$ into (148) and hence $\mathbf{C}_3$ reduces to the mode $\mathbf{D}_2$.

4.2.2 Propagation orthogonal to the c-axis

Whenever the propagation direction is orthogonal to the c-axis, say $\mathbf{m} = \cos\theta\mathbf{e}_1 + \sin\theta\mathbf{e}_2$ then we have, for all θ, a transverse wave which is directed as the c-axis:

$$\omega_1(\xi) = \xi\sqrt{\frac{\hat{\mathbb{C}}_{44}}{\rho}}, \quad \mathbf{a}_1 = \mathbf{e}_3 \tag{152}$$

and two waves which in general are neither transverse nor longitudinal:

$$\begin{aligned}\omega_{2,3}(\xi) &= \xi\sqrt{\frac{1}{2\rho}(a \pm \sqrt{b\cos^2 2\theta + c\sin^2 2\theta + 2d\sin 2\theta\cos 2\theta})},\\ \mathbf{a}_2 &= \cos\beta\mathbf{e}_1 + \sin\beta\mathbf{e}_2,\\ \mathbf{a}_3 &= -\sin\beta\mathbf{e}_1 + \cos\beta\mathbf{e}_2,\end{aligned} \tag{153}$$

with

$$\tan\beta = \frac{a - \sqrt{b\cos^2 2\theta + c\sin^2 2\theta + 2d\sin 2\theta\cos 2\theta}}{2\hat{\mathbb{C}}_{16}\cos 2\theta + (\hat{\mathbb{C}}_{66} + 2\hat{\mathbb{C}}_{12})\sin 2\theta}, \tag{154}$$

where, for the classes 4, $\bar{4}$ and $4/m$:

$$\begin{aligned}a &= \hat{\mathbb{C}}_{11} + \hat{\mathbb{C}}_{66}, \quad b = (\hat{\mathbb{C}}_{11} - \hat{\mathbb{C}}_{66})^2 + 4\hat{\mathbb{C}}_{16}^2,\\ c &= \hat{\mathbb{C}}_{16}^2 + (\hat{\mathbb{C}}_{66} + 2\hat{\mathbb{C}}_{12})^2, \; d = \hat{\mathbb{C}}_{16}(\hat{\mathbb{C}}_{11} + \hat{\mathbb{C}}_{66} + 4\hat{\mathbb{C}}_{12}).\end{aligned} \tag{155}$$

We may search for the angle θ such that $\mathbf{a}_2$ is a longitudinal wave with frequency ω_2 and $\mathbf{a}_3$ is a transverse wave with frequency ω_3, which is equivalent to require either that $\mathbf{a}_2 \times \mathbf{m} = \mathbf{0}$ and hence $\theta = \beta$, or that $A_{12} = 0$ which gives

$$\tan 2\theta = -\frac{2\hat{\mathbb{C}}_{16}}{\hat{\mathbb{C}}_{12} + 2\hat{\mathbb{C}}_{66}}. \tag{156}$$

For the classes $4mm$, 422, $4/mmm$ and $\bar{4}2m$ with $\hat{\mathbb{C}}_{16} = 0$ then

$$b = (\hat{\mathbb{C}}_{11} - \hat{\mathbb{C}}_{66})^2, \quad c = (\hat{\mathbb{C}}_{66} + 2\hat{\mathbb{C}}_{12})^2, \quad d = 0, \tag{157}$$

and then from (156) and (154) with $\beta = \theta$ it is easy to see that for $\theta \in \{0, \pi/4, \pi/2\}$ we have a longitudinal ($\mathbf{a}_2 = \mathbf{m}$) and a transverse ($\mathbf{a}_3 = \mathbf{e}_3 \times \mathbf{m}$) waves whose frequencies are given by $(153)_1$ when we use (157) in place of (155).

The microdistortion which corresponds to the transverse wave (152) is:

$$\mathbf{C}_1 = iL_m\xi\mathbb{B}^{-1}\mathbf{Q}^T(\theta)\mathbf{e}_3, \tag{158}$$

and by (110) and (92) we obtain again (150) with for instance, when $k = 1$:

$$
\begin{aligned}
2\alpha_1 &= \mathsf{Q}_{323}\,(\mathbb{B}^{-1}_{44} + \mathbb{B}^{-1}_{47}) + \mathsf{Q}_{332}\,(\mathbb{B}^{-1}_{47} - \mathbb{B}^{-1}_{45}) \\
&\quad + \mathsf{Q}_{331}\mathbb{B}^{-1}_{45} + \mathsf{Q}_{313}\mathbb{B}^{-1}_{55}, \\
2\beta_1 &= \mathsf{Q}_{331}\,(\mathbb{B}^{-1}_{47} + \mathbb{B}^{-1}_{55}) + \mathsf{Q}_{313}\,(\mathbb{B}^{-1}_{44} + \mathbb{B}^{-1}_{47}) \\
&\quad - \mathsf{Q}_{332}\mathbb{B}^{-1}_{45} + \mathsf{Q}_{323}\mathbb{B}^{-1}_{47}, \\
2\gamma_1 &= \mathsf{Q}_{323}\,(\mathbb{B}^{-1}_{44} - \mathbb{B}^{-1}_{47}) + \mathsf{Q}_{332}\,(\mathbb{B}^{-1}_{47} + \mathbb{B}^{-1}_{45}) \\
&\quad + \mathsf{Q}_{331}\mathbb{B}^{-1}_{45} - \mathsf{Q}_{313}\mathbb{B}^{-1}_{55}, \\
2\delta_1 &= \mathsf{Q}_{331}\,(\mathbb{B}^{-1}_{47} - \mathbb{B}^{-1}_{55}) + \mathsf{Q}_{313}\,(\mathbb{B}^{-1}_{44} - \mathbb{B}^{-1}_{47}) \\
&\quad - \mathsf{Q}_{332}\mathbb{B}^{-1}_{45} - \mathsf{Q}_{323}\mathbb{B}^{-1}_{47};
\end{aligned} \tag{159}
$$

here the components $\mathsf{Q}_{33\alpha}(\theta)$ and $\mathsf{Q}_{3\alpha 3}(\theta)$, $\alpha = 1, 2$ are given by (108) with $m_1 = \cos\theta$, $m_2 = \sin\theta$ and $m_3 = 0$.

Accordingly this transverse wave is associated to a shear between $\mathbf{m}$ and $\mathbf{e}_3$ and a rigid rotation about $\mathbf{e}_3 \times \mathbf{m}$.

When we turn our attention to the other two waves, which are neither transverse nor longitudinal, we have that the two associated microdistortions

$$
\mathbf{C}_\gamma = iL_m\xi\mathbb{B}^{-1}\mathbf{Q}^T(\theta)\,\mathbf{a}_\gamma, \quad \gamma = 2, 3, \tag{160}
$$

in view of (92) and (110) can be represented as:

$$
\begin{aligned}
\mathbf{C}_2 &= iL_m\xi\,(\cos\beta\;\mathbf{B}_1 + \sin\beta\;\mathbf{B}_2), \\
\mathbf{C}_3 &= iL_m\xi\,(-\sin\beta\;\mathbf{B}_1 + \cos\beta\;\mathbf{B}_2);
\end{aligned} \tag{161}
$$

the two tensors $\mathbf{B}_k = \mathbb{B}^{-1}\mathbf{Q}^T(\theta)\,\mathbf{e}_k$, $k = 1, 2$ have the common structure

$$
\mathbf{B}_k = \alpha_k\mathbf{W}_1 + \beta_k\mathbf{W}_2 + \gamma_k\mathbf{W}_3 + \delta_k\hat{\mathbf{W}}_6 + \epsilon_k\bar{\mathbf{W}}_6, \tag{162}
$$

where

$$
\begin{aligned}
\alpha_k &= \mathbb{B}^{-1}_{11}\mathsf{Q}_{k11} + \mathbb{B}^{-1}_{12}\mathsf{Q}_{k22} + \mathbb{B}^{-1}_{13}\mathsf{Q}_{k33} + \mathbb{B}^{-1}_{16}\mathsf{Q}_{k12} + \mathbb{B}^{-1}_{26}\mathsf{Q}_{k21}, \\
\beta_k &= \mathbb{B}^{-1}_{12}\mathsf{Q}_{k11} + \mathbb{B}^{-1}_{11}\mathsf{Q}_{k22} + \mathbb{B}^{-1}_{13}\mathsf{Q}_{k33} + \mathbb{B}^{-1}_{26}\mathsf{Q}_{k12} + \mathbb{B}^{-1}_{16}\mathsf{Q}_{k21}, \\
\gamma_k &= \mathbb{B}^{-1}_{13}\,(\mathsf{Q}_{k11} + \mathsf{Q}_{k22}) + \mathbb{B}^{-1}_{33}\mathsf{Q}_{k33} + \mathbb{B}^{-1}_{36}\,(\mathsf{Q}_{k12} - \mathsf{Q}_{k21}), \quad k = 1, 2, \\
\delta_k &= \frac{1}{2}(\mathbb{B}^{-1}_{16} - \mathbb{B}^{-1}_{26})\,(\mathsf{Q}_{k11} + \mathsf{Q}_{k22}) + \frac{1}{2}(\mathbb{B}^{-1}_{66} - \mathbb{B}^{-1}_{69})\,(\mathsf{Q}_{k12} + \mathsf{Q}_{k21}), \\
\epsilon_k &= \frac{1}{2}(\mathbb{B}^{-1}_{16} + \mathbb{B}^{-1}_{26})\,(\mathsf{Q}_{k11} + \mathsf{Q}_{k22}) + \frac{1}{2}(\mathbb{B}^{-1}_{66} - \mathbb{B}^{-1}_{69})\,(\mathsf{Q}_{k12} - \mathsf{Q}_{k21}) \\
&\quad + \mathbb{B}^{-1}_{36}\mathsf{Q}_{k33};
\end{aligned} \tag{163}
$$

Table 1 Modes in the long-wavelength (LW) and microvibration (M) approximations.

	m	$\mathbf{e}_1$	$\mathbf{e}_2$	$\mathbf{e}_3$	$\mathbf{D}_1$	$\mathbf{D}_2$	$\mathbf{D}_3$	$\mathbf{S}_{12}$	$\mathbf{S}_{3v}$	$\mathbf{R}_3$	$\mathbf{R}_\omega$
		·	·	$\Re_L$	·	$\Im$	·	·	·	$\Im$†	·
LW	$\Vert\mathbf{e}_3$	$\Re_T$	·	·	·	·	·	·	$\Im$	·	$\Im$
		·	$\Re_T$	·	·	·	·	·	$\Im$	·	$\Im$
		·	·	$\Re_T$	·	·	·	·	$\Im$	·	$\Im$
LW	$\perp\mathbf{e}_3$	$\Re$	·	·	[$\Im$]	($\Im$)	·	$\Im$	·	$\Im$	·
		·	$\Re$	·	[$\Im$]	($\Im$)	·	$\Im$	·	$\Im$	·
	(2)	·	·	·	$\Re$	·	·	$\Re$∘	·	$\Re$•	·
	(2)	·	·	·	$\Re$	·	·	$\Re$∘	·	$\Re$•	·
M		·	·	·	·	$\Re$	·	·	·	$\Re$	·
LS	(2), ♭	·	·	·	·	·	·	·	$\Re$	·	$\Re$
	(2), ♭	·	·	·	·	·	·	·	$\Re$	·	$\Re$
		·	·	·	·	·	$\Re$	·	·	·	·
	(2)	·	·	·	$\Re$	·	·	·	·	·	·
M		·	·	·	·	·	·	$\Re$	·	·	·
HS		·	·	·	·	·	·	·	·	$\Re$	·
	(2), ♭	·	·	·	·	·	·	·	$\Re$	·	$\Re$
	(2), ♭	·	·	·	·	·	·	·	$\Re$	·	$\Re$

HS = classes $4mm$, 422, $4/mmm$, $\bar{4}2m$, LS = classes 4, $\bar{4}$, $4/m$; $\Re$ = Real component, $\Im$ = Immaginary component; L = longitudinal wave, T = transverse wave; • = equal values components; ∘ = equal value, opposite sign components; (#) = number of independent eigentensors with same structure; ♭ = multiple eigenvalues; † vanishes for HS classes; [$\Im$] = only LS classes; ($\Im$) = only HS classes.

here the components of $Q(\theta)$ are obtained again from (108) with $m_1 = \cos\theta$, $m_2 = \sin\theta$ and $m_3 = 0$.

Interestingly enough for both the waves with amplitudes $\mathbf{a}_2$ and $\mathbf{a}_3$, the corresponding microdistortions are a combination of the modes $\mathbf{D}_1$, $\mathbf{S}_{12}$

and $\mathbf{R}_3$. We remark that this situation is maintained even when the propagation direction is given by (156) and the two waves becomes one transverse and the other longitudinal.

We finish by looking at the higher-symmetry classes $4mm$, 422, $4/mmm$, $\bar{4}2m$: for these classes, by (112) we have that formula (162) holds with $\alpha_k = \beta_k$ and $\delta_k = \epsilon_k$ and therefore the mode $\mathbf{D}_1$ changes into $\mathbf{D}_2$.

The results obtained in §§.4.1 and 4.2 for the two limit problems are represented in Table 1: it is remarkable that in the long-wavelength approximation the macroscopic displacements are real whereas the lattice microdistortions are purely immaginary; conversely, in the microvibration problem the matrix A reduces to a real one and the lattice microdistortions are real. In both cases we find that the lattice modes can be described by the means of three kind of dilatations, two shear deformations and two rigid rotations.

4.3 Micromorphic continua

4.3.1 Propagation along the tetragonal c-axis

4.3.1.1 Classes 4, $\bar{4}$ and $4/m$

We begin with the lower-symmetry classes; in this case the blocks of the matrix A have the following non-null components

$$A \equiv \begin{bmatrix} \bullet & \bullet & \cdot & \cdot & \cdot & \cdot & \bullet & \bullet & \cdot & \bullet & \bullet & \cdot \\ \bullet & \bullet & \cdot & \cdot & \cdot & \cdot & \bullet & \bullet & \cdot & \bullet & \bullet & \cdot \\ \cdot & \cdot & \bullet & \bullet & \bullet & \bullet & \cdot & \cdot & \bullet & \cdot & \cdot & \bullet \\ \cdot & \cdot & \bullet & \bullet & \bullet & \bullet & \cdot & \cdot & \bullet & \cdot & \cdot & \bullet \\ \cdot & \cdot & \bullet & \bullet & \bullet & \bullet & \cdot & \cdot & \bullet & \cdot & \cdot & \bullet \\ \cdot & \cdot & \bullet & \bullet & \bullet & \bullet & \cdot & \cdot & \bullet & \cdot & \cdot & \bullet \\ \bullet & \bullet & \cdot & \cdot & \cdot & \cdot & \bullet & \bullet & \cdot & \bullet & \bullet & \cdot \\ \bullet & \bullet & \cdot & \cdot & \cdot & \cdot & \bullet & \bullet & \cdot & \bullet & \bullet & \cdot \\ \cdot & \cdot & \bullet & \bullet & \bullet & \bullet & \cdot & \cdot & \bullet & \cdot & \cdot & \bullet \\ \bullet & \bullet & \cdot & \cdot & \cdot & \cdot & \bullet & \bullet & \cdot & \bullet & \bullet & \cdot \\ \bullet & \bullet & \cdot & \cdot & \cdot & \cdot & \bullet & \bullet & \cdot & \bullet & \bullet & \cdot \\ \cdot & \cdot & \bullet & \bullet & \bullet & \bullet & \cdot & \cdot & \bullet & \cdot & \cdot & \bullet \end{bmatrix}, \tag{164}$$

with the independent components given by (104), (114), (108) and (122) evaluated for $m_1 = m_2 = 0$ and $m_3 = 1$ (the line/column ordering in the matrix A is 1, 2, 3|1, 2, 3, 4, 5, 6, 7, 8, 9).

The matrix (164) is reduced by the pairs $\mathcal{M}_1$ and $\mathcal{M}_2$:

$$\mathcal{M}_1 \equiv \mathcal{V}_\parallel \oplus \mathcal{U}_1, \quad \mathcal{M}_2 \equiv \mathcal{V}_\perp \oplus \mathcal{U}_2, \tag{165}$$

were $\mathcal{U}_1$ and $\mathcal{U}_2$ are defined by (124).

We begin with the subspace $\mathcal{M}_1$ and define the normalized components:

$$
\begin{aligned}
a(\xi) &= \xi^2 A_{33}, \\
b(\xi) &= \xi^2 \frac{\mathsf{P}_{311} L_c^2}{L_m} + i\xi \mathbf{Q}_{311}, \quad c(\xi) = \xi^2 \frac{\mathsf{P}_{333} L_c^2}{L_m} + i\xi \mathbf{Q}_{333}, \\
d(\xi) &= \xi^2 \frac{\mathbb{A}_{66} L_c^2}{J_{11}} + \frac{\mathbb{B}_{66}}{\rho J_{11}}, \quad e(\xi) = \xi^2 \frac{\mathbb{A}_{69} L_c^2}{J_{11}} + \frac{\mathbb{B}_{69}}{\rho J_{11}}, \\
f(\xi) &= \xi^2 \frac{\mathbb{A}_{11} L_c^2}{J_{11}} + \frac{\mathbb{B}_{11}}{\rho J_{11}}, \quad g(\xi) = \xi^2 \frac{\mathbb{A}_{33} L_c^2}{J_{33}} + \frac{\mathbb{B}_{33}}{\rho J_{33}}, \\
h(\xi) &= \xi^2 \frac{\mathbb{A}_{12} L_c^2}{\sqrt{J_{11} J_{33}}} + \frac{\mathbb{B}_{12}}{\rho \sqrt{J_{11} J_{33}}}, \quad l(\xi) = \xi^2 \frac{\mathbb{A}_{13} L_c^2}{\sqrt{J_{11} J_{33}}} + \frac{\mathbb{B}_{13}}{\rho \sqrt{J_{11} J_{33}}}, \\
n(\xi) &= \xi^2 \frac{\mathbb{A}_{16} L_c^2}{J_{11}} + \frac{\mathbb{B}_{16}}{\rho J_{11}}, \quad p(\xi) = \xi^2 \frac{\mathbb{A}_{26} L_c^2}{J_{11}} + \frac{\mathbb{B}_{26}}{\rho J_{11}}, \\
m(\xi) &= \xi^2 \frac{\mathbb{A}_{36} L_c^2}{\sqrt{J_{11} J_{33}}} + \frac{\mathbb{B}_{36}}{\rho \sqrt{J_{11} J_{33}}}, \quad q(\xi) = \xi^2 \frac{\mathsf{P}_{312} L_c^2}{L_m} + i\xi \mathbf{Q}_{312};
\end{aligned}
\tag{166}
$$

then, since the characteristic equation can be factorized into a second- and a fourth-degree algebraic equations, we obtain by the means of the Cardano's formulae for the fourth-degree algebraic equations (Zwillinger, 2003), the explicit representation of the six eigenvalues

$$
\begin{aligned}
\omega_1^2 &= \frac{1}{2}\left(F - \sqrt{F^2 - 4G}\right), \\
\omega_2^2 &= \frac{1}{2}\left(F + \sqrt{F^2 - 4G}\right), \\
\omega_3^2 &= \frac{1}{4}A - \frac{1}{2}\sqrt{\Omega + W} - \frac{1}{2}\sqrt{2\Omega - W - \frac{S}{4\sqrt{\Omega + W}}}, \\
\omega_4^2 &= \frac{1}{4}A - \frac{1}{2}\sqrt{\Omega + W} + \frac{1}{2}\sqrt{2\Omega - W - \frac{S}{4\sqrt{\Omega + W}}}, \\
\omega_5^2 &= \frac{1}{4}A + \frac{1}{2}\sqrt{\Omega + W} - \frac{1}{2}\sqrt{2\Omega - W + \frac{S}{4\sqrt{\Omega + W}}}, \\
\omega_6^2 &= \frac{1}{4}A + \frac{1}{2}\sqrt{\Omega + W} + \frac{1}{2}\sqrt{2\Omega - W + \frac{S}{4\sqrt{\Omega + W}}},
\end{aligned}
\tag{167}
$$

where:

$$
\Omega = \frac{1}{4}A^2 + \frac{2}{3}B,
\tag{168}
$$

and

$$
\begin{aligned}
A &= a + d - e + f + g + h,\\
B &= c\bar{c} + 2(b\bar{b} + l^2 + m^2 + qq^*) + (n + p)^2\\
&\quad - (d - e)(a + f + g + h)\\
&\quad - (f + h)(a + g) - ag,\\
C &= 2b\bar{b}(d - e + g) + 2l^2(a + d - e) + 2m^2(a + f + h)\\
&\quad + 2qq^*(f + g + h)\\
&\quad + (a + g)((n + p)^2 - (f + h)(d - e))\\
&\quad + (c\bar{c} - ag)(d - e + f + h)\\
&\quad - 4((lm + q\bar{b})(n + p) + (lb + mq)\bar{c}),\\
D &= 2b\bar{b}(m^2 - g(d - e)) + 2qq^*(2l^2 - g(f + h))\\
&\quad + 2l(2b\bar{c} - al)(d - e)\\
&\quad + 2m(2cq^* - am)(f + h) + ((n + p)^2\\
&\quad - (f + h)(d - e))(c\bar{c} - ag)\\
&\quad + 4(n + p)(m(c\bar{b} + al) + q(g\bar{b} - \bar{c}l)) - 8lmq\bar{b}\\
F &= d + e + f - h,\\
G &= (d + e)(f - h) - (n - p)^2,
\end{aligned}
\tag{169}
$$

(here $\bar{b}$ and $\bar{c}$ represent the complex conjugates of b and c) and

$$
\begin{aligned}
P &= -2B^3 - 9ACB + 72DB + 27C^2 + 27A^2D,\\
Q &= B^2 + 3AC + 12D,\\
S &= A^3 + 4AB - 8C,\\
W &= \frac{1}{3}\left(\sqrt[3]{\frac{P + \sqrt{P^2 - 4Q^3}}{2}} + Q\frac{1}{\sqrt[3]{\frac{P + \sqrt{P^2 - 4Q^3}}{2}}}\right).
\end{aligned}
$$

By looking at (167), by (166) and (169) we notice that the frequencies $\omega_{1,2}$ depend solely on the components of $\mathbb{A}$ and $\mathbb{B}$, whereas the others depend also on the acoustic tensors **A**, P and Q.

The associated eigenvectors:

$$
\begin{aligned}
w_1 &= \{\mathbf{a}_o^1 = \alpha_1\mathbf{e}_3;\ \mathbf{V}_1 = \beta_1\mathbf{W}_1 + \gamma_1\mathbf{W}_2 + \delta_1\mathbf{W}_3\\
&\qquad + (\theta_1 + \epsilon_1)\hat{\mathbf{W}}_6 + (\epsilon_1 - \theta_1)\bar{\mathbf{W}}_6\},\\
w_2 &= \{\mathbf{a}_o^2 = \alpha_1\mathbf{e}_3;\ \mathbf{V}_2 = \gamma_1\mathbf{W}_1 + \beta_1\mathbf{W}_2 + \delta_1\mathbf{W}_3\\
&\qquad - (\theta_1 + \epsilon_1)\hat{\mathbf{W}}_6 + (\epsilon_1 - \theta_1)\bar{\mathbf{W}}_6\},
\end{aligned}
$$

$$
\begin{aligned}
w_3 &= \{\mathbf{a}_o^3 = \alpha_3 \mathbf{e}_3;\ \mathbf{V}_3 = \beta_3 (\mathbf{I} - \mathbf{W}_3) + \delta_3 \mathbf{W}_3 + 2\epsilon_3 \bar{\mathbf{W}}_6\}, \\
w_4 &= \{\mathbf{a}_o^4 = \alpha_4 \mathbf{e}_3;\ \mathbf{V}_4 = \beta_4 \mathbf{W}_1 + \gamma_4 \mathbf{W}_2 + \delta_4 \mathbf{W}_3 + (\beta_1 - \gamma_1) \hat{\mathbf{W}}_6 \\
&\qquad + (\beta_1 + \gamma_1) \bar{\mathbf{W}}_6\}, \\
w_5 &= \{\mathbf{a}_o^5 = \alpha_4 \mathbf{e}_3;\ \mathbf{V}_5 = \gamma_4 \mathbf{W}_1 + \beta_4 \mathbf{W}_2 + \delta_4 \mathbf{W}_3 - (\beta_1 - \gamma_1) \hat{\mathbf{W}}_6 \\
&\qquad + (\beta_1 + \gamma_1) \bar{\mathbf{W}}_6\}, \\
w_6 &= \{\mathbf{a}_o^6 = \alpha_6 \mathbf{e}_3;\ \mathbf{V}_6 = \beta_6 (\mathbf{I} - \mathbf{W}_3) + \delta_6 \mathbf{W}_3 + 2\epsilon_6 \bar{\mathbf{W}}_6\},
\end{aligned} \tag{170}
$$

represents a longitudinal wave combined either with the modes $\mathbf{D}_1$, $\mathbf{S}_{12}$, $\mathbf{R}_3$ or with the modes $\mathbf{D}_2$ and $\mathbf{R}_3$: the coefficients of (170) are given in §.A of the Appendix.

For $\xi \to 0$ since a, b, c and q vanish, then from (167) and the results of §.A we have that:

$$
\begin{aligned}
&(\omega_{1,2}(\xi),\ w_{1,2}) \to (\omega_{1,2},\ \{\mathbf{0},\ \mathbf{C}_{1,2}\}), \\
&(\omega_6(\xi),\ w_6) \to (\omega_3,\ \{\mathbf{0},\ \mathbf{C}_3\}), \\
&(\omega_{4,5}(\xi),\ w_{4,5}) \to (\omega_{4,5},\ \{\mathbf{0},\ \mathbf{C}_{4,5}\}), \\
&(\omega_3(\xi),\ w_3) \to (0,\ \{\mathbf{e}_3,\ \mathbf{0}\}),
\end{aligned} \tag{171}
$$

with the cut-off frequencies $\omega_{1,2}$ and $\omega_{3,4,5}$ given respectively by (129) and (130) and where the microdistortions $\mathbf{C}_{1-5}$ are given by (132); therefore in the subspace $\mathcal{M}_1$ there are one acoustic longitudinal wave and five optic waves with cut-off frequencies (129) and (130).

A qualitative graph of the dispersion relations $\omega = \omega(\xi)$ for the solutions in the subspace $\mathcal{M}_1$, which can be obtained for arbitrary values of the components, is given in Fig. 4.

In the subspace $\mathcal{M}_2$, provided we define the normalized components:

$$
\begin{aligned}
a(\xi) &= \xi^2 A_{11}, \quad b(\xi) = \xi^2 A_{12}, \\
c(\xi) &= \xi^2 \frac{\mathsf{P}_{113} L_c^2}{L_m} + i\xi \mathsf{Q}_{113}, \quad d(\xi) = \xi^2 \frac{\mathsf{P}_{131} L_c^2}{L_m} + i\xi \mathsf{Q}_{131}, \\
e(\xi) &= \xi^2 \frac{\mathbb{A}_{44} L_c^2}{J_{33}} + \frac{\mathbb{B}_{44}}{\rho J_{33}}, \quad f(\xi) = \xi^2 \frac{\mathbb{A}_{55} L_c^2}{J_{11}} + \frac{\mathbb{B}_{55}}{\rho J_{11}}, \\
g(\xi) &= \xi^2 \frac{\mathbb{A}_{47} L_c^2}{\sqrt{J_{11} J_{33}}} + \frac{\mathbb{B}_{47}}{\rho \sqrt{J_{11} J_{33}}}, \quad h(\xi) = \xi^2 \frac{\mathbb{A}_{45} L_c^2}{\sqrt{J_{11} J_{33}}} + \frac{\mathbb{B}_{45}}{\rho \sqrt{J_{11} J_{33}}}, \\
m(\xi) &= \xi^2 \frac{\mathsf{P}_{123} L_c^2}{L_m} + i\xi \mathsf{Q}_{123}, \quad n(\xi) = \xi^2 \frac{\mathsf{P}_{132} L_c^2}{L_m} + i\xi \mathsf{Q}_{132},
\end{aligned} \tag{172}
$$

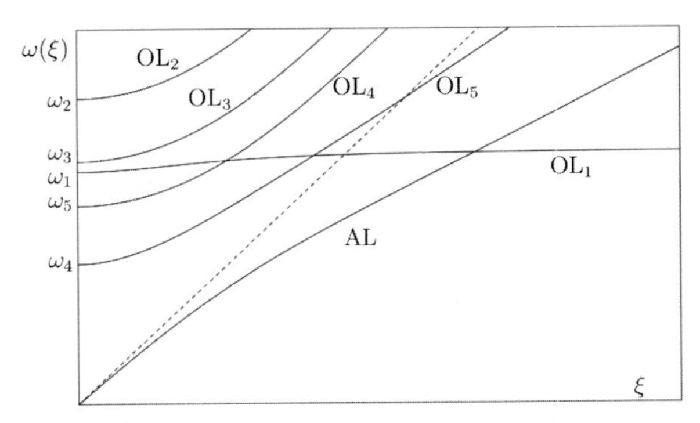

Fig. 4 Schematic of the dispersion relations in $\mathcal{M}_1$. The dotted line represent the linearly elastic longitudinal wave. AL = Acoustic longitudinal wave; OL$_{1,2,4,5}$ = optic longitudinal waves, modes $\mathbf{D}_1$, $\mathbf{S}_{12}$, $\mathbf{R}_3$; OL$_3$ = optic longitudinal wave, modes $\mathbf{D}_2$, $\mathbf{R}_3$. The frequencies on the ω-axis are the cut-off frequencies (129), (130).

we notice that the characteristic equation can be factorized into two cubic equations in ω^2. Accordingly the eigenvalues can be obtained by using twice the Cardano's formula:

$$
\begin{aligned}
\omega_7^2 &= \frac{1}{3}\left(A_1 + 2\sqrt{P_1}\cos\frac{\theta_1}{3}\right),\\
\omega_8^2 &= \frac{1}{3}\left(A_1 + 2\sqrt{P_1}\cos\frac{\theta_1 - 2\pi}{3}\right),\\
\omega_9^2 &= \frac{1}{3}\left(A_1 + 2\sqrt{P_1}\cos\frac{\theta_1 + 2\pi}{3}\right),\\
\omega_{10}^2 &= \frac{1}{3}\left(A_2 + 2\sqrt{P_2}\cos\frac{\theta_2}{3}\right),\\
\omega_{11}^2 &= \frac{1}{3}\left(A_2 + 2\sqrt{P_2}\cos\frac{\theta_2 - 2\pi}{3}\right),\\
\omega_{12}^2 &= \frac{1}{3}\left(A_2 + 2\sqrt{P_2}\cos\frac{\theta_2 + 2\pi}{3}\right),
\end{aligned}
\tag{173}
$$

where

$$
\begin{aligned}
A_{1,2} &= a \pm b + e + f, \\
B_{1,2} &= 2a^3 \pm 3a^2 (b \mp e \mp f) \\
&+ 3a\,(2b^2 + 3(c\bar{c} + d\bar{d}) - (e - f)^2 - 6(g^2 + h^2) \\
&+ 3(m\bar{m} + n\bar{n}) \\
&\mp 2b(e + f) + 4ef) \pm 2(b^3 \pm e^3 \pm f^3) \\
&\pm 3b(c\bar{c} + d\bar{d} - e^2 - f^2) + 9(g^2 + h^2)(\mp 2b + e + f) \\
&\pm 9m\bar{m}(b \pm e \mp 3f) \pm 9n\bar{n}(b \pm f \mp 3e) \\
&- 3e(b^2 + f^2 - 3c\bar{c} + 6d\bar{d}) \\
&- 3f(b^2 + 6c\bar{c} - 3d\bar{d} + e^2 + 4be) \\
&+ 54(g(c\bar{d} + m\bar{n}) + h(d\bar{m} - c\bar{n})) \\
C_{1,2} &= \mp c\bar{c} - (d\bar{d} + g^2 + h^2 + m\bar{m} + n\bar{n}) + (a \pm b)(e + f) + ef,
\end{aligned}
\tag{174}
$$

where $\bar{c}$, $\bar{d}$, $\bar{m}$, $\bar{n}$ are the complex conjugates of c, d, m, n and

$$
\theta_\alpha = \cos^{-1}\frac{B_\alpha}{2\sqrt[3]{P_\alpha}}, \qquad P_\alpha = A_\alpha^2 - 3C_\alpha, \qquad \alpha = 1,\, 2.
\tag{175}
$$

In this case all the frequencies depend on all the components of the block matrix A; as far as the corresponding eigenvectors are concerned, they are

$$
\begin{aligned}
w_k = &\{\mathbf{a}_o^k = \alpha_k \mathbf{e}_1 + \beta_k \mathbf{e}_2;\ \mathbf{V}_k = (\gamma_k + \epsilon_k)\hat{\mathbf{W}}_4 + (\delta_k + \theta_k)\hat{\mathbf{W}}_5 \\
&+ (\gamma_k - \epsilon_k)\bar{\mathbf{W}}_4 + (\delta_k - \theta_k)\bar{\mathbf{W}}_5\}, \quad k = 7, \ldots 12,
\end{aligned}
\tag{176}
$$

the coefficients being given in detail into the Appendix, §.A. The kinematics described by these eigenvectors is formed by a macroscopic transverse wave coupled with a combination of the modes $\mathbf{S}_{3\nu}$ and $\mathbf{R}_\omega$.

The behavior of the eigencouples (173), (176) for $\xi \to 0$, since in such a case $a = b = c = d = m = n = 0$, yields four optic waves (with multiplicity 2) with cut-off frequencies (133) and eigenvectors given by (134):

$$
w_7(0) = w_8(0) = \{\mathbf{0};\ \mathbf{C}_{6,7}\}, \qquad w_{10}(0) = w_{11}(0) = \{\mathbf{0};\ \mathbf{C}_{8,9}\}, \tag{177}
$$

and two acoustic waves with eigenvectors

$$w_9(0) = \{\mathbf{e}_1 - \mathbf{e}_2; \mathbf{0}\}, \quad w_{12}(0) = \{\mathbf{e}_1 + \mathbf{e}_2; \mathbf{0}\}. \tag{178}$$

As we did for $\mathcal{M}_1$ we give in Fig. 5 a representative graph for the dispersion relations in $\mathcal{M}_2$.

To summarize, for the Tetragonal classes 4, $\bar{4}$, $4/m$, for a propagation direction $\mathbf{m}$ along the tetragonal c-axis, we have three Acoustic and nine Optic waves which depend on 44 independent components of A (four components of $\mathbf{A}$, thirteen for both $\mathbb{A}$ and $\mathbb{B}$, seven for both P and Q): .

1. (AL) One Acoustic wave associated with a macroscopic displacement along c and a combination of the modes $\mathbf{D}_1$, $\mathbf{S}_{12}$ and $\mathbf{R}_3$, which for $\xi = 0$ reduced to a macroscopic longitudinal wave;
2. ($AT_{1,2}$) Two Acoustic waves associated with a macroscopic displacement orthogonal to c, coupled a combination of the modes $\mathbf{S}_{3\nu}$ and $\mathbf{R}_\omega$. For $\xi = 0$ these waves reduce to two macroscopic orthogonal transverse waves;
3. ($OL_{1,2,4,5}$) Four Optic waves associated with a macroscopic displacement along c and with a microdistortion which combines the modes $\mathbf{D}_1$, $\mathbf{S}_{12}$ and $\mathbf{R}_3$ which for $\xi = 0$ reduces to the pure microdistortions $(132)_{1,2,4,5}$;
4. (OL_3) One Optic wave associated with a macroscopic displacement along c and with a combination of the modes $\mathbf{D}_2$ and $\mathbf{R}_3$;
5. ($OT_{7,8,10,11}$) Four Optic waves associated with a macroscopic displacement orthogonal to c coupled with a combination of the $\mathbf{S}_{3\nu}$ and $\mathbf{R}_\omega$ modes, which for $\xi = 0$ reduce to the shear microdistortion $(134)_2$ and to the rigid rotation $(134)_1$.

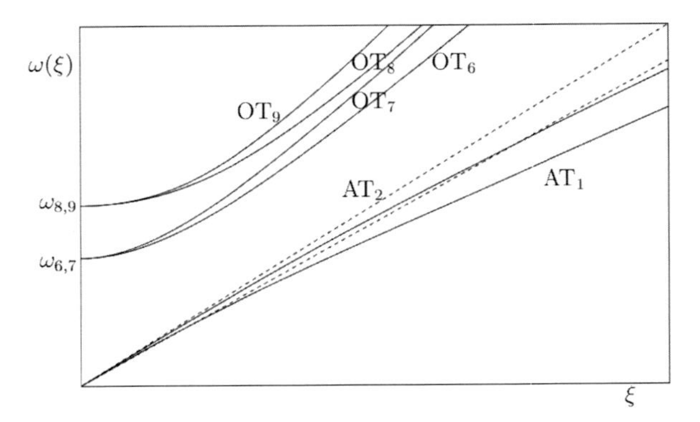

Fig. 5 Schematic of the dispersion relations in $\mathcal{M}_2$. The dotted line represent the linearly elastic transverse waves. $AT_{1,2}$ = Acoustic transverse wave; $OT_{7,8,10,11}$ = optic transverse waves, modes $\mathbf{S}_{3\nu}$, $\mathbf{R}_\omega$. The frequencies on the ω-axis are the cut-off frequencies (133).

4.3.1.2 Classes $4mm$, 422, $\bar{4}2m$ and $4/mmm$

When we deal with the high-symmetric tetragonal classes, for a propagation direction along the tetragonal c-axis with $\mathbf{m} = \mathbf{e}_3$, the matrix A has the following non-null components, the independent ones given by relations (105), (113), (117) and (123) of the Appendix, evaluted for $m_1 = m_2 = 0$ and $m_3 = 1$.

$$
A \equiv \begin{bmatrix}
\bullet & \bullet & \cdot & \cdot & \cdot & \cdot & \cdot & \bullet & \cdot & \cdot & \bullet & \cdot \\
\bullet & \bullet & \cdot & \cdot & \cdot & \cdot & \bullet & \cdot & \cdot & \bullet & \cdot & \cdot \\
\cdot & \cdot & \bullet & \bullet & \bullet & \bullet & \cdot & \cdot & \cdot & \cdot & \cdot & \cdot \\
\cdot & \cdot & \bullet & \bullet & \bullet & \bullet & \cdot & \cdot & \cdot & \cdot & \cdot & \cdot \\
\cdot & \cdot & \bullet & \bullet & \bullet & \bullet & \cdot & \cdot & \cdot & \cdot & \cdot & \cdot \\
\cdot & \cdot & \bullet & \bullet & \bullet & \bullet & \cdot & \cdot & \cdot & \cdot & \cdot & \cdot \\
\cdot & \bullet & \cdot & \cdot & \cdot & \cdot & \bullet & \cdot & \cdot & \bullet & \cdot & \cdot \\
\bullet & \cdot & \cdot & \cdot & \cdot & \cdot & \cdot & \bullet & \cdot & \cdot & \bullet & \cdot \\
\cdot & \cdot & \cdot & \cdot & \cdot & \cdot & \cdot & \cdot & \bullet & \cdot & \cdot & \bullet \\
\cdot & \bullet & \cdot & \cdot & \cdot & \cdot & \bullet & \cdot & \cdot & \bullet & \cdot & \cdot \\
\bullet & \cdot & \cdot & \cdot & \cdot & \cdot & \cdot & \bullet & \cdot & \cdot & \bullet & \cdot \\
\cdot & \cdot & \cdot & \cdot & \cdot & \cdot & \cdot & \cdot & \bullet & \cdot & \cdot & \bullet
\end{bmatrix} : \tag{179}
$$

accordingly A is reduced by the three subspaces:

$$
\mathcal{N}_1 \equiv \mathcal{V}_{\parallel} \oplus \mathcal{Z}_1, \quad \mathcal{N}_2 \equiv \mathcal{Z}_2, \quad \mathcal{N}_3 \equiv \mathcal{V}_{\perp} \oplus \mathcal{Z}_3; \tag{180}
$$

We begin our analysis with the subspace $\mathcal{N}_1$ and notice that the eigenvalues can be obtained from those in $\mathcal{M}_1$ when we set $d = e = 0$ and $m = n = p = q = 0$ into (166). Then the characteristic equation admits the root $\omega^2 = f - h$ and thus, by the means of the Cardano's formulae (*vid. e,g.* (Zwillinger, 2003)) we obtain the four eigenvalues (for once we write one of them in terms components and characteristic length):

$$
\begin{aligned}
\omega_1^2(\xi) &= \xi^2\left(\frac{\mathbb{A}_{11}}{J_{11}} + \frac{\mathbb{A}_{12}}{\sqrt{J_{11}J_{33}}}\right)L_c^2 + \frac{\mathbb{B}_{11}}{\rho J_{11}} + \frac{\mathbb{B}_{12}}{\rho\sqrt{J_{11}J_{33}}}, \\
\omega_2^2(\xi) &= \frac{1}{3}\left(A + 2\sqrt{P}\cos\frac{\theta}{3}\right), \\
\omega_3^2(\xi) &= \frac{1}{3}\left(A + 2\sqrt{P}\cos\frac{\theta + 2\pi}{3}\right), \\
\omega_4^2(\xi) &= \frac{1}{3}\left(A + 2\sqrt{P}\cos\frac{\theta - 2\pi}{3}\right),
\end{aligned} \tag{181}
$$

where

$$P = A^2 - 3B, \qquad Q = 2A^3 - 9AB - 27C,$$
$$\theta = \cos^{-1}\frac{Q}{2\sqrt[3]{P}}, \tag{182}$$

and A, B and C are obtained by setting $d = e = m = n = p = q = 0$ into (169). The corresponding (non-normalized) eigenvectors are

$$\begin{aligned}
w_1(\xi) &= \{\mathbf{0};\, \mathbf{V}_1 = \mathbf{W}_1 - \mathbf{W}_2\},\\
w_2(\xi) &= \{\alpha_2\mathbf{e}_3;\, \mathbf{V}_2 = \beta_2\mathbf{W}_1 + \gamma_2\mathbf{W}_2 + \delta_2\mathbf{W}_3\},\\
w_3(\xi) &= \{\alpha_2\mathbf{e}_3;\, \mathbf{V}_3 = \gamma_2\mathbf{W}_1 + \beta_2\mathbf{W}_2 + \delta_2\mathbf{W}_3\},\\
w_4(\xi) &= \{\alpha_4\mathbf{e}_3;\, \mathbf{V}_4 = \beta_4(\mathbf{I} - \mathbf{W}_3) + \delta_4\mathbf{W}_3\},
\end{aligned} \tag{183}$$

where the complex components of w_2, w_3 and w_4 are given by:

$$\begin{aligned}
\alpha_2 &= (b(g-\omega^2) - cl)((f-\omega^2) - h),\\
\beta_2 &= -((a-\omega^2)(l^2 - (g-\omega^2)h) - 2b\bar{c}l + c\bar{c}h\\
&\quad + b\bar{b}(g-\omega^2)),\\
\gamma_2 &= (a-\omega^2)(l^2 - (f-\omega^2)(g-\omega^2)) - 2b\bar{c}l\\
&\quad + b\bar{b}(g-\omega^2) + c\bar{c}(f-\omega^2),\\
\delta_2 &= ((a-\omega^2)l - b\bar{c})((f-\omega^2) - h),\\
\alpha_4 &= 2l^2 - (g-\omega^2)(h + (f-\omega^2)),\\
\beta_4 &= b(g-\omega^2) - cl,\\
\delta_4 &= c(h + (f-\omega^2)) - 2bl.
\end{aligned} \tag{184}$$

Looking at (181) and (183) we notice first of all that the frequencies $\omega_{2,3,4}(\xi)$ are associated with the mode $\mathbf{D}_2$ coupled with a macroscopic longitudinal wave; the frequency $\omega_1(\xi)$ is instead associated uniquely to the traceless real microdistortion $\mathbf{D}_3$.

In the limit $\xi \to 0$, since a, b and c vanish, then the three frequencies $\omega_1(\xi)$ and $\omega_{2,3}(\xi)$ reduce to (137) with eigenvectors

$$w_1(0) = \{\mathbf{0};\, \mathbf{C}_1\}, \quad w_2(0) = \{\mathbf{0};\, \mathbf{C}_2\}, \quad w_3 = \{\mathbf{0};\, \mathbf{C}_3\}, \tag{185}$$

with the three microdistortions given by (139): these are optic frequencies, the values (137) being the associated cut-off values; the frequency $\omega_4(\xi)$ vanishes instead for $\xi \to 0$ with

$$w_4(0) = \{\mathbf{e}_3;\, \mathbf{0}\}, \tag{186}$$

which represents a purely macroscopic acoustic longitudinal wave.

The solutions in the subspace $\mathcal{N}_2$ are are the same as those in $\mathcal{Z}_2$ with frequencies

$$\begin{aligned}
\omega_5^2(\xi) &= \xi^2 L_c^2 \frac{\mathbb{A}_{66} - \mathbb{A}_{69}}{J_{11}} + \frac{\mathbb{B}_{66} - \mathbb{B}_{69}}{\rho J_{11}}, \\
\omega_6^2(\xi) &= \xi^2 L_c^2 \frac{\mathbb{A}_{66} + \mathbb{A}_{69}}{J_{11}} + \frac{\mathbb{B}_{66} + \mathbb{B}_{69}}{\rho J_{11}},
\end{aligned} \tag{187}$$

and accordingly describe optic waves with purely microdistortion amplitudes and whose cut-off frequencies are given by (141); the corresponding real eigenvectors are:

$$w_5(0) = \{\mathbf{0};\ \mathbf{V}_5 = \mathbf{C}_4\}, \qquad w_6(0) = \{\mathbf{0};\ \mathbf{V}_6 = \mathbf{C}_5\}, \tag{188}$$

with the microdistortions given by (141).

We finish with the subspace $\mathcal{N}_3$ where the solutions are obtained by setting $m = n = h = 0$ into (172) which yield the six eigenvalues

$$\begin{aligned}
\omega_7^2 &= \frac{1}{3}\left(A_1 + 2\sqrt{P_1}\cos\frac{\theta_1}{3}\right), \\
\omega_8^2 &= \frac{1}{3}\left(A_1 + 2\sqrt{P_1}\cos\frac{\theta_1 + 2\pi}{3}\right), \\
\omega_9^2 &= \frac{1}{3}\left(A_1 + 2\sqrt{P_1}\cos\frac{\theta_1 - 2\pi}{3}\right), \\
\omega_{10}^2 &= \frac{1}{3}\left(A_2 + 2\sqrt{P_2}\cos\frac{\theta_2}{3}\right), \\
\omega_{11}^2 &= \frac{1}{3}\left(A_2 + 2\sqrt{P_2}\cos\frac{\theta_2 + 2\pi}{3}\right), \\
\omega_{12}^2 &= \frac{1}{3}\left(A_2 + 2\sqrt{P_2}\cos\frac{\theta_2 - 2\pi}{3}\right),
\end{aligned} \tag{189}$$

where A_α and P_α for $\alpha = 1, 2$ are obtained when we put $m = n = h = 0$ into (174). The eigenvector have the same representation (176) with the coefficients given by those for $\mathcal{M}_1$ with $m = n = h = 0$:

$$\begin{aligned}
w_7 &= \{\mathbf{a}_o^7 = \alpha_7\mathbf{e}_1 + \beta_7\mathbf{e}_2; \\
\mathbf{V}_7 &= (\gamma_7 + \epsilon_7)\hat{\mathbf{W}}_4 + (\delta_7 + \theta_7)\hat{\mathbf{W}}_5 + (\gamma_7 - \epsilon_7)\bar{\mathbf{W}}_4 \\
&\quad + (\delta_7 - \theta_7)\bar{\mathbf{W}}_5\},
\end{aligned}$$

$$
\begin{aligned}
w_8 &= \{\mathbf{a}_o^8 = \alpha_8\mathbf{e}_1 + \beta_8\mathbf{e}_2; \\
\mathbf{V}_8 &= (\gamma_8 + \epsilon_8)\hat{\mathbf{W}}_4 + (\delta_8 + \theta_8)\hat{\mathbf{W}}_5 + (\gamma_8 - \epsilon_8)\bar{\mathbf{W}}_4 \\
&\quad + (\delta_8 - \theta_8)\bar{\mathbf{W}}_5\}, \\
w_9 &= \{\mathbf{a}_o^9 = \beta_8\mathbf{e}_1 + \alpha_8\mathbf{e}_2; \\
\mathbf{V}_9 &= (\delta_8 + \theta_8)\hat{\mathbf{W}}_4 + (\gamma_8 + \epsilon_8)\hat{\mathbf{W}}_5 - (\delta_8 - \theta_8)\bar{\mathbf{W}}_4 \\
&\quad - (\gamma_8 - \epsilon_8)\bar{\mathbf{W}}_5\}, \\
w_{10} &= \{\mathbf{a}_o^{10} = \beta_7\mathbf{e}_1 + \alpha_7\mathbf{e}_2; \\
\mathbf{V}_{10} &= (\delta_7 + \theta_7)\hat{\mathbf{W}}_4 + (\gamma_7 + \epsilon_7)\hat{\mathbf{W}}_5 - (\delta_7 - \theta_7)\bar{\mathbf{W}}_4 \\
&\quad - (\gamma_7 - \epsilon_7)\bar{\mathbf{W}}_5\}, \\
w_{11} &= \{\mathbf{a}_o^{11} = \alpha_{11}\mathbf{e}_1 + \beta_{11}\mathbf{e}_2; \\
\mathbf{V}_{11} &= (\gamma_{11} + \epsilon_{11})\hat{\mathbf{W}}_4 + (\delta_{11} + \theta_{11})\hat{\mathbf{W}}_5 + (\gamma_{11} - \epsilon_{11})\bar{\mathbf{W}}_4 \\
&\quad + (\delta_{11} - \theta_{11})\bar{\mathbf{W}}_5\}, \\
w_{12} &= \{\mathbf{a}_o^{12} = \beta_{11}\mathbf{e}_1 + \alpha_{11}\mathbf{e}_2; \\
\mathbf{V}_{12} &= (\delta_{11} + \theta_{11})\hat{\mathbf{W}}_4 + (\gamma_{11} + \epsilon_{11})\hat{\mathbf{W}}_5 - (\delta_{11} - \theta_{11})\bar{\mathbf{W}}_4 \\
&\quad - (\gamma_{11} - \epsilon_{11})\bar{\mathbf{W}}_5\},
\end{aligned} \tag{190}
$$

These eigenvectors describe a kinematics formed by a macroscopic transverse wave coupled a combination of the modes $\mathbf{S}_{3\nu}$ and $\mathbf{R}_{\omega}$.

The eigenvalues (189) and their associated eigenvectors becomes, since for $\xi \to 0$ it is $a = b = c = d = 0$:

$$
w_{7,8,9,10}(0) = \{\mathbf{0};\ \mathbf{C}_{6,7,8,9}\}, \tag{191}
$$

which correspond to four optic waves (with multiplicity 2) with cut-off frequencies (133) with $\mathbb{B}_{45} = 0$ and eigenvectors given by (134) and two acoustic waves with eigenvectors:

$$
w_{11}(0) = \{\mathbf{e}_1 - \mathbf{e}_2;\ \mathbf{0}\}, \qquad w_{12}(0) = \{\mathbf{e}_1 + \mathbf{e}_2;\ \mathbf{0}\}. \tag{192}
$$

To summarize, in the high-symmetric tetragonal classes $4mm$, 422, $\bar{4}2m$ and $4/mmm$, for the propagation direction along the tetragonal c-axis we have the three Acoustic and nine Optic waves, which depend on the 25 independent components of A (three of $\mathbf{A}$, seven each of $\mathbb{A}$ and $\mathbb{B}$ and four each of P and Q):

1. (AL) One Acoustic wave associated with a macroscopic displacement along c and the mode $\mathbf{D}_2$ which for $\xi = 0$ yields a macroscopic longitudinal wave;
2. ($AT_{1,2}$) Two Acoustic waves associated with a macroscopic displacemente orthogonal to c, coupled with the $\mathbf{S}_{3\nu}$ and $\mathbf{D}_2$ modes. For $\xi = 0$ they reduce to two macroscopic orthogonal transverse waves;
3. ($OL_{1,2,3}$) Two Optic waves associated to a macrodisplacement along $\mathbf{e}_3$ coupled with the mode $\mathbf{D}_2$, which for $\xi = 0$ reduces to the pure microdistortions $(139)_{2,3}$;
4. (OD) One Optic wave associated to the traceless distortion $\mathbf{D}_3$ which is independent on the macroscopic displacement and which reduces, for $\xi \to 0$, to $(139)_1$;
5. (OS) One Optic wave associated to the the mode $\mathbf{S}_{12}$ which is independent on the macroscopic displacement and reduces for $\xi \to 0$ to $(141)_2$;
6. (OR) One Optic wave associated to the mode $\mathbf{R}_3$, which is independent on the macroscopic displacement and reduces for $\xi \to 0$ to $(141)_1$;
7. ($OT_{1,2,3,4}$) Four Optic waves associated with a macroscopic displacement orthogonal to c coupled with a combination of the modes $\mathbf{S}_{3\nu}$ and $\mathbf{R}_\omega$, which for $\xi = 0$ reduce to the shear microdistortions $(134)_2$ and to the rigid rotations $(134)_1$ with $\mathbb{B}_{45} = 0$.

As far as the eigenvalues ordering is concerned, we can only say that:

$$\begin{aligned}
&\omega_5 < \omega_6, \\
&\omega_3 < \omega_2 < \omega_4, \text{ (or } \omega_3 > \omega_2 > \omega_4), \\
&\omega_8 < \omega_7 < \omega_9, \text{ (or } \omega_8 > \omega_7 > \omega_9), \\
&\omega_{11} < \omega_{10} < \omega_{12}, \text{ (or } \omega_{11} > \omega_{10} > \omega_{12}).
\end{aligned} \tag{193}$$

To summarize the results obtained in this subsection for the coupled bulk and lattice waves propagating along the tetragonal c-axis, in all cases but three the modes are complex and fully-couples macroscopic displacements with lattice microdistortions: only for the high-symmetric classes we have three fully optic modes with real microdistortions without macroscopic displacements. These results are represented in the Table 2.

Table 2 Modes, wave propagation along the direction c.

		$\mathbf{e}_1$	$\mathbf{e}_2$	$\mathbf{e}_3$	$\mathbf{D}_1$	$\mathbf{D}_2$	$\mathbf{D}_3$	$\mathbf{S}_{12}$	$\mathbf{S}_{3v}$	$\mathbf{R}_3$	$\mathbf{R}_\omega$	$\xi \to 0$
	(2)	·	·	$\bullet\mathfrak{C}_L$	$\mathfrak{C}$	·	·	$\mathfrak{C}^\circ$	·	$\mathfrak{C}\bullet$	·	$\mathrm{OL}_{1,4}$
	(2)	·	·	$\bullet\mathfrak{C}_L$	$\mathfrak{C}$	·	·	$\mathfrak{C}^\circ$	·	$\mathfrak{C}\bullet$	·	$\mathrm{OL}_{2,5}$
LS		·	·	$\mathfrak{C}_L$	·	$\mathfrak{C}$	·	·	·	$\mathfrak{C}$	·	OL_3
		·	·	$\mathfrak{C}_L$	·	$\mathfrak{C}$	·	·	·	$\mathfrak{C}$	·	AL
	(4)	$\mathfrak{C}_T$	$\mathfrak{C}_T$	·	·	·	·	·	$\mathfrak{C}$	·	$\mathfrak{C}$	OT_{6-9}
	(2)	$\mathfrak{C}_T$	$\mathfrak{C}_T$	·	·	·	·	·	$\mathfrak{C}$	·	$\mathfrak{C}$	AT_{1-2}
		·	·	·	·	·	$\mathfrak{R}$	·	·	·	·	OL_1
		·	·	$\bullet\mathfrak{C}_L$	$\mathfrak{C}$	·	·	·	·	·	·	OL_2
		·	·	$\bullet\mathfrak{C}_L$	$\mathfrak{C}$	·	·	·	·	·	·	OL_3
		·	·	$\mathfrak{C}_L$	·	$\mathfrak{C}$	·	·	·	·	·	AL

HS		·	·	·	·	·	·	$\mathfrak{R}$	·	·	·	OS
		·	·	·	·	·	·	·	·	$\mathfrak{R}$	·	OR
	(2)	$\mathfrak{C}_T$	$\mathfrak{C}_T$	·	·	·	·	·	$\mathfrak{C}$	·	$\mathfrak{C}$	$OT_{6,9}$
	(2)	$\mathfrak{C}_T$	$\mathfrak{C}_T$	·	·	·	·	·	$\mathfrak{C}$	·	$\mathfrak{C}$	$OT_{7,8}$
	(2)	$\mathfrak{C}_T$	$\mathfrak{C}_T$	·	·	·	·	·	$\mathfrak{C}$	·	$\mathfrak{C}$	$AT_{1,2}$

HS = classes $4mm$, 422, $4/mmm$, $\bar{4}2m$, LS = classes 4, $\bar{4}$, $4/m$; $\mathfrak{R}$ = Real components, $\mathfrak{C}$ = Complex components; L = longitudinal wave, T = transverse wave; • = equal values components; ∘ = equal value, opposite sign components; (♯) = number of independent eigentensors with same structure.

4.3.2 Propagation orthogonal to the tetragonal c-axis

When the propagation direction is orthogonal to the tetragonal c-axis, namely

$$\mathbf{m} = \mathbf{m}_{\perp} = \cos\theta\mathbf{e}_1 + \sin\theta\mathbf{e}_2, \tag{194}$$

then we assume into the constitutive relations (104), (105), (108), (112), (114), (117), (122) and (123) that ($m_1 = \cos\theta$, $m_2 = \sin\theta$, $m_3 = 0$).

4.3.2.1 Classes 4 $\bar{4}$

For these two classes, if we look at (108) and (114) with $m_3 = 0$ we notice that the 3×9 sub matrix $[\xi^2\mathrm{P} + i\xi\mathrm{Q}]$ has 27 non-zero components. Accordingly the matrix A is not reduced by subspaces of $\mathcal{V} \oplus \mathrm{Lin}$ and the twelve eigencouples (ω_k, w_k), $k = 1, \ldots, 12$ the eigenvectors span all the space:

$$w_k = \{\mathbf{a}_o^k = \alpha_j^k\mathbf{e}_j;\ \mathbf{V}_k = \beta_h^k\mathbf{W}_h,\ j = 1, 2, 3,\ h = 1, \ldots, 9\}; \tag{195}$$

we cannot give an explicit expression of the eigenvalues in terms of the components of A and the best we can say, by the results of §.2.3, is that we shall have three acoustic waves and nine optic waves. These waves, in the limit for $\xi = 0$, yield one macroscopic transverse wave and other two macroscopic waves, neither longitudinal nor transverse, together with the nine optic waves (132) and (134) whose cut-off frequencies are given by the solution of the microvibrations problem on $\mathcal{U}_1$ and $\mathcal{U}_2$ of §.4.1.

4.3.2.2 Class 4/*m*

For the centrosymmetric classe $4/m$ the tensor P vanishes altogether since it depends on the components of an odd-tensor; then the matrix A has the structure:

$$A \equiv \begin{bmatrix}
\bullet & \bullet & \cdot & \bullet & \bullet & \bullet & \cdot & \cdot & \bullet & \cdot & \cdot & \bullet \\
\bullet & \bullet & \cdot & \bullet & \bullet & \bullet & \cdot & \cdot & \bullet & \cdot & \cdot & \bullet \\
\cdot & \cdot & \bullet & \cdot & \cdot & \cdot & \bullet & \bullet & \cdot & \bullet & \bullet & \cdot \\
\bullet & \bullet & \cdot & \bullet & \bullet & \bullet & \cdot & \cdot & \bullet & \cdot & \cdot & \bullet \\
\bullet & \bullet & \cdot & \bullet & \bullet & \bullet & \cdot & \cdot & \bullet & \cdot & \cdot & \bullet \\
\bullet & \bullet & \cdot & \bullet & \bullet & \bullet & \cdot & \cdot & \bullet & \cdot & \cdot & \bullet \\
\cdot & \cdot & \bullet & \cdot & \cdot & \cdot & \bullet & \bullet & \cdot & \bullet & \cdot & \cdot \\
\cdot & \cdot & \bullet & \cdot & \cdot & \cdot & \bullet & \bullet & \cdot & \cdot & \bullet & \cdot \\
\bullet & \bullet & \cdot & \bullet & \bullet & \bullet & \cdot & \cdot & \bullet & \cdot & \cdot & \bullet \\
\cdot & \cdot & \bullet & \cdot & \cdot & \cdot & \bullet & \cdot & \cdot & \bullet & \bullet & \cdot \\
\cdot & \cdot & \bullet & \cdot & \cdot & \cdot & \cdot & \bullet & \cdot & \bullet & \bullet & \cdot \\
\bullet & \bullet & \cdot & \bullet & \bullet & \bullet & \cdot & \cdot & \bullet & \cdot & \cdot & \bullet
\end{bmatrix}, \tag{196}$$

which is reduced by the pairs

$$\mathcal{L}_1 \equiv \mathcal{V}_{\parallel} \oplus \mathcal{U}_2, \quad \mathcal{L}_2 \equiv \mathcal{V}_{\perp} \oplus \mathcal{U}_1, \tag{197}$$

with $\dim \mathcal{L}_1 = 5$ and $\dim \mathcal{L}_1 = 7$.

In the subspace $\mathcal{L}_1$ the fifth-degree characteristic equation in ω^2 can not be factorized into lower-degree equations and hence we do not have an explicit expression for the eigencouples in terms of the 15 independent components of the submatrix of A.

However, the eigenvectors w shall have the general representation:

$$\begin{aligned} w_k = \{\mathbf{a}_o^k = \alpha_1^k \mathbf{e}_3;\ \mathbf{V}_k \\ = \beta_4^k \mathbf{W}_4 + \beta_5^k \mathbf{W}_5 + \beta_7^k \mathbf{W}_7 + \beta_8^k \mathbf{W}_8\}, \quad k = 1, \ldots, 5; \end{aligned} \tag{198}$$

such a kinematics represents a macroscopic transverse wave directed as $\mathbf{e}_3$ plus a combination of the shear mode $\mathbf{S}_{3\nu}$ with the rigid rotation $\mathbf{R}_\omega$. In the limit $\xi \to 0$ the eigenvectors reduce to a transverse acoustic wave and to the micro-deformations (134).

Likewise, in the subspace $\mathcal{L}_2$ we shall have non-factorizable 7th degree algebraic equation for ω^2 and the seven eigenvectors shall have the general representation

$$\begin{aligned} w_k = \{\alpha_1^k \mathbf{e}_1 + \alpha_2^k \mathbf{e}_2;\ \mathbf{V}_k = \beta_j^k \mathbf{W}_j + \beta_6 \mathbf{W}_6 + \beta_9^k \mathbf{W}_9, \\ j = 1, 2, 3\}, \quad k = 1, \ldots, 7, \end{aligned} \tag{199}$$

that represents a macroscopic wave (nor longitudinal neither transverse) in the plane orthogonal to $\mathbf{e}_3$ coupled with a combination of the three dilatations $\mathbf{D}_k$ the shear $\mathbf{S}_{12}$ and the rigid rotation $\mathbf{R}_3$. In the limit $\xi \to 0$ these eigenvectors reduce to two acoustic waves (nor longitudinal, neither transverse) and to the microdeformations (132).

To summarize, for the Tetragonal class $4/m$, for a propagation direction $\mathbf{m}_\perp$ orthogonal to the tetragonal c-axis, we have three Acoustic and nine Optic waves which depend on 56 independent components of A (four components of $\mathbf{A}$, 26 for $\mathbb{A}$ and 13 for $\mathbb{B}$, 14 for Q):

1. (AT) One Acoustic wave associated with a macroscopic displacement along c and a combination of the five modes $\mathbf{D}_1$, $\mathbf{D}_2$, $\mathbf{D}_3$, $\mathbf{S}_{12}$ and $\mathbf{R}_3$, which for $\xi = 0$ reduced to a macroscopic longitudinal wave;
2. ($A_{1,2}$) Two Acoustic waves associated with a macroscopic displacement orthogonal to c, coupled a combination of the two modes $\mathbf{S}_{3\nu}$ and $\mathbf{R}_\omega$. For $\xi = 0$ these waves reduce to two macroscopic waves, nor transverse neither longitudinal;

3. ($O_{1,2,3,4,5}$) Five Optic waves associated with a macroscopic displacement orthogonal to the direction c and with a microdistortion which combines the five modes $\mathbf{D}_1$, $\mathbf{D}_2$, $\mathbf{D}_3$, $\mathbf{S}_{12}$ and $\mathbf{R}_3$ which for $\xi = 0$ reduces to the pure microdistortions (132);
4. ($OL_{6,7,8,9}$) Four Optic waves associated with a macroscopic displacement in the direction c coupled with a combination of the two $\mathbf{S}_{3\nu}$ and $\mathbf{R}_\omega$ modes, which for $\xi = 0$ reduce to the shear microdistortion $(134)_2$ and to the rigid rotation $(134)_1$.

For the classes 4 and $\bar{4}$ the acoustic waves AT and $AL_{1,2}$ depends on all the seven modes (19)-(25), as well as the optic waves $O_{1,\ldots,9}$ that, in the limit for $\xi = 0$ reduce to (132) and (134).

4.3.2.3 Classes $4/mmm$, $\bar{4}2m$, 422 polar

Also for these classes, if we look at (112) and (117) with $m_3 = 0$, the 3×9 sub matrix $[\xi^2 P + i\xi Q]$ has 27 non-zero components.

We are therefore in the same situation of the classes 4 and $\bar{4}$ and the consideration we did for those classes applies once more, with the eigenvector given by (195). There are three acoustic and nine optic waves which, in the limit for $\xi = 0$, yield one macroscopic transverse wave and two macroscopic waves that are neither longitudinal nor transverse, together with the nine optic waves (134), (139) and (141) whose cut-off frequencies are given by the solution of the microvibrations problem on $\mathcal{Z}_1$, $\mathcal{Z}_2$ and $\mathcal{Z}_3$ of §.4.1.

However, in the two special cases $\theta = 0$ *i.e.* $\mathbf{m} = \mathbf{e}_1$ and $\theta = \pi/2$ *i.e.* $\mathbf{m} = \mathbf{e}_2$ the matrix A admits a simpler structure. If we consider for instance the propagation direction with $\mathbf{m} = \mathbf{e}_2$ then we have

$$A \equiv \begin{bmatrix} \bullet & \cdot & \cdot & \cdot & \cdot & \cdot & \cdot & \bullet & \bullet & \cdot & \bullet & \bullet \\ \cdot & \bullet & \cdot & \bullet & \bullet & \bullet & \bullet & \cdot & \cdot & \bullet & \cdot & \cdot \\ \cdot & \cdot & \bullet & \bullet & \bullet & \bullet & \bullet & \cdot & \cdot & \cdot & \cdot & \cdot \\ \cdot & \bullet & \bullet & \bullet & \bullet & \bullet & \cdot & \cdot & \cdot & \cdot & \cdot & \cdot \\ \cdot & \bullet & \bullet & \bullet & \bullet & \bullet & \cdot & \cdot & \cdot & \cdot & \cdot & \cdot \\ \cdot & \bullet & \bullet & \bullet & \bullet & \bullet & \cdot & \cdot & \cdot & \cdot & \cdot & \cdot \\ \cdot & \bullet & \bullet & \cdot & \cdot & \cdot & \bullet & \cdot & \cdot & \bullet & \cdot & \cdot \\ \bullet & \cdot & \cdot & \cdot & \cdot & \cdot & \cdot & \bullet & \cdot & \cdot & \bullet & \cdot \\ \bullet & \cdot & \cdot & \cdot & \cdot & \cdot & \cdot & \cdot & \bullet & \cdot & \cdot & \bullet \\ \cdot & \bullet & \cdot & \cdot & \cdot & \cdot & \bullet & \cdot & \cdot & \bullet & \cdot & \cdot \\ \bullet & \cdot & \cdot & \cdot & \cdot & \cdot & \cdot & \bullet & \cdot & \cdot & \bullet & \cdot \\ \bullet & \cdot & \cdot & \cdot & \cdot & \cdot & \cdot & \cdot & \bullet & \cdot & \cdot & \bullet \end{bmatrix}, \tag{200}$$

and hence A is reduced by the two seven- and five-dimensional subspaces:

$$\begin{aligned} \mathcal{K}_1 &\equiv \operatorname{span}\{\mathbf{e}_1, \mathbf{e}_3\} \oplus \mathcal{Z}_1 \oplus \operatorname{span}\{\mathcal{W}_4, \mathcal{W}_7\}, \\ \mathcal{K}_2 &\equiv \operatorname{span}\{\mathbf{e}_2\} \oplus \mathcal{Z}_2 \oplus \operatorname{span}\{\mathcal{W}_5, \mathcal{W}_8\}. \end{aligned} \tag{201}$$

The characteristic equations cannot be factorized and we are not able to give a explicit expression for the eigenvalues: however in the subspace $\mathcal{K}_1$ the eigenvector have the general representation

$$w_k = \{\mathbf{a}_o^k = \alpha_1^k \mathbf{e}_1 + \alpha_3^k \mathbf{e}_3;\ \mathbf{V}_k = \alpha_j^k \mathbf{W}_j + \alpha_4^k \mathbf{W}_4 + \alpha_7^k \mathbf{W}_7,\\ j = 1, 2, 3\}, \quad k = 1, \dots, 7, \tag{202}$$

whereas in $\mathcal{K}_2$ they are

$$w_k = \{\mathbf{a}_o^k = \alpha_2^k \mathbf{e}_2;\ \mathbf{V}_k = \alpha_5^k \mathbf{W}_5 + \alpha_6^k \mathbf{W}_6 + \alpha_8^k \mathbf{W}_8 + \alpha_9^k \mathbf{W}_9\},\\ k = 1, \dots, 5. \tag{203}$$

Therefore, for the direction of propagation $\mathbf{m} = \mathbf{e}_2$ we get:

1. (AL) One Acoustic wave associated with a macroscopic longitudinal wave and a combination of the shear $\mathbf{S}_{13}$ between $\mathbf{e}_1$ and $\mathbf{e}_3$ and the rigid rotation $\mathbf{R}_2$ about the direction of propagation $\mathbf{e}_2$, which for $\xi = 0$ reduced to a macroscopic longitudinal wave.
2. ($AT_{1,2}$) Two Acoustic waves associated with a macroscopic displacement orthogonal to $\mathbf{e}_2$, coupled a combination of the shear $\mathbf{S}_{1\nu*}$ between $\mathbf{e}_1$ and the direction $\mathbf{v} = \alpha\mathbf{e}_2 + \beta\mathbf{e}_3$ and the rigid rotation $\mathbf{R}_{\omega*}$ about a direction $\boldsymbol{\omega}^* = \omega_2^*\mathbf{e}_2 - \omega_3^*\mathbf{e}_3$. For $\xi = 0$ these waves reduce to two macroscopic transverse waves.
3. ($O_{1,2,3,4,5}$) Five Optic waves associated with a transverse macroscopic displacement and with a microdistortion which for $\xi = 0$ reduces to the microdistortions (139), the shear $\mathbf{S}_{13}$ and the rotation $\mathbf{R}_2$.
4. ($OL_{6,7,8,9}$) Four Optic waves associated with a longitudinal macroscopic displacement coupled with a combination of $\mathbf{S}_{1\nu*}$ and $\mathbf{R}_{\omega*}$ that for $\xi = 0$ reduces to (141), the shear $\mathbf{S}_{23}$ and the rotation $\mathbf{R}_1$.

When the direction of propagation is $\mathbf{m} = \mathbf{e}_1$ the roles of ($\mathbf{e}_2$, $\mathbf{W}_5$, $\mathbf{W}_8$) and ($\mathbf{e}_1$, $\mathbf{W}_4$, $\mathbf{W}_7$) are swapped.

4.3.2.4 Classes 4*mm*, 422 axial

For the centrosymmetric class $4mm$ for the axial tensors of the class 422 we have $\mathrm{P} = \mathbf{0}$. Accordingly the matrix A admits the same structure as (164) which is reduced once more by the pairs $\mathcal{M}_1$ and $\mathcal{M}_2$.

Accordingly all the results obtained in §.4.3.1 for the propagation along the c-axis for the classes 4, $\bar{4}$ and $4/m$ still hold provided in (166) and (172) we set:

$$\mathbb{B}_{16} = \mathbb{B}_{26} = \mathbb{B}_{36} = \mathbb{B}_{45} = \mathbb{B}_{47} = 0, \qquad \text{all } \mathsf{P}_{ijk} = 0. \tag{204}$$

The eigenvectors w have the same interpretation as in §.4.3.1, the only difference being that in the limit $\xi = 0$ one acoustic wave is transverse, rather than longitudinal and the other two acoustic waves are neither orthogonal nor transverse.

For all the Tetragonal classes but three, the propagation in a direction orthogonal to the tetragonal c-axis of coupled bulk and lattice waves is described by complex and fully-coupled macroscopic and lattice waves. For the centrosymmetric class $4/m$ the matrix A is reduced by the five-dimensional subspace $\mathcal{L}_1$ and the seven-dimensional subspace $\mathcal{L}_2$ whereas for the centrosymmetric class $4/m$ and for the axial tensors of the class 422 the same solution obtained in the subspaces $\mathcal{M}_1$ and $\mathcal{M}_2$ for the propagation along c still holds with different values of the components. In the special cases of propagation along either $\mathbf{e}_1$ or $\mathbf{e}_2$ for the classes $4/mmm$, $\bar{4}2m$ and the polar tensors of the class 422 the matrix A is reduced by a seven- and a five-dimensional subspaces different from $\mathcal{L}_{1,2}$.

As a final remark, we have seen that the matrix $\mathbb{B}$ reduces Lin in two or three subspaces, depending on the tetragonal classes we are considering.

Likewise, the matrix A reduces the space $\mathcal{V} \oplus \text{Lin}$ in two or three different subspaces, depending on both the direction of propagation $\mathbf{m}$ and the tetragonal classes: further, these subspaces are related to those reduced by $\mathbb{B}$ in the limits for $\zeta_{1,2} \to 0$.

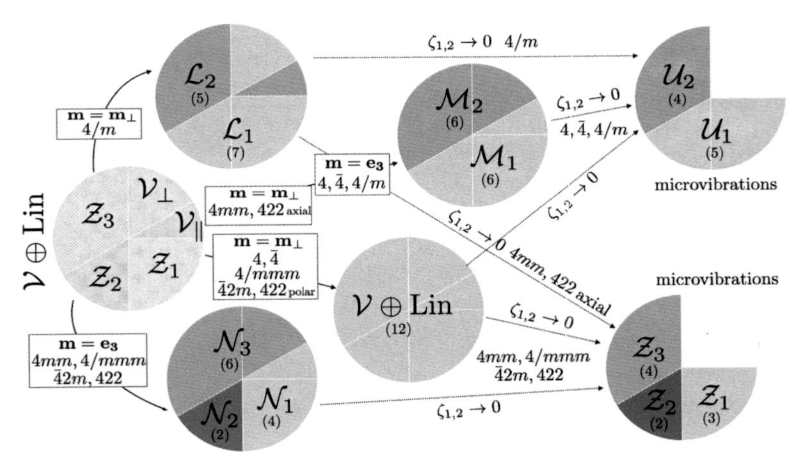

Fig. 6 The subspaces $\mathcal{M}_j$, $\mathcal{L}_j$, $\mathcal{N}_k \subset \mathcal{V} \oplus \text{Lin}$ reduced by A and their relation with the direction of propagation **m** and with the subspaces $\mathcal{U}_j$, $\mathcal{Z}_k \subset \text{Lin}$ reduced by $\mathbb{B}$.

The relations between all these subspaces, the crystal classes and the direction of propagation are graphically depicted in Fig. 6.

Acknowledgments

The research leading to these results is within the scope of CERN R&D Experiment 18 "Crystal Clear Collaboration" and the PANDA Collaboration at GSI-Darmstadt. This work was done under the auspices of the Italian Gruppo Nazionale per la Fisica Matematica (GNFM) and Istituto Nazionale di Alta Matematica (INDAM).

Appendix A. The components of the eigentensors $\mathbf{C}_k$ and the eigenvectors $\mathbf{w}_k$ in the subspaces $\mathcal{U}_1$, $\mathcal{M}_1$ and $\mathcal{M}_2$

In this Appendix we shall list the coefficients of the linear combinations between the elements of the bases $\{\mathbf{e}_k\}$ and $\{\mathbf{W}_h\}$, $k = 1, 2, 3$ and $h = 1, \ldots, 9$ which describes the eigenvectors of the two propagation conditions (70) and (86) for the various subspaces of Lin and $\mathcal{V} \oplus$ Lin respectively reduced by $\mathbb{B}$ and A. These components are calculated by the means of the free web application dCode (dCode.fr).

Subspace $\mathcal{U}_1$

Here the real coefficients a, b, c, d, e and f are defined by (126).

- $\mathbf{C}_1$ and $\mathbf{C}_2$:

$$
\begin{aligned}
\alpha_1 = {} & (hd - (a-\omega^2)g)(l^2 + (b-\omega^2)f) \\
& + (h(a-\omega^2) - gd)(l^2 - (b-\omega^2)(c-\omega^2)) \\
& + hg(2el - (b-\omega^2)g) \\
& + e^2(f - (c-\omega^2))(g \\
& + h) + el(g^2 - 3h^2 \\
& + (f + (c-\omega^2))((a-\omega^2) - d)) + (b-\omega^2)h^3, \\
\beta_1 = {} & -(h(a-\omega^2) + gd)(l^2 + (b-\omega^2)f) \\
& - (hd + (a-\omega^2)g)(l^2 - (b-\omega^2)(c-\omega^2)) \\
& + e^2(g+h)(f - (c-\omega^2)) + (b-\omega^2)g(h^2 - g^2) \\
& + el(2gh - h^2 + 3g^2 \\
& + (f + (c-\omega^2))(d - (a-\omega^2))), \\
\gamma_1 = {} & -l((a-\omega^2) + d)(h^2 + g^2) - egh(h+g) \\
& + e(d - (a-\omega^2))(f - (c-\omega^2))(g-h) \\
& + e(h^3 + g^3) + l(f + (c-\omega^2))((a-\omega^2)^2 - d\bar{d}),
\end{aligned}
$$

$$
\begin{aligned}
\sqrt{2}\delta_1 &= ((a-\omega^2)^2 - d\bar{d})(2l^2 + (b-\omega^2)(f-c)) \\
&+ (h^2+g^2)((b-\omega^2)(a+d) - 2e^2) \\
&+ 2((a-\omega^2) - d)(h(g(b-\omega^2) - el) - e^2(c-\omega^2)) \\
&+ 2((a-\omega^2) + d)e^2 f, \\
\sqrt{2}\epsilon_1 &= (b-\omega^2)(((a-\omega^2)^2 - d\bar{d})(f+c) \\
&+ (h+g)^2(d - (a-\omega^2))) \\
&+ 2(h(el - g(b-\omega^2)) \\
&+ 2e^2(c-\omega^2))(d-(a-\omega^2)) - 2e^2 f((a-\omega^2) + d).
\end{aligned}
$$

- $\mathbf{C}_3$:

$$
\begin{aligned}
\alpha_3 &= e((c-\omega^2) - f) - l(h+g), \\
\gamma_3 &= (h+g)^2 + (d + (a-\omega^2))(f - (c-\omega^2)), \\
\sqrt{2}\delta_3 &= l(d + (a-\omega^2)) - e(h+g).
\end{aligned}
\tag{205}
$$

- $\mathbf{C}_4$ and $\mathbf{C}_5$:

$$
\begin{aligned}
\alpha_4 &= ((b-\omega^2)d - e^2)(f^2 - (c-\omega^2)^2) \\
&+ (l^2 - (b-\omega^2)f)(h^2+g^2) \\
&+ 2((b-\omega^2)(c-\omega^2) - l^2)gh \\
&+ 2l(dl - e(g+h))(f + (c-\omega^2)), \\
\beta_4 &= ((a-\omega^2)(b-\omega^2) - e^2)((c-\omega^2)^2 - f^2) \\
&+ (l^2 - (b-\omega^2)(c-\omega^2))(h^2+g^2) \\
&+ 2((b-\omega^2)f - l^2)gh - 2l((a-\omega^2)l - e(g+h)) \\
&\quad (f + (c-\omega^2)), \\
\gamma_4 &= e(h-g)^2((c-\omega^2) - f) - l(h^3 + g^3 - gh(h+g)) \\
&+ e((a-\omega^2) - d)(f^2 - (c-\omega^2)^2) \\
&+ l(g+h)((a-\omega^2) - d)(f + (c-\omega^2)), \\
\sqrt{2}\delta_4 &= (h-g)((2l^2 + (b-\omega^2)(f - (c-\omega^2)))((a-\omega^2) \\
&\quad + d) \\
&+ 2e^2((c-\omega^2) - f) \\
&+ (b-\omega^2)(h^2-g^2) + 4l(h+g)), \\
\sqrt{2}\epsilon_4 &= ((b-\omega^2)(h-g) - 2el)(h^2 - g^2 \\
&\quad + ((a-\omega^2) - d)(f + (c-\omega^2))),
\end{aligned}
$$

Subspace $\mathcal{M}_1$

The components of A are defined by (166).

- w_1 and w_2:

$$
\begin{aligned}
\alpha_1 = & -(((2l^2 - \bar{g}h - (f - \omega^2)\bar{g})(p - n)^2 \\
& + 2(\bar{d} + e)(h - (f - \omega^2))l^2 + \bar{g}(\bar{d} + e)((f - \omega^2)^2 - h^2))q \\
& + c(-lp^3 + (ln + hm + (f - \omega^2)m)p^2 \\
& + (ln\bar{n} - 2mn(h + (f - \omega^2)) \\
& + (\bar{d} + e)((f - \omega^2) - h)l)p \\
& - ln^3 + (hm + (f - \omega^2)m)n\bar{n} \\
& + ((-\bar{d} - e)h + (\bar{d} + e)(f - \omega^2))ln \\
& + (\bar{d} + e)h^2m + (-\bar{d} - e)(f - \omega^2)^2m) \\
& + b(\bar{g}p^3 + (-\bar{g}n - 2lm)p^2 \\
& + (-\bar{g}n\bar{n} + 4lmn + (\bar{d} + e)\bar{g}h + (-\bar{d} - e)(f - \omega^2)\bar{g})p \\
& + \bar{g}n^3 - 2lmn\bar{n} + ((\bar{d} + e)\bar{g}h + (-\bar{d} - e)(f - \omega^2)\bar{g})n \\
& + l((-2\bar{d} - 2e)hm + (2\bar{d} + 2e)(f - \omega^2)m))), \\
\beta_1 = & -(((2l^2 - \bar{g}h - (f - \omega^2)\bar{g})p \\
& + (-2l^2 + \bar{g}h + (f - \omega^2)\bar{g})n)q\bar{q} \\
& + (c(-3lp^2 + (2ln + 2hm + 2(f - \omega^2)m)p + ln\bar{n} \\
& + (-2hm - 2(f - \omega^2)m)n + ((-\bar{d} - e)h \\
& + (\bar{d} + e)(f - \omega^2))l) + b(3\bar{g}p^2 + (-2\bar{g}n - 4lm) \\
& \quad p - \bar{g}n\bar{n} + 4lmn \\
& + (\bar{d} + e)\bar{g}h + (-\bar{d} - e)(f - \omega^2)\bar{g}))q \\
& + (a - \omega^2)(-\bar{g}p^3 + 3lmp^2 + (\bar{g}n\bar{n} - 2lmn \\
& + (f - \omega^2)(\bar{d}\bar{g} - m\bar{m}) + h(-m\bar{m} - e\bar{g}) \\
& + (e - \bar{d})l^2)p - lmn\bar{n} + (h(m\bar{m} - \bar{d}\bar{g}) + (f - \omega^2)(m\bar{m} \\
& \quad + e\bar{g}) + (\bar{d} - e)l^2)n + l((\bar{d} + e)hm + (-\bar{d} - e)(f \\
& \quad - \omega^2)m)) \\
& + c\bar{c}(p^3 + (-n\bar{n} + eh - \bar{d}(f - \omega^2))p + (\bar{d}h - e(f - \omega^2))n) \\
& \quad + bc(-3mp^2 + (2mn + (2\bar{d} - 2e)l)p \\
& + mn\bar{n} + (2e - 2\bar{d})ln + (-\bar{d} - e)hm + (\bar{d} + e)(f - \omega^2)m) \\
& \quad + b\bar{b}((2m\bar{m} + (e - \bar{d})\bar{g})p + ((\bar{d} - e)\bar{g} - 2m\bar{m})n)),
\end{aligned}
$$

$$
\begin{aligned}
\gamma_1 = {} & (((2l^2 - \bar{g}h - (f - \omega^2)\bar{g})p + (-2l^2 + \bar{g}h + (f - \omega^2)\bar{g})n)q\bar{q} \\
& + (c(-lp^2 + (-2ln + 2hm + 2(f - \omega^2)m)p + 3ln\bar{n} \\
& - (2hm + 2(f - \omega^2)m)n + ((\bar{d} + e)h + (-\bar{d} - e)(f - \omega^2))l) \\
& + b(\bar{g}p^2 + (2\bar{g}n - 4lm)p - 3\bar{g}n\bar{n} + 4lmn \\
& + (-\bar{d} - e)\bar{g}h + (\bar{d} + e)(f - \omega^2)\bar{g}))q + (a - \omega^2)((lm \\
& - \bar{g}n)p^2 + (2lmn + h(\bar{d}\bar{g} - m\bar{m}) + (f - \omega^2)(-m\bar{m} - e\bar{g}) \\
& + (e - \bar{d})l^2)p \\
& + \bar{g}n^3 - 3lmn\bar{n} + ((f - \omega^2)(m\bar{m} - \bar{d}\bar{g}) + h(m\bar{m} + e\bar{g}) + (\bar{d} \\
& - e)l^2)n + l((-\bar{d} - e)hm + (\bar{d} + e)(f - \omega^2)m)) \\
& + c\bar{c}(np^2 + (e(f - \omega^2) - \bar{d}h)p - n^3 + (\bar{d}(f - \omega^2) - eh)n) \\
& + bc(-mp^2 + ((2\bar{d} - 2e)l - 2mn)p + 3mn\bar{n} \\
& + (2e - 2\bar{d})ln + (\bar{d} + e)hm + (-\bar{d} - e)(f - \omega^2)m) \\
& + b\bar{b}((2m\bar{m} + (e - \bar{d})\bar{g})p + ((\bar{d} - e)\bar{g} - 2m\bar{m})n)), \\
\delta_1 = {} & -((b(-2lp^2 + 4lnp - 2ln\bar{n} + ((-2\bar{d} - 2e)h + (2\bar{d} + 2e)(f \\
& - \omega^2))l) + c((h + (f - \omega^2))p^2 + (-2h - 2(f - \omega^2))np \\
& + (h + (f - \omega^2))n\bar{n} \\
& + (\bar{d} + e)h^2 + (-\bar{d} - e)(f - \omega^2)^2))q + (a - \omega^2)(lp^3 + (-ln \\
& - hm - (f - \omega^2)m)p^2 + (-ln\bar{n} + (2hm + 2(f - \omega^2)m)n \\
& + ((\bar{d} + e)h \\
& - (\bar{d} + e)(f - \omega^2))l)p + ln^3 + (-hm - (f - \omega^2)m)n\bar{n} + ((\bar{d} \\
& + e)h + (-\bar{d} - e)(f - \omega^2))ln + (-\bar{d} - e)h^2m + (\bar{d} + e)(f \\
& - \omega^2)^2m) \\
& + bc(-p^3 + np^2 + (n\bar{n} + (-\bar{d} - e)h + (\bar{d} + e)(f - \omega^2))p \\
& - n^3 + ((-\bar{d} - e)h + (\bar{d} + e)(f - \omega^2))n) + b\bar{b}(2mp^2 \\
& - 4mnp + 2mn\bar{n} + (2\bar{d} + 2e)hm + (-2\bar{d} - 2e)(f - \omega^2)m)), \\
\epsilon_1 = {} & (((2h - 2(f - \omega^2))l^2 - \bar{g}h^2 + (f - \omega^2)^2\bar{g})q\bar{q} + (c((2(f - \omega^2) \\
& - 2h)lp + (2(f - \omega^2) - 2h)ln + 2h^2m - 2(f - \omega^2)^2m) \\
& + (2\bar{g}h - 2(f - \omega^2)\bar{g})n + l(4(f - \omega^2)m - 4hm)))q + (a \\
& - \omega^2)((l^2 - \bar{g}h)p^2 + ((2(f - \omega^2)\bar{g} - 2l^2)n + l(2hm - 2(f \\
& - \omega^2)m))p + (l^2 - \bar{g}h)n\bar{n} \\
& + l(2hm - 2(f - \omega^2)m)n + (f - \omega^2)^2(m\bar{m} + e\bar{g}) + h^2(-m\bar{m} \\
& - e\bar{g}) + (2eh - 2e(f - \omega^2))l^2) + bc(-2lp^2 + (4ln - 2hm \\
& + 2(f - \omega^2)m)p
\end{aligned}
$$

$$
\begin{aligned}
&+ (2(f-\omega^2)m - 2hm)n + (4e(f-\omega^2) - 4eh)l) \\
&+ c\bar{c}(hp^2 - 2(f-\omega^2)np + hn\bar{n} + eh^2 - e(f-\omega^2)^2) \\
&\quad + b\bar{b}(\bar{g}p^2 \\
&+ \bar{g}n\bar{n} + h(2m\bar{m} + 2e\bar{g}) + (f-\omega^2)(-2m\bar{m} - 2e\bar{g}))) \\
&\quad + b((2\bar{g}h - 2(f-\omega^2)\bar{g})p - 2ln\bar{n} - 2\bar{g}np,
\end{aligned}
$$

$$
\begin{aligned}
\theta_1 = {} & (((2h - 2(f-\omega^2))l^2 - \bar{g}h^2 + (f-\omega^2)^2\bar{g})q\bar{q} + (c((2(f \\
&\quad - \omega^2) - 2h)lp + (2(f-\omega^2) - 2h)ln + 2h^2m - 2(f \\
&\quad - \omega^2)^2m) \\
&+ (2\bar{g}h - 2(f-\omega^2)\bar{g})n + l(4(f-\omega^2)m - 4hm)))q + (a \\
&\quad - \omega^2)((((f-\omega^2)\bar{g} - l^2)p^2 + ((2l^2 - 2\bar{g}h)n + l(2hm \\
&\quad - 2(f-\omega^2)m))p \\
&+ l(2hm - 2(f-\omega^2)m)n + (f-\omega^2)^2(m\bar{m} - \bar{d}\bar{g}) + h^2(\bar{d}\bar{g} \\
&\quad - m\bar{m}) + (2\bar{d}(f-\omega^2) - 2\bar{d}h)l^2) + bc(2lp^2 + (-4ln \\
&\quad - 2hm + 2(f-\omega^2)m)p \\
&+ (2(f-\omega^2)m - 2hm)n + (4\bar{d}h - 4\bar{d}(f-\omega^2))l) \\
&\quad + b\bar{b}(-\bar{g}p^2 + 2\bar{g}np - \bar{g}n\bar{n} + h(2m\bar{m} - 2\bar{d}\bar{g}) + (f \\
&\quad - \omega^2)(2\bar{d}\bar{g} - 2m\bar{m})) + 2ln\bar{n} \\
&+ c\bar{c}(-(f-\omega^2)p^2 + 2hnp - (f-\omega^2)n\bar{n} - \bar{d}h^2 + \bar{d}(f \\
&\quad - \omega^2)^2)) + b((2\bar{g}h - 2(f-\omega^2)\bar{g})p \\
&+ ((f-\omega^2)\bar{g} - l^2)n\bar{n};
\end{aligned}
$$

- w_3:

$$
\begin{aligned}
\alpha_3 &= 2(ql + bm)(p + n) - c(p + n)^2 + (c(\bar{d} - e) - 2qm)((f \\
&\qquad - \omega^2) + h) - 2bl(\bar{d} - e), \\
\beta_3 &= (qc - (a - \omega^2)m)(p + n) + 2q(bm - lq) + ((a - \omega^2)l \\
&\qquad + bc)(e - \bar{d}), \\
\delta_3 &= 2q\bar{q}(h + (f-\omega^2)) - 4bq(p + n) + (a - \omega^2)(p + n)^2 \\
&\qquad + (a - \omega^2)(e - \bar{d})(h + (f-\omega^2)) + 2b\bar{b}(\bar{d} - e), \\
\epsilon_3 &= ((a - \omega^2)m - qc)((f-\omega^2) + h) + (bc - (a - \omega^2)l)(p \\
&\qquad + n) - 2b(bm + ql).
\end{aligned}
$$

- w_4 and w_5:

$$
\begin{aligned}
\alpha_4 = & -((\bar{g}p^3 + (-\bar{g}n - 2lm)p^2 + (-\bar{g}n\bar{n} + 4lmn + (\bar{d} + e)\bar{g}h \\
& + (-\bar{d} - e)(f - \omega^2)\bar{g})p + \bar{g}n^3 - 2lmn\bar{n} \\
& + ((\bar{d} + e)\bar{g}h - (\bar{d} + e)(f - \omega^2)\bar{g})n + l((-2\bar{d} - 2e)hm \\
& + (2\bar{d} + 2e)(f - \omega^2)m))q + c(-mp^3 + (mn + (\bar{d} \\
& - e)l)p^2 \\
& + (mn\bar{n} + (2e - 2\bar{d})ln + (-\bar{d} - e)hm + (\bar{d} + e)(f \\
& - \omega^2)m)p - mn^3 + (\bar{d} - e)ln\bar{n} + ((-\bar{d} - e)hm + (\bar{d} \\
& + e)(f - \omega^2)m)n \\
& + ((\bar{d}^2 - e^2)h + (e^2 - \bar{d}^2)(f - \omega^2))l) + b((2m\bar{m} + (e \\
& - \bar{d})\bar{g})p^2 + ((2\bar{d} - 2e)\bar{g} - 4m\bar{m})np + (2m\bar{m} + (e \\
& - \bar{d})\bar{g})n\bar{n}, \\
& + h((2\bar{d} + 2e)m\bar{m} + (e^2 - \bar{d}^2)\bar{g}) + (f - \omega^2)((-2\bar{d} \\
& - 2e)m\bar{m} + (\bar{d}^2 - e^2)\bar{g}))) + (\bar{d} - e)ln\bar{n} + (\bar{d} + e)((f \\
& - \omega^2) - h)mn \\
& + ((\bar{d}^2 - e^2)(h - (f - \omega^2)))l) + b((2m\bar{m} + (e - \bar{d})\bar{g})p^2 \\
& + ((2\bar{d} - 2e)\bar{g} - 4m\bar{m})np \\
& + (2m\bar{m} + (e - \bar{d})\bar{g})n\bar{n} + h((2\bar{d} + 2e)m\bar{m} + (e^2 - \bar{d}^2)\bar{g}) \\
& + (f - \omega^2)((-2\bar{d} - 2e)m\bar{m} + (\bar{d}^2 - e^2)\bar{g}))), \\
\beta_4 = & (\bar{g}q\bar{q}(p - n)^2 + 2(\bar{d} + e)(\bar{g}h - l^2)q\bar{q} + (c(-2mp^2 \\
& + (4mn + (2\bar{d} + 2e)l)p - 2mn\bar{n} + (2\bar{d} + 2e)ln \\
& + (-4\bar{d} - 4e)hm) + b((-2\bar{d} - 2e)\bar{g}p + (-2\bar{d} - 2e)\bar{g}n \\
& + (4\bar{d} + 4e)lm))q + (a - \omega^2)((m\bar{m} + e\bar{g})p^2 \\
& + ((2\bar{d}\bar{g} - 2m\bar{m})n + (-2\bar{d} - 2e)lm)p + (m\bar{m} + e\bar{g})n\bar{n} \\
& + (-2\bar{d} - 2e)lmn + h((2\bar{d} + 2e)m\bar{m} + (e^2 - \bar{d}^2)\bar{g}) \\
& + (\bar{d}^2 - e^2)l^2) + c\bar{c}(-ep^2 - 2\bar{d}np - en\bar{n} + (\bar{d}^2 - e^2)h) \\
& + bc((2\bar{d} + 2e)mp + (2\bar{d} + 2e)mn \\
& + (2e^2 - 2\bar{d}^2)l) + b\bar{b}((-2\bar{d} - 2e)m\bar{m} + (\bar{d}^2 - e^2)\bar{g}),
\end{aligned}
$$

$$\begin{aligned}
\gamma_4 = {} & (\bar{g}q\bar{q}(p-n)^2 + 2(\bar{d}+e)(l^2 - (f-\omega^2)\bar{g})q\bar{q} + (c(-2mp^2 \\
& + (4mn + (-2\bar{d} - 2e)l)p - 2mn\bar{n} + (-2\bar{d} - 2e)ln \\
& + (4\bar{d} + 4e)(f-\omega^2)m) + b((2\bar{d} + 2e)\bar{g}p + (2\bar{d} + 2e)\bar{g}n \\
& + (-4\bar{d} - 4e)lm))q + (a - \omega^2)((m\bar{m} - \bar{d}\bar{g})p^2 \\
& + (2\bar{d} + 2e)lmn + (f-\omega^2)((-2\bar{d} - 2e)m\bar{m} + (\bar{d}^2 \\
& - e^2)\bar{g}) + (e^2 - \bar{d}^2)l^2) + c\bar{c}(\bar{d}p^2 + 2enp + \bar{d}n\bar{n} + (e^2 \\
& - \bar{d}^2)(f-\omega^2)) \\
& + ((-2m\bar{m} - 2e\bar{g})n + (2\bar{d} + 2e)lm)p + (m\bar{m} - \bar{d}\bar{g})n\bar{n} \\
& + bc((-2\bar{d} - 2e)mp \\
& + (-2\bar{d} - 2e)mn + (2\bar{d}^2 - 2e^2)l) + b\bar{b}((2\bar{d} + 2e)m\bar{m} \\
& + (e^2 - \bar{d}^2)\bar{g}), \\
\delta_4 = {} & -(2l(p-n)^2q\bar{q} + 2(\bar{d}+e)(h - (f-\omega^2))lq\bar{q} + (c(-p^3 \\
& + np^2 + (n\bar{n} + (-\bar{d} - e)h + (\bar{d} + e)(f-\omega^2))p - n^3 \\
& + ((-\bar{d} - e)h \\
& + (\bar{d} + e)(f-\omega^2))n) + b(-2mp^2 + 4mnp - 2mn\bar{n} \\
& + (-2\bar{d} - 2e)hm + (2\bar{d} + 2e)(f-\omega^2)m))q + (a \\
& - \omega^2)(mp^3 + ((e - \bar{d})l - mn)p^2 \\
& + (-mn\bar{n} + (2\bar{d} - 2e)ln + (\bar{d} + e)hm + (-\bar{d} - e)(f \\
& - \omega^2)m)p + mn^3 + (e - \bar{d})ln\bar{n} + ((\bar{d} + e)hm + (-\bar{d} \\
& - e)(f-\omega^2)m)n \\
& + ((e^2 - \bar{d}^2)h + (\bar{d}^2 - e^2)(f-\omega^2))l) + bc((\bar{d} - e)p^2 \\
& + (2e - 2\bar{d})np + (\bar{d} - e)n\bar{n} + (\bar{d}^2 - e^2)h + (e^2 \\
& - \bar{d}^2)(f-\omega^2))),
\end{aligned}$$

- w_6:

$$\begin{aligned}
\alpha_6 = {} & \bar{g}(p+n)^2 - 4lm(p+n) \\
& + (h + (f-\omega^2))(2m\bar{m} + (e - \bar{d})\bar{g}) + 2l^2(\bar{d} - e),
\end{aligned}$$

$$\begin{aligned}
\beta_6 &= l(2mq + c(e - \bar{d})) + (p + n)(cm - \bar{g}q) \\
&\quad + b((\bar{d} - e)\bar{g} - 2m^2), \\
\delta_6 &= -(f + h)(2mq + c(e - \bar{d})) - 2(p + n)(lq + bm) \\
&\quad + c(p + n)^2 + 2bl(\bar{d} - e), \\
\epsilon_6 &= (h + (f - \omega^2))(\bar{g}q - cm) \\
&\quad + (cl - b\bar{g})(p + n) + 2l(bm - lq),
\end{aligned}$$

Subspace $\mathcal{M}_2$

The components of A are defined by (172).

- w_7

$$\begin{aligned}
\alpha_7 &= ((e - \omega^2)hn^3 + (-2ghm + c(2h^2 + (e - \omega^2)(f - \omega^2)) \\
&\quad - (e - \omega^2)dg)n\bar{n} + ((f - \omega^2)hm\bar{m} + (d(2g^2 - 2h^2) \\
&\quad - 2c(f - \omega^2)g)m \\
&\quad + b(-gh^2 - g^3 + (e - \omega^2)(f - \omega^2)g) - 4cdgh + 3c\bar{c}(f \\
&\quad - \omega^2)h + (e - \omega^2)d\bar{d}h)n + (c(f - \omega^2)^2 - d(f \\
&\quad - \omega^2)g)m\bar{m} \\
&\quad + 2d\bar{d}gh - 2cd(f - \omega^2)h)m + d(b(((e - \omega^2)(f - \omega^2) \\
&\quad - g^2)h - h^3) + (a - \omega^2)(-gh^2 - g^3 + (e - \omega^2)(f \\
&\quad - \omega^2)g) - 3c\bar{c}(f - \omega^2)g) \\
&\quad + cd\bar{d}(2g^2 + (e - \omega^2)(f - \omega^2)) - (e - \omega^2)d^3g + c^3(f \\
&\quad - \omega^2)^2) + (a - \omega^2)(h^3 + (g^2 - (e - \omega^2)(f - \omega^2))h) \\
&\quad + (b((f - \omega^2)h^2 + (f - \omega^2)g^2 - (e - \omega^2)(f - \omega^2)^2) + (a \\
&\quad - \omega^2)c((f - \omega^2)h^2 + (f - \omega^2)g^2 - (e - \omega^2)(f \\
&\quad - \omega^2)^2), \\
\beta_7 &= ((e - \omega^2)gn^3 + ((-2g^2 - (e - \omega^2)(f - \omega^2))m + 2cgh \\
&\quad + (e - \omega^2)dh)n\bar{n} + (3(f - \omega^2)gm\bar{m} + (-4dgh - 2c(f \\
&\quad - \omega^2)h)m
\end{aligned}$$

$$
\begin{aligned}
&+ (a-\omega^2)(gh^2+g^3-(e-\omega^2)(f-\omega^2)g) + cd(2h^2 \\
&- 2g^2) + c\bar{c}(f-\omega^2)g + (e-\omega^2)d\bar{d}g)n - (f-\omega^2)^2m^3 \\
&+ 3d(f-\omega^2)hm\bar{m} \\
&+ d\bar{d}(-2h^2-(e-\omega^2)(f-\omega^2)) + 2cd(f-\omega^2)g - c\bar{c}(f \\
&- \omega^2)^2)m + d((a-\omega^2)(h^3+(g^2-(e-\omega^2)(f \\
&- \omega^2))h) + b(gh^2+g^3 \\
&+ bc(-(f-\omega^2)h^2-(f-\omega^2)g^2+(e-\omega^2)(f-\omega^2)^2) \\
&- 2cd\bar{d}gh + (e-\omega^2)d^3h) + b(((e-\omega^2)(f-\omega^2) \\
&- g^2)h - h^3) \\
&+ ((a-\omega^2)(-(f-\omega^2)h^2-(f-\omega^2)g^2+(e-\omega^2)(f \\
&- \omega^2)^2) - (e-\omega^2)(f-\omega^2)g) + c\bar{c}(f-\omega^2)h),
\end{aligned}
$$

$$
\begin{aligned}
\gamma_7 = &- (b(h^2-g^2)n\bar{n} + (2b(f-\omega^2)gm - 4bdgh + 2bc(f \\
&- \omega^2)h)n - b(f-\omega^2)^2m\bar{m} + 2bd(f-\omega^2)hm + bd\bar{d}(g^2 \\
&- h^2) \\
&- 2bcd(f-\omega^2)g + bc\bar{c}(f-\omega^2)^2),
\end{aligned}
$$

$$
\begin{aligned}
\delta_7 = &- ((e-\omega^2)mn^3 + (-2gm\bar{m} + 2chm - (e-\omega^2)bh - (e \\
&- \omega^2)(a-\omega^2)g + (e-\omega^2)cd)n\bar{n} + ((f-\omega^2)m^3 \\
&- 2dhm\bar{m} \\
&+ c\bar{c}(f-\omega^2) + (e-\omega^2)d\bar{d})m + c(b(-h^2-g^2-(e \\
&- \omega^2)(f-\omega^2)) - 2(a-\omega^2)gh) + d(2c\bar{c}h + 2(e \\
&- \omega^2)bg))n + (b(f-\omega^2)h \\
&+ (d(b(-h^2-g^2-(e-\omega^2)(f-\omega^2)) + 2(a-\omega^2)gh) \\
&- 2cd\bar{d}h + 2bc(f-\omega^2)g)m + d((a-\omega^2)c(h^2+3g^2 \\
&- (e-\omega^2)(f-\omega^2)) \\
&+ (a-\omega^2)^2(-gh^2-g^3+(e-\omega^2)(f-\omega^2)g) + c\bar{c}(-b(f \\
&- \omega^2)h - (a-\omega^2)(f-\omega^2)g) + d\bar{d}((e-\omega^2)bh - 2c\bar{c}g \\
&- (e-\omega^2)(a-\omega^2)g)
\end{aligned}
$$

$$
\begin{aligned}
& + ((a-\omega^2)(h^2+3g^2-(e-\omega^2)(f-\omega^2)) - 4cdg - (a \\
& \quad -\omega^2)(f-\omega^2)g + cd(f-\omega^2))m\bar{m} + c^3(f-\omega^2)) \\
& \quad + b\bar{b}(gh^2+g^3) \\
& - (e-\omega^2)(f-\omega^2)g + (e-\omega^2)cd^3), \\
\epsilon_7 = {} & -((e-\omega^2)cn^3 + ((-2cg-(e-\omega^2)d)m + 2c\bar{c}h + (e \\
& \quad -\omega^2)(a-\omega^2)h + (e-\omega^2)bg)n\bar{n} + ((2dg+c(f \\
& \quad -\omega^2))m\bar{m} \\
& + (a-\omega^2)c(3h^2+g^2-(e-\omega^2)(f-\omega^2)) + d(2(e \\
& \quad -\omega^2)bh - 2c\bar{c}g) + c^3(f-\omega^2) + (e-\omega^2)cd\bar{d})n - d(f \\
& \quad -\omega^2)m^3 \\
& + (d((a-\omega^2)(-3h^2-g^2+(e-\omega^2)(f-\omega^2)) - c\bar{c}(f \\
& \quad -\omega^2)) - 2bc(f-\omega^2)h + 2cd\bar{d}g - (e-\omega^2)d^3)m \\
& + cd(b(h^2+g^2+(e-\omega^2)(f-\omega^2)) - 2(a-\omega^2)gh) \\
& \quad + c\bar{c}((a-\omega^2)(f-\omega^2)h - b(f-\omega^2)g) + d\bar{d}((e \\
& \quad -\omega^2)(a-\omega^2)h - (e-\omega^2)bg)) \\
& + (b(-h^2-g^2-(e-\omega^2)(f-\omega^2)) - 2(a-\omega^2)gh \\
& \quad - 4cdh)m + ((a-\omega^2)(f-\omega^2)h + 2d\bar{d}h + b(f \\
& \quad -\omega^2)g)m\bar{m} \\
& + (a-\omega^2)^2(h^3+(g^2-(e-\omega^2)(f-\omega^2))h) + b\bar{b}(((e \\
& \quad -\omega^2)(f-\omega^2) - g^2)h - h^3), \\
\theta_7 = {} & ((e-\omega^2)n^4 + (2ch-2gm)n^3 + ((f-\omega^2)m\bar{m} - 2dhm \\
& \quad + (a-\omega^2)(h^2+g^2-2(e-\omega^2)(f-\omega^2)) - 2bgh \\
& \quad - 2cdg + c\bar{c}(f-\omega^2) \\
& + ((2b(f-\omega^2)h + 2(a-\omega^2)(f-\omega^2)g - 2d\bar{d}g)m \\
& \quad + bd(2g^2-2h^2) + c(-2(a-\omega^2)(f-\omega^2)h - 2b(f \\
& \quad -\omega^2)g) + 2cd\bar{d}h)n + (d\bar{d}(f-\omega^2) \\
& + (d(2(a-\omega^2)(f-\omega^2)h - 2b(f-\omega^2)g) - 2d^3h \\
& \quad + 2bc(f-\omega^2)^2)m + d\bar{d}((a-\omega^2)(h^2+g^2-2(e \\
& \quad -\omega^2)(f-\omega^2)) + 2bgh + c\bar{c}(f-\omega^2))
\end{aligned}
$$

$$+ (a-\omega^2)^2(-(f-\omega^2)h^2 - (f-\omega^2)g^2 + (e-\omega^2)(f-\omega^2)^2) + cd(2(a-\omega^2)(f-\omega^2)g - 2b(f-\omega^2)h) - 2cd^3g - (a-\omega^2)c\bar{c}(f-\omega^2)^2 + b\bar{b}((f-\omega^2)h^2 + (f-\omega^2)g^2 - (e-\omega^2)(f-\omega^2)^2) - (a-\omega^2)(f-\omega^2)^2)m\bar{m} + 2(e-\omega^2)d\bar{d})n\bar{n} + (e-\omega^2)d^4);$$

- w_8

$$\begin{aligned}
\alpha_8 = {} & ((e-\omega^2)^2n^3 + ((e-\omega^2)ch - 3(e-\omega^2)gm)n\bar{n} + ((-2h^2 + 2g^2 + (e-\omega^2)(f-\omega^2))m\bar{m} - 2cghm + (a-\omega^2)((e-\omega^2)h^2 \\
& + (e-\omega^2)g^2 - (e-\omega^2)^2(f-\omega^2)) + c\bar{c}((e-\omega^2)(f-\omega^2) - 2h^2) + (a-\omega^2)(gh^2 - g^3 + (e-\omega^2)(f-\omega^2)g) \\
& - 2(e-\omega^2)bgh - 2(e-\omega^2)cdg + (e-\omega^2)^2d\bar{d})n - (f-\omega^2)gm^3 + (2dgh - c(f-\omega^2)h)m\bar{m} + (b((g^2 + (e-\omega^2)(f-\omega^2))h - h^3) \\
& + 2cdg^2 - c\bar{c}(f-\omega^2)g - (e-\omega^2)d\bar{d}g)m + c((a-\omega^2)((g^2 + (e-\omega^2)(f-\omega^2))h - h^3) + b(-gh^2 + g^3 - (e-\omega^2)(f-\omega^2)g)) \\
& - 2(e-\omega^2)(a-\omega^2)gh) - c^3(f-\omega^2)h + (e-\omega^2)cd\bar{d}h) \\
& + d(b(-(e-\omega^2)h^2 - (e-\omega^2)g^2 + (e-\omega^2)^2(f-\omega^2)), \\
\beta_8 = {} & -(((e-\omega^2)hm + (e-\omega^2)cg - (e-\omega^2)^2d)n\bar{n} + ((2(e-\omega^2)dg - 2cg^2)m + b((e-\omega^2)h^2 + (e-\omega^2)g^2 - (e-\omega^2)^2(f-\omega^2)) \\
& + 2c\bar{c}gh - 2(e-\omega^2)(a-\omega^2)gh)n + (e-\omega^2)(f-\omega^2)g) \\
& - 2cdgh - c\bar{c}(f-\omega^2)h + (e-\omega^2)d\bar{d}h)m \\
& + c(b((g^2 + (e-\omega^2)(f-\omega^2))h - h^3) + (a-\omega^2)(-gh^2 + g^3 - (e-\omega^2)(f-\omega^2)g)) + c^3(f-\omega^2)g + 3(e-\omega^2)cd\bar{d}g - (e-\omega^2)^2d^3) \\
& + d((a-\omega^2)(-(e-\omega^2)h^2 - (e-\omega^2)g^2 + (e-\omega^2)^2(f-\omega^2)) + c\bar{c}(2h^2 - 2g^2 - (e-\omega^2)(f-\omega^2)) - 2(e-\omega^2)bgh),
\end{aligned}$$

$$
\begin{aligned}
\gamma_8 = & -((e-\omega^2)mn^3 + (-2gm\bar{m} + (e-\omega^2)bh - (e-\omega^2)(a \\
& -\omega^2)g + (e-\omega^2)cd)n\bar{n} + ((f-\omega^2)m^3 \\
& + ((a-\omega^2)(-h^2+3g^2-(e-\omega^2)(f-\omega^2)) - 2bgh - 4cdg \\
& + c\bar{c}(f-\omega^2) + (e-\omega^2)d\bar{d})m) + bcn(g^2 - h^2 + (e \\
& -\omega^2)(f-\omega^2) - 2(e-\omega^2)bdg) \\
& + (b(f-\omega^2)h - (a-\omega^2)(f-\omega^2)g + cd(f-\omega^2))m\bar{m} \\
& + (bd(-h^2+g^2+(e-\omega^2)(f-\omega^2)) - 2bc(f-\omega^2)g)m \\
& + (a-\omega^2)^2(gh^2 - g^3 + (e-\omega^2)(f-\omega^2)g) + b\bar{b}(-gh^2+g^3 \\
& -(e-\omega^2)(f-\omega^2)g) + d(c((a-\omega^2)(-h^2+3g^2-(e \\
& -\omega^2)(f-\omega^2)) + 2bgh) \\
& + c^3(f-\omega^2)) + bc(-h^2+g^2+(e-\omega^2)(f-\omega^2) - 2(e \\
& -\omega^2)bdg)n + d\bar{d}(-(e-\omega^2)bh - 2c\bar{c}g - (e-\omega^2)(a \\
& -\omega^2)g) + (e-\omega^2)cd^3), \\
\delta_8 = & ((e-\omega^2)^2bn\bar{n} + (2hm^3 + (2c\bar{c}h - 2(e-\omega^2)(a-\omega^2)h - 2(e \\
& -\omega^2)bg)m)n \\
& + (b(h^2+g^2) - 2(a-\omega^2)gh + 2cdh)m\bar{m} + c\bar{c}(b(-h^2-g^2) \\
& -2(a-\omega^2)gh) \\
& + d(c(2(e-\omega^2)bg - 2(e-\omega^2)(a-\omega^2)h) + 2c^3h) - 2(e \\
& -\omega^2)b\bar{b}gh + 2(e-\omega^2)(a-\omega^2)^2gh - (e-\omega^2)^2bd\bar{d}), \\
\epsilon_8 = & (((e-\omega^2)m\bar{m} + (e-\omega^2)c\bar{c} - (e-\omega^2)^2(a-\omega^2))n\bar{n} \\
& + (-2gm^3 + 2chm\bar{m} + (2(e-\omega^2)bh - 2c\bar{c}g + 2(e-\omega^2)(a \\
& -\omega^2)g)m \\
& + c(2(e-\omega^2)bg - 2(e-\omega^2)(a-\omega^2)h) + 2c^3h + (bc(2h^2 \\
& -2g^2) \\
& -2(e-\omega^2)^2bd)n + (f-\omega^2)m^4 - 2dhm^3 + ((a-\omega^2)(h^2 \\
& + g^2 - 2(e-\omega^2)(f-\omega^2)) - 2bgh - 2cdg + 2c\bar{c}(f-\omega^2) \\
& + (e-\omega^2)d\bar{d})m\bar{m} \\
& + d(-2c\bar{c}h + 2(e-\omega^2)(a-\omega^2)h + 2(e-\omega^2)bg))m + c\bar{c}((a \\
& -\omega^2)(h^2+g^2-2(e-\omega^2)(f-\omega^2)) + 2bgh) \\
& + b\bar{b}((e-\omega^2)h^2 + (e-\omega^2)g^2 - (e-\omega^2)^2(f-\omega^2)) + ((e \\
& -\omega^2)c\bar{c} - (e-\omega^2)^2(a-\omega^2))d\bar{d}) \\
& + (a-\omega^2)^2(-(e-\omega^2)h^2 - (e-\omega^2)g^2 + (e-\omega^2)^2(f \\
& -\omega^2)) + d(c(2(e-\omega^2)(a-\omega^2)g - 2(e-\omega^2)bh) \\
& -2c^3g) + c^4(f-\omega^2),
\end{aligned}
$$

$$\begin{aligned}
\theta_8 = {} & -((e-\omega^2)cn^3 + (2hm\bar{m} + (-2cg - (e-\omega^2)d)m + 2c\bar{c}h \\
& - (e-\omega^2)(a-\omega^2)h + (e-\omega^2)bg)n\bar{n} + ((2dg + c(f \\
& - \omega^2))m\bar{m} \\
& + (b(3h^2 - g^2 - (e-\omega^2)(f-\omega^2)) - 2(a-\omega^2)gh)m \\
& + (a-\omega^2)c(-h^2 + g^2 - (e-\omega^2)(f-\omega^2)) + d(-2(e \\
& - \omega^2)bh - 2c\bar{c}g) \\
& + c^3(f-\omega^2) + (e-\omega^2)cd\bar{d})n + 2bc(f-\omega^2)h + 2cd\bar{d}g \\
& - (e-\omega^2)d^3)m \\
& - d(f-\omega^2)m^3 + (-(a-\omega^2)(f-\omega^2)h + 2d\bar{d}h + b(f \\
& - \omega^2)g)m\bar{m} + (d((a-\omega^2)(h^2 - g^2 + (e-\omega^2)(f \\
& - \omega^2)) - c\bar{c}(f-\omega^2)) \\
& + b\bar{b}(h^3 + (-g^2 - (e-\omega^2)(f-\omega^2))h) + (a-\omega^2)^2((g^2 \\
& + (e-\omega^2)(f-\omega^2))h - h^3) + cd(b(-3h^2 + g^2 + (e \\
& - \omega^2)(f-\omega^2)) - 2(a-\omega^2)gh) \\
& + c\bar{c}(-(a-\omega^2)(f-\omega^2)h - b(f-\omega^2)g) + d\bar{d}(2c\bar{c}h - (e \\
& - \omega^2)(a-\omega^2)h - (e-\omega^2)bg));
\end{aligned}$$

- w_9

$$\begin{aligned}
\alpha_9 = {} & -(((e-\omega^2)hm + (e-\omega^2)cg - (e-\omega^2)^2d)n\bar{n} + b(-(e \\
& - \omega^2)h^2 - (e-\omega^2)g^2 + (e-\omega^2)^2(f-\omega^2)) \\
& + (-2ghm\bar{m} + (c(2h^2 - 2g^2) + 2(e-\omega^2)dg)m + ((a \\
& - \omega^2)(h^3 + (g^2 - (e-\omega^2)(f-\omega^2))h) + b(gh^2 + g^3 \\
& - (e-\omega^2)(f-\omega^2)g) \\
& + 2c\bar{c}gh - 2(e-\omega^2)cdh)n + (f-\omega^2)hm^3 + (d(-2h^2 - (e \\
& - \omega^2)(f-\omega^2)) + c(f-\omega^2)g)m\bar{m} \\
& - 4cdgh + c\bar{c}(f-\omega^2)h + 3(e-\omega^2)d\bar{d}h)m + c(b(((e \\
& - \omega^2)(f-\omega^2) - g^2)h - h^3) + (a-\omega^2)(gh^2 + g^3 - (e \\
& - \omega^2)(f-\omega^2)g)) \\
& + d((a-\omega^2)(-(e-\omega^2)h^2 - (e-\omega^2)g^2 + (e-\omega^2)^2(f \\
& - \omega^2)) + c\bar{c}(-2g^2 - (e-\omega^2)(f-\omega^2))) + c^3(f \\
& - \omega^2)g + 3(e-\omega^2)cd\bar{d}g - (e-\omega^2)^2d^3),
\end{aligned}$$

$$
\begin{aligned}
\beta_9 = & -((e-\omega^2)^2n^3 + (3(e-\omega^2)ch - 3(e-\omega^2)gm)n\bar{n} + ((2g^2 \\
& + (e-\omega^2)(f-\omega^2))m\bar{m} + (-4cgh - 2(e-\omega^2)dh)m + (a \\
& - \omega^2)((e-\omega^2)h^2 \\
& + (e-\omega^2)g^2 - (e-\omega^2)^2(f-\omega^2)) + c\bar{c}(2h^2 + (e-\omega^2)(f \\
& - \omega^2)) - 2(e-\omega^2)cdg \\
& + (e-\omega^2)^2d\bar{d})n - (f-\omega^2)gm^3 + (2dgh + c(f-\omega^2)h)m\bar{m} \\
& + (b(((e-\omega^2)(f-\omega^2) - g^2)h - h^3) \\
& + (a-\omega^2)(-gh^2 - g^3 + (e-\omega^2)(f-\omega^2)g) + cd(2g^2 \\
& - 2h^2) - c\bar{c}(f-\omega^2)g - (e-\omega^2)d\bar{d}g)m + c((a-\omega^2)(h^3 \\
& + (g^2 - (e-\omega^2)(f-\omega^2))h) \\
& + b(-gh^2 - g^3 + (e-\omega^2)(f-\omega^2)g)) + d(b((e-\omega^2)h^2 \\
& + (e-\omega^2)g^2 - (e-\omega^2)^2(f-\omega^2)) - 2c\bar{c}gh) + c^3(f \\
& - \omega^2)h + (e-\omega^2)cd\bar{d}h), \\
\gamma_9 = & ((e-\omega^2)cn^3 + ((-2cg - (e-\omega^2)d)m + 2c\bar{c}h + (e-\omega^2)(a \\
& - \omega^2)h - (e-\omega^2)bg)n\bar{n} + ((2dg + c(f-\omega^2))m\bar{m} \\
& + (b(h^2 + g^2 + (e-\omega^2)(f-\omega^2)) - 2(a-\omega^2)gh - 4cdh)m \\
& + ((a-\omega^2)(f-\omega^2)h + 2d\bar{d}h - b(f-\omega^2)g)m\bar{m} \\
& + (a-\omega^2)c(3h^2 + g^2 - (e-\omega^2)(f-\omega^2)) + d(-2(e \\
& - \omega^2)bh - 2c\bar{c}g) + c^3(f-\omega^2) + (e-\omega^2)cd\bar{d})n - d(f \\
& - \omega^2)m^3 \\
& + (d((a-\omega^2)(-3h^2 - g^2 + (e-\omega^2)(f-\omega^2)) - c\bar{c}(f \\
& - \omega^2)) + 2bc(f-\omega^2)h + 2cd\bar{d}g - (e-\omega^2)d^3)m \\
& + (a-\omega^2)^2(h^3 + (g^2 - (e-\omega^2)(f-\omega^2))h) + b\bar{b}(((e \\
& - \omega^2)(f-\omega^2) - g^2)h - h^3) \\
& + cd(b(-h^2 - g^2 - (e-\omega^2)(f-\omega^2)) - 2(a-\omega^2)gh) \\
& + c\bar{c}((a-\omega^2)(f-\omega^2)h + b(f-\omega^2)g) + d\bar{d}((e-\omega^2)(a \\
& - \omega^2)h + (e-\omega^2)bg)), \\
\delta_9 = & (((e-\omega^2)m\bar{m} + (e-\omega^2)c\bar{c} - (e-\omega^2)^2(a-\omega^2))n\bar{n} \\
& + (-2gm^3 + 2chm\bar{m} + (-2(e-\omega^2)bh - 2c\bar{c}g + 2(e \\
& - \omega^2)(a-\omega^2)g)m \\
& + c(-2(e-\omega^2)(a-\omega^2)h - 2(e-\omega^2)bg) + b\bar{b}((e-\omega^2)h^2 \\
& + (e-\omega^2)g^2 - (e-\omega^2)^2(f-\omega^2))
\end{aligned}
$$

$$
\begin{aligned}
&+ 2c^3h + 2(e-\omega^2)^2bd)n + (f-\omega^2)m^4 - 2dhm^3 + ((a\\
&\quad -\omega^2)(h^2+g^2-2(e-\omega^2)(f-\omega^2)) + 2bgh - 2cdg\\
&\quad + 2c\bar{c}(f-\omega^2) + (e-\omega^2)d\bar{d})m\bar{m}\\
&+ (bc(2g^2-2h^2) + d(-2c\bar{c}h + 2(e-\omega^2)(a-\omega^2)h - 2(e\\
&\quad -\omega^2)bg))m + c\bar{c}((a-\omega^2)(h^2+g^2-2(e-\omega^2)(f\\
&\quad -\omega^2)) - 2bgh)\\
&+ (a-\omega^2)^2(-(e-\omega^2)h^2 - (e-\omega^2)g^2 + (e-\omega^2)^2(f\\
&\quad -\omega^2)) + d(c(2(e-\omega^2)bh + 2(e-\omega^2)(a-\omega^2)g)\\
&\quad - 2c^3g)\\
&+ c^4(f-\omega^2) + ((e-\omega^2)c\bar{c} - (e-\omega^2)^2(a-\omega^2))d\bar{d}),
\end{aligned}
$$

$$
\begin{aligned}
\epsilon_9 = {}& ((e-\omega^2)^2bn\bar{n} + (2(e-\omega^2)bch - 2(e-\omega^2)bgm)n + b(g^2\\
&\quad - h^2)m\bar{m} + (2(e-\omega^2)bdh - 4bcgh)m + bc\bar{c}(h^2-g^2)\\
&+ 2(e-\omega^2)bcdg - (e-\omega^2)^2bd\bar{d}),
\end{aligned}
$$

$$
\begin{aligned}
\theta_9 = {}& -((e-\omega^2)mn^3 + (-2gm\bar{m} + 2chm - (e-\omega^2)bh - (e\\
&\quad -\omega^2)(a-\omega^2)g + (e-\omega^2)cd)n\bar{n} + ((f-\omega^2)m^3\\
&\quad - 2dhm\bar{m}\\
&+ ((a-\omega^2)(h^2+3g^2-(e-\omega^2)(f-\omega^2)) - 4cdg + c\bar{c}(f\\
&\quad -\omega^2) + (e-\omega^2)d\bar{d})m\\
&+ c(b(-h^2-g^2-(e-\omega^2)(f-\omega^2)) - 2(a-\omega^2)gh)\\
&\quad + d(2c\bar{c}h + 2(e-\omega^2)bg))n + (b(f-\omega^2)h - (a-\omega^2)(f\\
&\quad -\omega^2)g + cd(f-\omega^2))m\bar{m}\\
&+ (d(b(-h^2-g^2-(e-\omega^2)(f-\omega^2)) + 2(a-\omega^2)gh)\\
&\quad - 2cd\bar{d}h + 2bc(f-\omega^2)g)m + d((a-\omega^2)c(h^2+3g^2-(e\\
&\quad -\omega^2)(f-\omega^2))\\
&+ c^3(f-\omega^2)) + b\bar{b}(gh^2+g^3-(e-\omega^2)(f-\omega^2)g) + (e\\
&\quad -\omega^2)cd^3)\\
&+ (a-\omega^2)^2(-gh^2-g^3+(e-\omega^2)(f-\omega^2)g) + c\bar{c}(-b(f\\
&\quad -\omega^2)h - (a-\omega^2)(f-\omega^2)g) + d\bar{d}((e-\omega^2)bh - 2c\bar{c}g\\
&\quad - (e-\omega^2)(a-\omega^2)g);
\end{aligned}
$$

- w_{10}

$$
\begin{aligned}
\alpha_{10} = {}& -((e-\omega^2)gn^3 + ((-2h^2-2g^2-(e-\omega^2)(f-\omega^2))m\\
&\quad + (e-\omega^2)dh)n\bar{n} + (3(f-\omega^2)gm\bar{m} - 2c(f-\omega^2)hm
\end{aligned}
$$

$$
\begin{aligned}
& + b((-g^2 - (e-\omega^2)(f-\omega^2))h - h^3) + (a-\omega^2)(gh^2 \\
& \quad + g^3 - (e-\omega^2)(f-\omega^2)g) - 2cd(h^2+g^2) \\
& + c\bar{c}(f-\omega^2)g + (e-\omega^2)d\bar{d}g)n - (f-\omega^2)^2 m^3 + d(f \\
& \quad - \omega^2)hm\bar{m} + ((a-\omega^2)((f-\omega^2)h^2 - (f-\omega^2)g^2 + (e \\
& \quad - \omega^2)(f-\omega^2)^2) \\
& + 2b(f-\omega^2)gh + 2cd(f-\omega^2)g - c\bar{c}(f-\omega^2)^2 - (e \\
& \quad - \omega^2)d\bar{d}(f-\omega^2))m + d((a-\omega^2)((-g^2 - (e-\omega^2)(f \\
& \quad - \omega^2))h - h^3) \\
& + b(-gh^2 - g^3 + (e-\omega^2)(f-\omega^2)g) - c\bar{c}(f-\omega^2)h) \\
& \quad + c(b(-(f-\omega^2)h^2 + (f-\omega^2)g^2 - (e-\omega^2)(f-\omega^2)^2) \\
& \quad + 2(a-\omega^2)(f-\omega^2)gh) \\
& + (e-\omega^2)d^3 h),
\end{aligned}
$$

$$
\begin{aligned}
\beta_{10} = \; & ((e-\omega^2)hn^3 + ((e-\omega^2)c(f-\omega^2) - (e-\omega^2)dg)n\bar{n} \\
& \quad + (-(f-\omega^2)hm\bar{m} + (d(2h^2+2g^2) - 2c(f-\omega^2)g)m \\
& + (a-\omega^2)((-g^2 - (e-\omega^2)(f-\omega^2))h - h^3) + b(gh^2 \\
& \quad + g^3 - (e-\omega^2)(f-\omega^2)g) + c\bar{c}(f-\omega^2)h + (e \\
& \quad - \omega^2)d\bar{d}h)n \\
& + (c(f-\omega^2)^2 - d(f-\omega^2)g)m\bar{m} + (b((f-\omega^2)h^2 - (f \\
& \quad - \omega^2)g^2 + (e-\omega^2)(f-\omega^2)^2) + 2(a-\omega^2)(f-\omega^2)gh \\
& \quad - 2cd(f-\omega^2)h)m \\
& + d(b((-g^2 - (e-\omega^2)(f-\omega^2))h - h^3) + (a-\omega^2)(-gh^2 \\
& \quad - g^3 + (e-\omega^2)(f-\omega^2)g) - 3c\bar{c}(f-\omega^2)g) \\
& + c((a-\omega^2)(-(f-\omega^2)h^2 + (f-\omega^2)g^2 - (e-\omega^2)(f \\
& \quad - \omega^2)^2) + 2b(f-\omega^2)gh) + cd\bar{d}(2h^2 + 2g^2 + (e \\
& \quad - \omega^2)(f-\omega^2)) \\
& - (e-\omega^2)d^3 g + c^3(f-\omega^2)^2),
\end{aligned}
$$

$$
\begin{aligned}
\gamma_{10} = \; & ((e-\omega^2)n^4 - 2gmn^3 + ((f-\omega^2)m\bar{m} + (a-\omega^2)(-h^2 \\
& \quad + g^2 - 2(e-\omega^2)(f-\omega^2)) - 2cdg + c\bar{c}(f-\omega^2) + 2(e \\
& \quad - \omega^2)d\bar{d})n\bar{n} \\
& + ((2(a-\omega^2)(f-\omega^2)g - 2d\bar{d}g)m + bd(-2h^2 - 2g^2) \\
& \quad + 2bc(f-\omega^2)g)n
\end{aligned}
$$

$$
\begin{aligned}
&+ (d\bar{d}(f-\omega^2) - (a-\omega^2)(f-\omega^2)^2)m\bar{m} + (2bd(f \\
&\quad - \omega^2)g - 2bc(f-\omega^2)^2)m + (a-\omega^2)^2((f-\omega^2)h^2 \\
&\quad - (f-\omega^2)g^2 + (e-\omega^2)(f-\omega^2)^2) \\
&+ b\bar{b}(-(f-\omega^2)h^2 + (f-\omega^2)g^2 - (e-\omega^2)(f-\omega^2)^2) \\
&+ d\bar{d}((a-\omega^2)(-h^2 + g^2 - 2(e-\omega^2)(f-\omega^2)) + c\bar{c}(f \\
&\quad - \omega^2)) + 2(a-\omega^2)cd(f-\omega^2)g - 2cd^3g - (a \\
&\quad - \omega^2)c\bar{c}(f-\omega^2)^2 + (e-\omega^2)d^4),
\end{aligned}
$$

$$
\begin{aligned}
\delta_{10} = {} & ((e-\omega^2)cn^3 + (-2hm\bar{m} + (-2cg - (e-\omega^2)d)m + (e \\
&\quad - \omega^2)(a-\omega^2)h - (e-\omega^2)bg)n\bar{n} + ((2dg + c(f \\
&\quad - \omega^2))m\bar{m} \\
&+ (b(h^2 + g^2 + (e-\omega^2)(f-\omega^2)) + 2(a-\omega^2)gh \\
&\quad - 4cdh)m + (a-\omega^2)c(-h^2 + g^2 - (e-\omega^2)(f-\omega^2)) \\
&+ d(-2(e-\omega^2)bh - 2c\bar{c}g) + c^3(f-\omega^2) + (e \\
&\quad - \omega^2)cd\bar{d})n - d(f-\omega^2)m^3 + ((a-\omega^2)(f-\omega^2)h \\
&\quad - b(f-\omega^2)g)m\bar{m} \\
&+ (d((a-\omega^2)(h^2 - g^2 + (e-\omega^2)(f-\omega^2)) - c\bar{c}(f \\
&\quad - \omega^2)) + 2bc(f-\omega^2)h + 2cd\bar{d}g - (e-\omega^2)d^3)m \\
&+ b\bar{b}(h^3 + (g^2 + (e-\omega^2)(f-\omega^2))h) + (a-\omega^2)^2((-g^2 \\
&\quad - (e-\omega^2)(f-\omega^2))h - h^3) + cd(b(-h^2 - g^2 - (e \\
&\quad - \omega^2)(f-\omega^2)) + 2(a-\omega^2)gh) \\
&+ c\bar{c}((a-\omega^2)(f-\omega^2)h + b(f-\omega^2)g) + d\bar{d}(-2c\bar{c}h + (e \\
&\quad - \omega^2)(a-\omega^2)h + (e-\omega^2)bg)),
\end{aligned}
$$

$$
\begin{aligned}
\epsilon_{10} = {} & -((e-\omega^2)mn^3 + (-2gm\bar{m} + 2chm + (e-\omega^2)bh - (e \\
&\quad - \omega^2)(a-\omega^2)g + (e-\omega^2)cd)n\bar{n} + ((f-\omega^2)m^3 \\
&\quad - 2dhm\bar{m} \\
&+ ((a-\omega^2)(h^2 + 3g^2 - (e-\omega^2)(f-\omega^2)) - 4cdg \\
&\quad + c\bar{c}(f-\omega^2) + (e-\omega^2)d\bar{d})m + (d(b(h^2 + g^2 + (e \\
&\quad - \omega^2)(f-\omega^2)) + 2(a-\omega^2)gh) \\
&+ c(b(h^2 + g^2 + (e-\omega^2)(f-\omega^2)) - 2(a-\omega^2)gh) \\
&\quad + d(2c\bar{c}h - 2(e-\omega^2)bg))n + (-b(f-\omega^2)h - (a \\
&\quad - \omega^2)(f-\omega^2)g + cd(f-\omega^2))m\bar{m}
\end{aligned}
$$

$$
\begin{aligned}
& - 2cd\bar{d}h - 2bc(f - \omega^2)g)m + d((a - \omega^2)c(h^2 + 3g^2 - (e \\
& \quad - \omega^2)(f - \omega^2)) + c^3(f - \omega^2)) + b\bar{b}(gh^2 + g^3 - (e \\
& \quad - \omega^2)(f - \omega^2)g) \\
& + (a - \omega^2)^2(-gh^2 - g^3 + (e - \omega^2)(f - \omega^2)g) + c\bar{c}(b(f \\
& \quad - \omega^2)h - (a - \omega^2)(f - \omega^2)g) + d\bar{d}(-(e - \omega^2)bh \\
& \quad - 2c\bar{c}g - (e - \omega^2)(a - \omega^2)g) \\
& + (e - \omega^2)cd^3),
\end{aligned}
$$

$$
\begin{aligned}
\theta_{10} = & \ (2hmn^3 + (b(h^2 + g^2) - 2(a - \omega^2)gh + 2cdh)n\bar{n} + (-2(a \\
& \quad - \omega^2)(f - \omega^2)h + 2d\bar{d}h - 2b(f - \omega^2)g)mn \\
& + b(f - \omega^2)^2 m\bar{m} + d\bar{d}(b(-h^2 - g^2) - 2(a - \omega^2)gh) \\
& + cd(2b(f - \omega^2)g - 2(a - \omega^2)(f - \omega^2)h) - 2b\bar{b}(f - \omega^2)gh \\
& \quad + 2(a - \omega^2)^2(f - \omega^2)gh + 2cd^3h - bc\bar{c}(f - \omega^2)^2);
\end{aligned}
$$

- w_{11}

$$
\begin{aligned}
\alpha_{11} = & \ ((2h^3m + (a - \omega^2)(gh^2 + g^3 - (e - \omega^2)(f - \omega^2)g) + c(gh^2 \\
& \quad - g^3 + (e - \omega^2)(f - \omega^2)g) - 2(e - \omega^2)dh^2)n \\
& + (c((f - \omega^2)h^2 + (f - \omega^2)g^2 - (e - \omega^2)(f - \omega^2)^2) \\
& \quad + d(c(h^3 + (-g^2 - (e - \omega^2)(f - \omega^2))h) \\
& + (a - \omega^2)((f - \omega^2)h^2 - (f - \omega^2)g^2 + (e - \omega^2)(f - \omega^2)^2) \\
& \quad - 2dgh^2)m + b(h^4 - g^4 + 2(e - \omega^2)(f - \omega^2)g^2 - (e \\
& \quad - \omega^2)^2(f - \omega^2)^2) \\
& + (a - \omega^2)((-g^2 - (e - \omega^2)(f - \omega^2))h - h^3)) + 2(a \\
& \quad - \omega^2)c(f - \omega^2)gh + 2(e - \omega^2)d\bar{d}gh),
\end{aligned}
$$

$$
\begin{aligned}
\beta_{11} = & \ (((e - \omega^2)h^2 + (e - \omega^2)g^2 - (e - \omega^2)^2(f - \omega^2))n\bar{n} + ((2(e \\
& \quad - \omega^2)(f - \omega^2)g - 2g^3)m + 2cg^2h - 2(e - \omega^2)dgh)n \\
& + (-(f - \omega^2)h^2 + (f - \omega^2)g^2 - (e - \omega^2)(f - \omega^2)^2)m\bar{m} \\
& \quad + c\bar{c}((f - \omega^2)h^2 + (f - \omega^2)g^2 - (e - \omega^2)(f - \omega^2)^2) \\
& + (d(2h^3 + 2(e - \omega^2)(f - \omega^2)h) - 2c(f - \omega^2)gh)m + (a \\
& \quad - \omega^2)(-h^4 + g^4 - 2(e - \omega^2)(f - \omega^2)g^2 + (e - \omega^2)^2(f \\
& \quad - \omega^2)^2) \\
& + d\bar{d}(-(e - \omega^2)h^2 + (e - \omega^2)g^2 - (e - \omega^2)^2(f - \omega^2)) \\
& \quad + cd(2(e - \omega^2)(f - \omega^2)g - 2g^3)),
\end{aligned}
$$

$$
\begin{aligned}
\gamma_{11} = {} & ((e-\omega^2)hn^3 + (-2ghm + cg^2 + (a-\omega^2)((e-\omega^2)(f-\omega^2) \\
& - g^2) - (e-\omega^2)dg)n\bar{n} + ((f-\omega^2)hm\bar{m} + (2dg^2 - 2c(f \\
& - \omega^2)g)m \\
& + (a-\omega^2)((g^2 - (e-\omega^2)(f-\omega^2))h - h^3) + b((f-\omega^2)h^2 \\
& - (f-\omega^2)g^2 + (e-\omega^2)(f-\omega^2)^2)m \\
& + b(-gh^2 + g^3 - (e-\omega^2)(f-\omega^2)g) - 2cdgh + c\bar{c}(f-\omega^2)h \\
& + (e-\omega^2)d\bar{d}h)n + (c(f-\omega^2)^2 - d(f-\omega^2)g)m\bar{m} \\
& + d(b((g^2 - (e-\omega^2)(f-\omega^2))h - h^3) + (a-\omega^2)(gh^2 - g^3 \\
& + (e-\omega^2)(f-\omega^2)g) - c\bar{c}(f-\omega^2)g - 2(a-\omega^2)c(f \\
& - \omega^2)g) \\
& + d\bar{d}((a-\omega^2)(h^2 + (e-\omega^2)(f-\omega^2)) + c(2g^2 - h^2)) + (a \\
& - \omega^2)^2(-(f-\omega^2)h^2 + (f-\omega^2)g^2 - (e-\omega^2)(f-\omega^2)^2) \\
& - (e-\omega^2)d^3g + (a-\omega^2)c\bar{c}(f-\omega^2)^2), \\
\delta_{11} = {} & -((e-\omega^2)^2n^3 + ((e-\omega^2)(a-\omega^2)h - 3(e-\omega^2)gm)n\bar{n} \\
& + ((2h^2 + 2g^2 + (e-\omega^2)(f-\omega^2))m\bar{m} + (-cgh + (a \\
& - \omega^2)gh - 2(e-\omega^2)dh)m \\
& + (a-\omega^2)(-(e-\omega^2)h^2 + (e-\omega^2)g^2 - (e-\omega^2)^2(f \\
& - \omega^2)) - 2(e-\omega^2)bgh + (a-\omega^2)cg^2 + (b(h^3 + (g^2 + (e \\
& - \omega^2)(f-\omega^2))h) \\
& + c\bar{c}((e-\omega^2)(f-\omega^2) - g^2) + d(-(e-\omega^2)cg - (e-\omega^2)(a \\
& - \omega^2)g) + (e-\omega^2)^2 d\bar{d})n - (f-\omega^2)gm^3 + c(f-\omega^2)hm\bar{m} \\
& + d(c(h^2 + 2g^2 - (e-\omega^2)(f-\omega^2)) + (a-\omega^2)(h^2 + (e \\
& - \omega^2)(f-\omega^2))) + (a-\omega^2)(-gh^2 - g^3 + (e-\omega^2)(f \\
& - \omega^2)g) \\
& - (a-\omega^2)c(f-\omega^2)g - (e-\omega^2)d\bar{d}g)m + (a-\omega^2)^2((-g^2 \\
& - (e-\omega^2)(f-\omega^2))h - h^3) + (a-\omega^2)c\bar{c}(f-\omega^2)h - (e \\
& - \omega^2)cd\bar{d}h) \\
& + d(b(-(e-\omega^2)h^2 + (e-\omega^2)g^2 - (e-\omega^2)^2(f-\omega^2)) \\
& - c\bar{c}gh + (a-\omega^2)cgh + 2(e-\omega^2)(a-\omega^2)gh) + bc(-gh^2 \\
& - g^3 + (e-\omega^2)(f-\omega^2)g), \\
\epsilon_{11} = {} & -(((e-\omega^2)hm + (e-\omega^2)cg - (e-\omega^2)^2d)n\bar{n} + (-2ghm\bar{m} \\
& + (c(2h^2 - g^2 - (e-\omega^2)(f-\omega^2)) + (a-\omega^2)((e \\
& - \omega^2)(f-\omega^2) - g^2)
\end{aligned}
$$

$$
\begin{aligned}
& + 2(e-\omega^2)dg)m + b((e-\omega^2)h^2 + (e-\omega^2)g^2 - (e \\
& \quad - \omega^2)^2(f-\omega^2)) + ((a-\omega^2)(h^3 + (g^2 - (e-\omega^2)(f \\
& \quad - \omega^2))h) \\
& + d(-(e-\omega^2)ch - (e-\omega^2)(a-\omega^2)h) + c\bar{c}gh + (a \\
& \quad - \omega^2)cgh)n + (f-\omega^2)hm^3 + (d(-2h^2 - (e-\omega^2)(f \\
& \quad - \omega^2)) + c(f-\omega^2)g)m\bar{m} \\
& + b(-gh^2 - g^3 + (e-\omega^2)(f-\omega^2)g) - d(3cgh + (a \\
& \quad - \omega^2)gh) + (a-\omega^2)c(f-\omega^2)h + 3(e-\omega^2)d\bar{d}h)m \\
& \quad + bc(h^3 + (g^2 - (e-\omega^2)(f-\omega^2))h) \\
& + d((a-\omega^2)(-(e-\omega^2)h^2 - (e-\omega^2)g^2 + (e-\omega^2)^2(f \\
& \quad - \omega^2)) + c\bar{c}(h^2 - (e-\omega^2)(f-\omega^2)) + (a-\omega^2)c(-h^2 \\
& \quad - 2g^2)) \\
& + (a-\omega^2)^2(gh^2 + g^3 - (e-\omega^2)(f-\omega^2)g) + d\bar{d}(2(e \\
& \quad - \omega^2)cg + (e-\omega^2)(a-\omega^2)g) + (a-\omega^2)c\bar{c}(f-\omega^2)g \\
& \quad - (e-\omega^2)^2d^3),
\end{aligned}
$$

$$
\begin{aligned}
\theta_{11} = {} & ((e-\omega^2)gn^3 + ((2h^2 - 2g^2 - (e-\omega^2)(f-\omega^2))m + cgh \\
& \quad + (a-\omega^2)gh - (e-\omega^2)dh)n\bar{n} + (3(f-\omega^2)gm\bar{m} \\
& \quad + (-2dgh - c(f-\omega^2)h \\
& + (a-\omega^2)(f-\omega^2)h)m - (a-\omega^2)c(f-\omega^2)^2)m + d((a \\
& \quad - \omega^2)((g^2 + (e-\omega^2)(f-\omega^2))h - h^3) + (a-\omega^2)(f \\
& \quad - \omega^2)g) \\
& + b(h^3 - (g^2 + (e-\omega^2)(f-\omega^2))h) + d(c(h^2 - g^2) - (a \\
& \quad - \omega^2)(h^2 + g^2)) + (a-\omega^2)(-gh^2 + g^3 - (e-\omega^2)(f \\
& \quad - \omega^2)g) + (a-\omega^2)c(f-\omega^2)g \\
& + (e-\omega^2)d\bar{d}g)n - (f-\omega^2)^2m^3 + d(f-\omega^2)hm\bar{m} + d\bar{d}((a \\
& \quad - \omega^2)gh - cgh) - 2(a-\omega^2)^2(f-\omega^2)gh - (e-\omega^2)d^3h) \\
& + ((a-\omega^2)(-(f-\omega^2)h^2 - (f-\omega^2)g^2 + (e-\omega^2)(f \\
& \quad - \omega^2)^2) + d\bar{d}(2h^2 - (e-\omega^2)(f-\omega^2)) + 2b(f-\omega^2)gh \\
& \quad + d(c(f-\omega^2)g \\
& + b(-gh^2 + g^3 - (e-\omega^2)(f-\omega^2)g) + (a-\omega^2)c(f \\
& \quad - \omega^2)h) + bc(-(f-\omega^2)h^2 - (f-\omega^2)g^2 + (e-\omega^2)(f \\
& \quad - \omega^2)^2),
\end{aligned}
$$

- w_{12}

$$
\begin{aligned}
\alpha_{12} = {} & (((e-\omega^2)h^2 - (e-\omega^2)g^2 + (e-\omega^2)^2(f-\omega^2))n\bar{n} + ((2g^3 \\
& - 2(e-\omega^2)(f-\omega^2)g)m + c(2h^3 + 2(e-\omega^2)(f-\omega^2)h) \\
& - 2(e-\omega^2)dgh)n + (-(f-\omega^2)h^2 - (f-\omega^2)g^2 + (e \\
& - \omega^2)(f-\omega^2)^2)m\bar{m} + (e-\omega^2)(f-\omega^2)^2) + d\bar{d}(-(e \\
& - \omega^2)h^2 - (e-\omega^2)g^2 \\
& + (2dg^2h - 2c(f-\omega^2)gh)m + (a-\omega^2)(h^4 - g^4 + 2(e \\
& - \omega^2)(f-\omega^2)g^2 - (e-\omega^2)^2(f-\omega^2)^2) + c\bar{c}((f-\omega^2)h^2 \\
& - (f-\omega^2)g^2 \\
& + (e-\omega^2)^2(f-\omega^2)) + cd(2g^3 - 2(e-\omega^2)(f-\omega^2)g)), \\
\beta_{12} = {} & (2(e-\omega^2)ghn\bar{n} + ((-2g^2 - 2(e-\omega^2)(f-\omega^2))hm + 2cgh^2 \\
& + 2(e-\omega^2)dh^2)n + 2(f-\omega^2)ghm\bar{m} + (-2dgh^2 - 2c(f \\
& - \omega^2)h^2)m \\
& + b(-h^4 + g^4 - 2(e-\omega^2)(f-\omega^2)g^2 + (e-\omega^2)^2(f \\
& - \omega^2)^2) + 2cdh^3), \\
\gamma_{12} = {} & ((e-\omega^2)gn^3 + ((-2g^2 - (e-\omega^2)(f-\omega^2))m + (e \\
& - \omega^2)dh)n\bar{n} + (3(f-\omega^2)gm\bar{m} - 2dghm + b((g^2 - (e \\
& - \omega^2)(f-\omega^2))h - h^3) \\
& + (a-\omega^2)(-gh^2 + g^3 - (e-\omega^2)(f-\omega^2)g) + 2cd(f \\
& - \omega^2)g - c\bar{c}(f-\omega^2)^2 - (e-\omega^2)d\bar{d}(f-\omega^2))m \\
& - 2cdg^2 + c\bar{c}(f-\omega^2)g + (e-\omega^2)d\bar{d}g)n - (f-\omega^2)^2m^3 \\
& + d(f-\omega^2)hm\bar{m} + ((a-\omega^2)((f-\omega^2)h^2 - (f-\omega^2)g^2 \\
& + (e-\omega^2)(f-\omega^2)^2) \\
& + d((a-\omega^2)((g^2 - (e-\omega^2)(f-\omega^2))h - h^3) + b(gh^2 - g^3 \\
& + (e-\omega^2)(f-\omega^2)g) + c\bar{c}(f-\omega^2)h) \\
& + bc(-(f-\omega^2)h^2 + (f-\omega^2)g^2 - (e-\omega^2)(f-\omega^2)^2) \\
& - 2cd\bar{d}gh + (e-\omega^2)d^3h), \\
\delta_{12} = {} & -(((e-\omega^2)hm - (e-\omega^2)cg + (e-\omega^2)^2d)n\bar{n} + ((c(2h^2 \\
& + 2g^2) - 2(e-\omega^2)dg)m + b(-(e-\omega^2)h^2 + (e-\omega^2)g^2 \\
& - (e-\omega^2)^2(f-\omega^2)) \\
& - 2(e-\omega^2)(a-\omega^2)gh + 2(e-\omega^2)cdh)n - (f-\omega^2)hm^3 \\
& + ((e-\omega^2)d(f-\omega^2) - c(f-\omega^2)g)m\bar{m} + ((a-\omega^2)(h^3 \\
& + (g^2 + (e-\omega^2)(f-\omega^2))h)
\end{aligned}
$$

$$
\begin{aligned}
&+ b(-gh^2 - g^3 + (e-\omega^2)(f-\omega^2)g) - c\bar{c}(f-\omega^2)h - (e \\
&\quad - \omega^2)d\bar{d}h)m + c(b((-g^2 - (e-\omega^2)(f-\omega^2))h - h^3) \\
&+ (a-\omega^2)(-gh^2 - g^3 + (e-\omega^2)(f-\omega^2)g)) + d((a \\
&\quad - \omega^2)(-(e-\omega^2)h^2 + (e-\omega^2)g^2 - (e-\omega^2)^2(f-\omega^2)) \\
&\quad + c\bar{c}(2h^2 + 2g^2 \\
&+ (e-\omega^2)(f-\omega^2)) + 2(e-\omega^2)bgh) - c^3(f-\omega^2)g - 3(e \\
&\quad - \omega^2)cd\bar{d}g + (e-\omega^2)^2 d^3),
\end{aligned}
$$

$$
\begin{aligned}
\epsilon_{12} \;=\; &-((e-\omega^2)^2 n^3 + (3(e-\omega^2)ch - 3(e-\omega^2)gm)n\bar{n} + ((2g^2 \\
&\quad + (e-\omega^2)(f-\omega^2))m\bar{m} + (-4cgh - 2(e-\omega^2)dh)m \\
&+ (a-\omega^2)((e-\omega^2)h^2 + (e-\omega^2)g^2 - (e-\omega^2)^2(f-\omega^2)) \\
&\quad + c\bar{c}(2h^2 + (e-\omega^2)(f-\omega^2)) - 2(e-\omega^2)cdg + (e \\
&\quad - \omega^2)^2 d\bar{d})n \\
&- (f-\omega^2)gm^3 + (2dgh + c(f-\omega^2)h)m\bar{m} + (b(h^3 + (g^2 \\
&\quad - (e-\omega^2)(f-\omega^2))h) + (a-\omega^2)(-gh^2 - g^3 + (e \\
&\quad - \omega^2)(f-\omega^2)g) \\
&- c\bar{c}(f-\omega^2)g - (e-\omega^2)d\bar{d}g)m + c((a-\omega^2)(h^3 + (g^2 - (e \\
&\quad - \omega^2)(f-\omega^2))h) + b(gh^2 + g^3 - (e-\omega^2)(f-\omega^2)g)) \\
&+ d(b(-(e-\omega^2)h^2 - (e-\omega^2)g^2 + (e-\omega^2)^2(f-\omega^2)) \\
&\quad - 2c\bar{c}gh) + cd(2g^2 - 2h^2 + c^3(f-\omega^2)h + (e-\omega^2)cd\bar{d}h),
\end{aligned}
$$

$$
\begin{aligned}
\theta_{12} \;=\; &((e-\omega^2)hn^3 + (c(2h^2 - (e-\omega^2)(f-\omega^2)) + (e \\
&\quad - \omega^2)dg)n\bar{n} + (-(f-\omega^2)hm\bar{m} + (2c(f-\omega^2)g - 2dg^2)m \\
&+ (a-\omega^2)(h^3 + (-g^2 - (e-\omega^2)(f-\omega^2))h) + (b(-(f \\
&\quad - \omega^2)h^2 - (f-\omega^2)g^2 + (e-\omega^2)(f-\omega^2)^2) \\
&+ b(-gh^2 + g^3 - (e-\omega^2)(f-\omega^2)g) + 2cdgh - c\bar{c}(f \\
&\quad - \omega^2)h + (e-\omega^2)d\bar{d}h)n + (d(f-\omega^2)g - c(f-\omega^2)^2)m\bar{m} \\
&+ 2(a-\omega^2)(f-\omega^2)gh - 2d\bar{d}gh)m + d(b((g^2 + (e-\omega^2)(f \\
&\quad - \omega^2))h - h^3) + (a-\omega^2)(-gh^2 + g^3 - (e-\omega^2)(f \\
&\quad - \omega^2)g) \\
&- 2b(f-\omega^2)gh) + cd\bar{d}(2h^2 - 2g^2 - (e-\omega^2)(f-\omega^2)) + (e \\
&\quad - \omega^2)d^3 g - c^3(f-\omega^2)^2) + 3c\bar{c}(f-\omega^2)g) \\
&+ c((a-\omega^2)(-(f-\omega^2)h^2 - (f-\omega^2)g^2 + (e-\omega^2)(f \\
&\quad - \omega^2)^2).
\end{aligned}
$$

References

Achenbach, J. D. (1973). *Wave propagation in elastic solids, volume 16 of N. H. Series in applied mathematics and mechanics*. Amsterdam, New York, Oxford: North-Holland Publishing Company.

Annenkova, P. L. A. A., & Korzhik, M. V. (2002). Lead tungstate scintillation material. *Nuclear Instrument and Methods in Physics Research-A, 490*, 30–50.

Authier, A. (2003). (editor.) *International tables for crystallography, Volume D: Physical properties of crystals*. Dordrecht: Kluwer Academic Publ.

Barbagallo, G., Madeo, A., d'Agostino, M. V., Abreu, R., Ghiba, I. D., & Neff, P. (2017). Transparent anisotropy for the relaxed micromorphic model: macroscopic consistency conditions and long wave length asymptotics. *International Journal of Solids and Structuress, 120*, 7–30.

Berezovsky, M. B. A., & Engelbrecht, J. (2011). Waves in microstructured solids: a unified viewpoint of modeling. *Acta Mechanica, 220*(1), 349–363.

Berezovsky, A., Engelbrecht, J., Salupere, A., Tamm, K., Peets, T., & Berezovsky, M. (2013). Dispersive waves in microstructured solids. *International Journal of Solids and Structures, 50*(11–12), 1981–1990.

Born, M., & Huang, K. (1954). *Dynamical theory of crystal lattices. International series of monographs on physics*. Clarendon Press.

Capriz, G., & Podio-Guidugli, P. (1976). Discrete and continuous bodies with affinestructure. *Annali di Matematica Pura ed Applicata, 115*, 195–217.

Capriz, W. W. G., & Podio-Guidugli, P. (1982). On balance equations for materials with affine structure. *Meccanica, 17*, 80–84.

Capriz, G. (1989). *Continua with microstructure, volume 35 of Springer Tracts in Natural Philosophy*. New York: Springer-Verlag.

d'Agostino, M. V., Barbagallo, G., Ghiba, I.-D., Eidel, B., Neff, P., & Madeo, A. (2020). Effective description of anisotropic wave dispersion in mechanical band-gap metamaterials via the relaxed micromorphic model. *Journal of Elasticity, 139*, 299–329.

Dormenev, V. I., Yu, G., Drobyshev, Korzhik, M. V., Lopatik, A. R., Peigneux, J.-P., & Sillou, D. (2005). Studying the kinetics of radiation damage in pwo crystals for the cms electromagnetic calorimeter (cern). *Instruments and Experimental Techniques, 48*, 303–307.

Dove, M. (1993). *Introduction to lattice dynamics*. Cambridge University Press.

Eringen, A. (1968). *Mechanics of micromorphic continua. Mechanics of generalized continua. IUTAM symposia.*. Berlin, Heidelberg: Springer-Verlag.

Eringen, A. (1999). *Microcontinuum field theories. I: Foundations and solids*. Springer-Verlag.

Erni, W., Keshelashvili, I., Krusche, B., Steinacher, M., Heng, Y., Liu, Z., ... Zmeskal, J. (2013). Technical design report for the PANDA (AntiProton Annihilations at Darmstadt) Straw Tube Tracker. *The European Physical Journal A, 49*, 13.

Fieschi, F. G. F. R. (1953). High-order tensors in symmetrical systems. *Il Nuovo Cimento, X* (7, serie nona), 865–882.

Gurtin, M. E. (1972). *The linear theory of elasticity, volume VIa/2 of Handbook of physics*. Berlin: Springer Verlag.

Halmos, P. (1987). *Finite-dimensional vector spaces*. New York Berlin Heidelberg: Springer-Verlag.

Hussein, M. I., Leamy, M. J., & Ruzzene, M. (2014). Dynamics of phononic materials andstructures: Historical origins, recent progress, and future outlook. *Applied Mechanics Reviews, 66*(4).

Hussein (Ed.), M. I. (2018). Advances in crystals and elastic metamaterials. *Part 1: Advances in Applied Mechanics, 51*, 1–164.

Hussein (Ed.), M. I. (2019). Advances in crystals and elastic metamaterials. *Part 2: Advances in Applied Mechanics, 52*, 1–181.

Lancaster, P. (1964). On eigenvalues of matrices dependent on a parameter. *Numerische Mathematik, 6*, 377–387.

Lecoq, P., Annekov, A., Getkin, A., Korzhik, M., & Pedrini, C. (2006). *Inorganic scintillators for detector systems*. Berlin, Heidelberg, New York: Springer.

Madeo, A., Neff, P., Ghiba, I. D., Placidi, L., & Rosi, G. (2015). Wave propagation in relaxed micromorphic continua: modelling metamaterials with frequency band-gaps. *Continuum Mechanics and Thermodynamics, 27*(4), 551–570.

Madeo, M. D. A., Neff, P., & Barbagallo, G. (2016). Complete band gaps including non-local effects occur only in the relaxed micromorphic model. *Comptes Rendus Mécanique, 344*, 784–796.

Madeo, P. N. A. (2017). *Dispersion of waves in micromorphic media and metamaterial. Handbook of nonlocal continuum mechanics for materials*. Springer Int. Publ. AG,.

Mengucci, P., André, G., Auffray, E., Barucca, G., Cecchi, C., Chipaux, R., ... Santecchia, E. (2015). Structural, mechanical and light yield characterisation of heat treated LYSO:Ce single crystals for medical imaging applications. *Nuclear Instrument and Methods in Physics Research-A, 785*, 110–116.

Mindlin, R. D., & Tiersten, H. F. (1963). Effects of couple-stresses in linear elasticity. *Arch. Rat. Mech. Anal. 11*, 415–448.

Mindlin, R. D. (1964). Micro-structure in linear elasticity. *Archive for Rational Mechanics and Analysis, 16*, 51–77.

Mindlin, R. D. (1965). On the equations of elastic materials with micro-structure. *International Journal of Solids and Structures, 1*, 73–78.

Montalto, L., Daví, F., Dormenev, V., Paone, N., & Rinaldi, D. (2023). PbWO4 acoustic properties measurement by laser ultrasonics with the aim of optical damage recovery. *Crystals, 13*, 556.

Moosavian, H. M. S. H. (2020). Mindlin-eringen anisotropic micromorphic elasticity and lattice dynamics representation. *Philosophical Magazine, Part A: Materials Science, 100*(2), 1671998.

Neff, P., Ghiba, I. D., Madeo, A., Placidi, L., & Rosi, G. (2014). A unifying perspective: the relaxed linear micromorphic continuum. *Continuum Mech. Thermodyn. 26*(5), 639–681.

Neff, P., Madeo, A., Barbagallo, G., d'Agostino, M. V., Abreu, R., & Ghiba, I. D. (2017). Real wave propagation in the isotropic relaxed micromorphic model. *Proc. Royal Soc. A, 473*, 20160790.

Novotny, R. W., Bremer, D., Dormenev, V., Drexler, P., Eissner, T., Kuske, T., & Moritz, M. (2011). Bremer and the PANDA collaboration. High-quality pwo crystals for the panda-emc. IV International Conference on Calorimetry in High Energy Physics (CALOR 2010) *Journal of Physics: Conference Series, 293*, 012003 IV International Conference on Calorimetry in High Energy Physics (CALOR 2010).

Olive, N. A. M. (2013). Symmetry classes for even-order tensors. *Mathematics and Mechanics of Complex Systems, 1*(2), 177–210.

Olive, N. A. M. (2014). Symmetry classes for odd-order tensors. *Journal of Applied Mathematics and Mechanics/Zeitschrift für Angewandte Mathematik und Mechanik, 94*(5), 421–447.

Sadaki, Z. (1941). Elastic waves in crystals. *Proceedings of the Physico-Mathematical Society of Japan, 23*(3), 539–547.

Shodja, H. M. H. M. (2020). Weakly nonlocal micromorphic elasticity for diamond structures vis-a-vis lattice dynamics. *Mechanics of Materials, 147*, 103365.

Singh, D. (2010). Structure and optical properties of high light output halide scintillators. *Physical Review B, 82*, 155145.

Zwillinger (Ed.), D. (2003). *Standard mathematical tables and formulae* (31st ed.). Boca Raton, FL: CRC Press Company.

CHAPTER THREE

Numerical modeling of highly nonlinear phenomena in heterogeneous materials and domains

Modesar Shakoor[a,*]
[a]Centre for Materials and Processes, IMT Nord Europe, Institut Mines-Télécom, University of Lille, Lille, France
*Corresponding author. e-mail address: modesar.shakoor@imt-nord-europe.fr

Contents

Advances in Applied Mechanics, Volume 57
ISSN 0065-2156, https://doi.org/10.1016/bs.aams.2023.09.003

Abstract

Recent research activities on numerical modeling of highly nonlinear phenomena in heterogeneous materials and domains are presented. The two studied phenomena are single-phase or two-phase flows where the heterogeneity is due to obstacles of a very small size compared to the simulation domain, and damage and fracture of structures where it is the mesostructure and/or the microstructure of the material that is heterogeneous.

In both cases, numerical modeling relies on a representation of the heterogeneity and an additional interface, which is the liquid/gas interface in two-phase flows, and the crack in fracture problems. These simulations are computationally demanding because these phenomena are nonlinear and also because there are interfaces at a very fine scale. This is the reason why it is suitable to accelerate simulations by model order reduction and multiscale modeling.

The originality of this work mainly consists in the development of advanced numerical methods. Heterogeneity discretization is achieved either by finite element mesh generation and adaption or voxel meshes and the fast Fourier transform based numerical method. Interfaces are modeled in flows using level-set functions and a quadratic finite element interpolation, while cracks are modeled in fracture problems using a phase-field approach. Model order reduction methods are developed by borrowing techniques from data science and deep learning, but some fundamental principles from mechanics are still enforced. Flow and fracture problems that are rarely dealt with in the literature are tackled thanks to multiscale modeling.

Nomenclature

Abbreviation	Description.
***n*D**	*n* Dimension(s).
ANR	Artificial Neural Network.
CFL	Courant-Friedrichs-Lewy.
CNN	Convolutional Neural Network.
DMA	Dynamic Mechanical Analysis.
FE	Finite Element.
FFT	Fast Fourier Transform.
FOM	Full Order Model.
LS	Level-Set.
PMVP	Principle of Multiscale Virtual Power.
POD	Proper Orthogonal Decomposition.
RBVMS	Residual-Based Variational MultiScale.
RMSE	Root Mean Square Error.
ROM	Reduced Order Model.
RVE	Representative Volume Element.
SCA	Self-consistent Clustering Analysis.
SPR	Superconvergent Patch Recovery.

1. Introduction

A presentation of recent research activities on numerical modeling of nonlinear mechanical phenomena in heterogeneous materials and domains is proposed. The common aspect of all works summarized herein is the significant ratio between the size of the studied structure or domain and the characteristic size of the heterogeneity. For instance, a composite structure of several meters such as a wind turbine blade contains fibers of a diameter of a few micrometers.

The video in Reference (Lagardère et al., 2019) illustrates quite well the different manufacturing steps of a composite structure. The reinforcement composed of several layers of continuous fibers that can be woven (Fig. 1A) is placed in a mold (Fig. 1B). The polymer resin is then injected in the mold (Fig. 1C and D) to form the final composite structure (Fig. 1E). In order to optimize the production rate while maintaining a good quality, it is essential to understand and control the impregnation, *i.e.*, the filling of the mold by the resin despite the flow resistance due to the presence of the reinforcement.

As shown in Fig. 2A, the continuous woven fibers reinforcement is composed of yarns of a diameter close to 100 μm. Those yarns are composed of fibers of a diameter close to 1 μm. Resin flow occurs simultaneously between yarns and between fibers. As illustrated in Fig. 2B, the resin front develops a complex morphology, which can lead to the formation of bubbles or voids between or inside yarns. The size of those voids can vary between 1 and 100 μm.

An example of image of a composite structure acquired by tomography is shown in Fig. 3. Voids of different sizes are clearly visible both inside and between yarns. Additional voids with an elongated shape are also visible between layers. All these voids have a significant influence on the mechanical performance of composite structures. For instance, they can reduce mechanical properties such as strength and fatigue life (Mehdikhani et al., 2019).

Composites structures are a very good application for the numerical models developed in this work as nonlinear mechanical phenomena play a key part during both their manufacturing and use. These phenomena are highly influenced by the heterogeneity due to fibers, yarns and voids. In this context, this work's research activities address:

- The development of advanced numerical methods for modeling the interfaces due to the heterogeneity. This consists mainly in the finite element method with interfaces modeling thanks to mesh generation

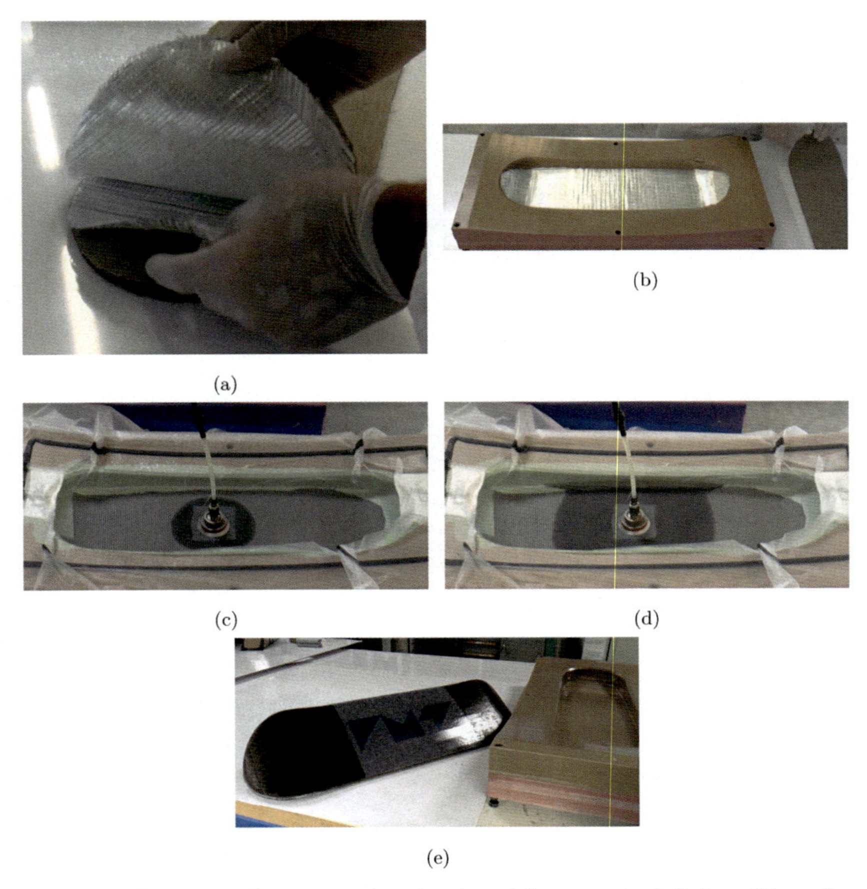

Fig. 1 Manufacturing of a composite skateboard by vacuum infusion: (A) preform composed of layers of flax (brown) and glass (white) fibers, (B) placement of the preform in the mold, (C,D) vacuum infusion of the resin, (E) final composite structure. Full video available from Lagardère, M., et al. (2019). Skate v2. ⟨https://youtu.be/QWRjgJEl1ao⟩.

and adaption, and the use of signed distance functions. These methods are presented in Section 2 with the objective of modeling two-phase flows with obstacles.

- Fracture modeling with a phase-field approach implemented with the fast Fourier transform based numerical method, as presented in Section 3.
- The acceleration of these mechanical computations with approaches based on data and model order reduction. The approaches presented in Section 4 rely on data clustering methods and deep learning with a convolutional neural network.

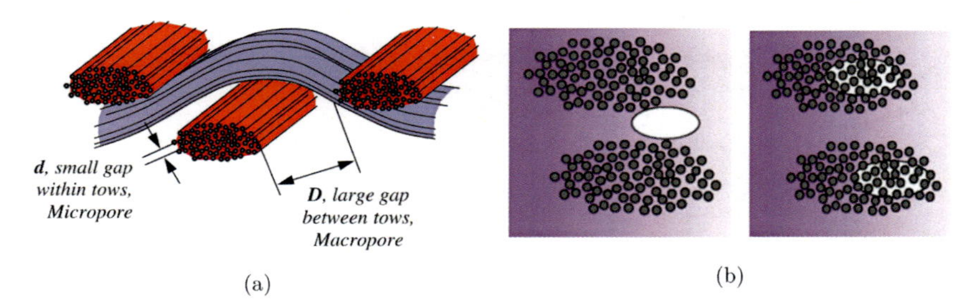

Fig. 2 Woven composite: (A) architecture and (B) formation of inter-yarn and intra-yarn voids during resin advancement. *Reprinted from Park, C. H., Lebel, A., Saouab, A., Bréard, J., Lee, W. I. (2011). Modeling and simulation of voids and saturation in liquid composite molding processes. Composites: Part A, 42, 658–668. with permission from Elsevier.*

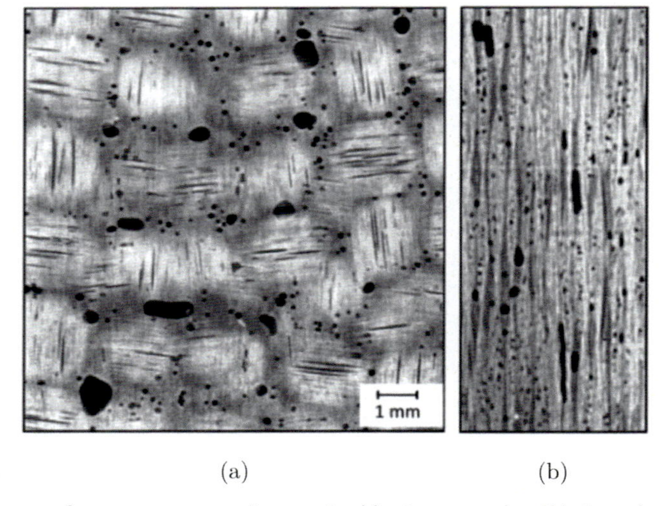

Fig. 3 Images of a woven composite acquired by tomography: (A) view along the lay-up direction (B) view across the lay-up direction. *Reprinted from Nikishkov, Y., Airoldi, L., & Makeev, A. (2013). Measurement of voids in composites by X-ray computed tomography. Composites Science and Technology, 89, 89–97 (Nikishkov et al., 2013) with permission from Elsevier.*

- The modeling of flow and fracture phenomena for large structures thanks to multiscale methods. As presented in Section 5, model order reduction is particularly relevant as it can take advantage of the partial redundancy that is inherent to these calculations.

2. Two-phase flows modeling with level-set functions and adaptive finite elements

2.1 Introduction

The materials considered in this work have a heterogeneous fine scale structure. Numerically modeling these materials necessarily raises the problem of discretizing interfaces. Numerous numerical methods have been proposed in the literature to deal with this problem, but no unique and universal approach has emerged. Depending on the quantity of interfaces, the complexity of their morphology, and the nature of their evolution during the simulation, different methods can be considered as more suitable.

The main objective of this section is to model two-phase flows in a domain containing obstacles, although the methods presented in this section certainly have other applications. This target problem is inspired from composites manufacturing processes such as resin transfer molding, where the resin flows within the mold, and is slowed down in its progression by the reinforcement. Modeling the interaction between the resin front and the obstacle formed by the reinforcement is essential in order to predict bubble entrapment during the manufacturing process. This is also important for the use of the produced parts as those bubbles are in fact voids that can significantly deteriorate the final properties of composite structures (Mehdikhani et al., 2019; Park et al., 2011).

Numerical methods for modeling two-phase flows in a domain containing obstacles are based on several components:

- A discretization of the simulation domain, this discretization potentially being adaptive.
- An approximation and solution method for flow equations.
- A modeling approach for obstacles.
- A modeling approach for the interface between the two phases, its evolution and surface tension.

The choice of the simulation domain discretization is related to the approximation and solution method for flow equations. The two most widely used methods in computational fluid mechanics are the volume of fluid method and the Finite Element (FE) method, and their variants. Depending on the chosen method, different meshes and element types may be used.

The considered problem being non-stationary, numerous adaptive approaches based on finite volumes or FEs have been proposed in the literature (Sussman, 2005; Zheng et al., 2005). They consist in dynamically adapting the mesh throughout the simulation in order to reduce the computational cost by coarsening elements where the approximation error is small, and improve the accuracy by refining elements where the approximation error is large.

Among these approaches, metric-driven FE mesh adaption methods are attractive because they consist in splitting the problem into two parts: an error estimator defining the remeshing metric, which prescribes the local edge size and element orientation, and a remeshing algorithm, which modifies the mesh. The literature on metric-driven mesh adaption methods is substantial but it mainly focuses on linear FE interpolation (Abgrall et al., 2014; Dapogny et al., 2014; Quan et al., 2014; Zhao et al., 2016). Although error estimators for higher interpolation orders are well established, the formulation of metrics based on those estimators has received less attention (Coulaud & Loseille, 2016; Huang, 2005).

Modeling obstacles, in the case where they are rigid, is a fluid-structure interaction problem that has been widely considered in the literature (Dowell & Hall, 2001; Hou et al., 2012). In the specific application of composites manufacturing, the difficulty comes from the geometry. In fact, woven composites have a complex mesostructure which involves interpenetration problems as well as artificial voids during discretization (Liu et al., 2017). These problems are even harder to deal with when the geometry is provided by a 3D image acquired by tomography. The robustness of existing mesh generation tools is still quite limited for these woven composites (Liu et al., 2017). This often restricts researchers to use voxelized meshes, as presented in Section 3.

The liquid/gas interface is undoubtedly the most important aspect of the problem. Tracking this interface with an explicit discretization as it undergoes deformations and even major topological changes requires advanced numerical techniques (Shakoor et al., 2017a). This is the reason why the two most widespread approaches for modeling a liquid/gas interface rely on an implicit representation. The interface is hence represented through a local volume fraction in the volume of fluid method Hirt and Nichols (1981) and through a signed distance function in the Level-Set (LS) method (Osher and Sethian, 1988). The use of a signed distance function is particularly relevant for modeling surface tension (Brackbill et al., 1992), as it gives a direct access to geometrical properties such as the normal vector or the mean curvature (Osher & Sethian, 1988; Sussman et al., 1994).

The LS method for two-phase flows modeling entails two major drawbacks which are the presence of parasitic currents and a poor mass conservation (Denner et al., 2017; Di Pietro et al., 2006). Parasitic currents are, among other reasons, caused by an imbalance between the discretization used for the LS function and that used for the pressure field (Francois et al., 2006), and by an inaccurate calculation of the mean curvature (Denner et al., 2017; Francois et al., 2006). The development of balanced FE approaches with a higher-order FE interpolation for the LS function has not been considered in the literature. Raising the order of the interpolation for the pressure field, indeed, raises the issue of the inf-sup condition. The question of the appropriate coupling between the velocity, the pressure and the LS function in such an approach is also worth investigating.

In addition, adaptive methods are relevant for two-phase flows as they enable maintaining a fine mesh close to the interface instead of refining the mesh uniformly over the whole simulation domain (Bui et al., 2012; Sussman, 2005; Zheng et al., 2005). The use of an implicit interface representation, moreover, does not necessarily imply that the mesh has to be fixed. Depending on the time step and the flow velocity, moving mesh and arbitrary Lagrangian–Eulerian approaches have also been developed in the literature (Enright et al., 2005; Hysing et al., 2009). The interest of a Lagrangian approach lies in the improved mass conservation that it provides (Hysing et al., 2009). The development of Lagrangian and arbitrary Lagrangian-Eulerian approaches with a quadratic interpolation for two-phase flows could hence be particularly interesting.

Numerical methods for mesh adaption with a quadratic interpolation and mesh generation with an explicit interface discretization are presented in Subsection 2.2. Some applications beyond two-phase flows are also tackled.

Contributions on numerical modeling of two-phase flows are presented in Subsection 2.3. These results rely on an LS method with quadratic interpolation and mesh adaption. The proposed approach is flexible as it is compatible with both Eulerian and Lagrangian meshes, and with both a strong and weak coupling between the velocity-pressure system and the LS function.

These numerical developments have required a substantial implementation effort. This section is hence directly related to the development of an FE code of a significant breadth. The management of this code is integrated in this research work, as well as its parallel implementation. This part of the work is presented in Subsection 2.4.

2.2 Mesh generation and adaption

One of the very first steps in any numerical modeling approach is the discretization. For heterogeneous materials, it means discretizing geometries that can originate from two sources. Experimental images can be acquired by optical or electronic microscopy, or tomography. Virtual geometries can also be generated from statistics. In the following, an approach based on signed distance functions to generate meshes for those geometries is presented. This approach was proposed initially in Reference (Shakoor, 2021) and has been implemented in the free software *FEMS* (Shakoor, 2022).

This approach relies on a preliminary step presented in Paragraph 2.2.1, which is the adaption of the mesh. Mesh generation in itself is presented in Paragraph 2.2.2. Its application to the computation of mechanical properties of long fiber composites is presented in Paragraph 2.2.3.

2.2.1 Mesh adaption method

Whatever the problem to be solved by the FE method, it is easy to define a sensor variable $\mathbf{s}$, which can even be a vector regrouping multiple quantities of interest. For instance, for solid mechanics $\mathbf{s} = \mathbf{u}$ and for incompressible flows $\mathbf{s} = \mathbf{v}$.

For a given FE approximation $\mathbf{s}_h$ on a current mesh, two alternative objectives can be addressed using unstructured mesh adaption. The current mesh can be modified to obtain a mesh either satisfying a prescribed approximation error tolerance with a complexity as low as possible, or reducing the approximation error as much as possible for a prescribed complexity. Here the complexity is considered directly proportional to the numbers of nodes and elements in the FE mesh.

Modifications may include removing, adding and moving nodes, as well as removing and adding elements. Criteria for determining which modifications to perform on the current mesh are defined through error estimators and then metric tensor fields, while modifications are operated through a remeshing algorithm. Modifications are not easy to implement for all element types, and mesh adaption is thus restricted in this paper to simplexes *i.e.*, linear triangles in 2D and linear tetrahedra in 3D.

The simulation domain is denoted $\Omega \subset \mathbb{R}^d$ in the following, with $d = 2$ or 3 being the spatial dimension. The mesh nodes set is denoted $\mathcal{N}$. A unique global number $n \in \mathcal{N}$ is associated to each node, and the coordinates of each node are given by $\mathbf{A}_n \in \Omega$. The mesh elements set is denoted $\mathcal{T}$, and an element K is defined by the set of its nodes $\mathcal{N}(K)$. A local number $n_K \in \mathcal{N}(K)$ identifies each node of an element. The connectivity operator Π_K is defined to map each local number to the associated global number. For instance, $n = \Pi_K(n_K)$ is the global number of the node which has local number n_K within element K.

2.2.1.1 Error estimators and remeshing metrics

Mesh adaption is a multi-objective optimization process targeting element qualities and edge lengths. In isotropic mesh adaption, a scalar mesh size field has to be defined on the FE mesh to determine the length prescribed locally for each edge. An optimal simplex has a volume as large as possible with the lengths of its edges as close as possible to this local mesh size. This contradiction between the two objectives requires to define a compromise. This will be addressed by the remeshing algorithm in the sequel.

For anisotropic mesh adaption, a metric tensor field **M** has to be defined on the FE mesh. This second order tensor defines locally d orthogonal directions and d independent scalar metrics in these directions (Arsigny et al., 2006). The optimization remains identical to that of isotropic mesh adaption, but this distortion of the Euclidean metric is embedded in the definitions of element volume $\forall K \in \mathcal{T}$,

$$|K|_{\mathbf{M}} = \int_K \sqrt{\det(\mathbf{M}(\mathbf{x}))}\,\mathrm{d}\mathbf{x}, \tag{1}$$

and edge length $\forall m_K, n_K \in \mathcal{N}(K),\ m_K \neq n_K,\ m = \Pi_K(m_K),\ n = \Pi_K(n_K)$,

$$||\mathbf{A}_m\mathbf{A}_n||_{\mathbf{M}} = \int_{\mathbf{A}_m\mathbf{A}_n} \sqrt{\mathbf{M}(\mathbf{x}).\ (\mathbf{A}_m - \mathbf{A}_n).\ (\mathbf{A}_m - \mathbf{A}_n)}\,\mathrm{d}\mathbf{x}. \tag{2}$$

As can be seen in Equation (2), a valid metric tensor $\mathbf{M}(\mathbf{x})$, $\mathbf{x} \in \Omega$ is a symmetric positive definite matrix. It can hence be expressed in diagonal form, for instance for $d = 3$ as

$$\mathbf{M}(\mathbf{x}) = \mathbf{R}(\mathbf{x}) \begin{pmatrix} \frac{1}{h_1^2(\mathbf{x})} & 0 & 0 \\ 0 & \frac{1}{h_2^2(\mathbf{x})} & 0 \\ 0 & 0 & \frac{1}{h_3^2(\mathbf{x})} \end{pmatrix} \mathbf{R}(\mathbf{x})^T \tag{3}$$

where each column $i = 1\ldots d$ of matrix $\mathbf{R}(\mathbf{x})$ is a direction vector along which mesh size $h_i(\mathbf{x})$ is prescribed. As shown in Fig. 4, the metric tensor gives direct control over element shape. It is determined by an error estimator. The two estimators used in this section are presented in the following.

First, metric-driven mesh adaption can be used to naturally adapt the mesh to a geometry and obtain meshes refined close to internal interfaces, and in particular in regions with large maximum principal curvature. This

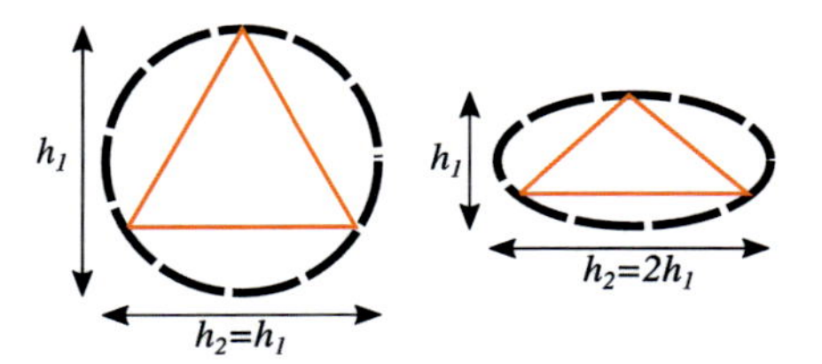

Fig. 4 Influence of the metric field on the final shape of a triangle with the isotropic case on the left, and the anisotropic case on the right, assuming the first direction vector is vertical and the second one is horizontal. *Reprinted from Shakoor, M. (2021). FEMS - A mechanics-oriented finite element modeling software. Computer Physics Communications, 260, 107729 with permission from Elsevier.*

is particularly interesting when the internal interface $\Gamma_{1,2}$ is defined through a signed distance function, or LS function, ϕ:

$$\phi(\mathbf{x}, t) = \begin{cases} +dist(\mathbf{x}, \Gamma_{1,2}(t)), & \mathbf{x} \in \Omega_1(t), \\ -dist(\mathbf{x}, \Gamma_{1,2}(t)), & \mathbf{x} \in \Omega_2(t), \\ 0, & \mathbf{x} \in \Gamma_{1,2}(t), \end{cases} \tag{4}$$

where it is assumed that at an instant t the domain $\Omega(t)$ is decomposed into two parts $\Omega_1(t)$ and $\Omega_2(t)$ which are separated by the interface $\Gamma_{1,2}(t)$. This function can be computed for simple geometrical entities and also for geometries segmented from 2D or 3D images. The sensor variable can then be defined as $\mathbf{s} = \phi$.

Although anisotropic curvature-based mesh adaption could be implemented quite easily in *FEMS* based on the literature (Abgrall et al., 2014; Quan et al., 2014), an isotropic criterion (Shakoor et al., 2018a) is preferred as this adaption process may be used as a first step for mesh generation as done in Paragraph 2.2.2. Matrix $\mathbf{R}(\mathbf{x})$, $\mathbf{x} \in \Omega$ is thus the identity matrix, while

$$\forall\ i = 1...d,\ h_i(\mathbf{x}) = \max\left(h_{min}, \min\left(h_{max}, \tilde{h}(\mathbf{x})\right)\right), \tag{5}$$

$$\tilde{h}(\mathbf{x}) = \frac{h_c}{\lambda_s(\mathbf{x})} + \left(h_{max} - \frac{h_c}{\lambda_s(\mathbf{x})}\right)\min\left(\frac{|\mathbf{s}(\mathbf{x})|}{h_{max}}, 1\right), \tag{6}$$

where $\lambda_s(\mathbf{x})$ is the maximum eigenvalue of the Hessian matrix $\nabla\nabla\ \mathbf{s}(\mathbf{x})$ of $\mathbf{s} = \phi$ at point $\mathbf{x}$. As the eigenvalues of this Hessian matrix include the principal curvatures of $\Gamma_{1,2}$, it can be seen that at the interface, where $\phi(\mathbf{x}) \approx 0$, mesh size is prescribed to be inversely proportional to the curvature, h_c being a control parameter. It is hence prescribed to be very small for singularities of the interface ($\lambda_s(\mathbf{x}) \rightarrow \infty$), and very large for flat regions of the interface ($\lambda_s(\mathbf{x}) \rightarrow 0$). At a distance larger than h_{max} from $\Gamma_{1,2}$, mesh size is prescribed to be equal to h_{max}, with a linear transition from $\phi(\mathbf{x}) \approx 0$ to $\phi(\mathbf{x}) \approx h_{max}$. Overall, the prescribed mesh size is bounded between parameters h_{min} and h_{max}.

Note that the metric tensor field is to be defined at mesh nodes in order to be interpolated at quadrature points to evaluate Equations (1) and (2). In Equation (5), sensor variable $\mathbf{s}$ can be replaced by its approximation $\mathbf{s}_h = \phi_h$ which is defined at mesh nodes, but not λ_s, which depends on the second derivatives of $\mathbf{s}$. The latter are not available at mesh nodes for Lagrange FEs, and are recovered in *FEMS* using an operation called Superconvergent Patch Recovery (SPR). SPR consists in recovering a higher-order and higher-regularity approximation of $\mathbf{s}$ around each mesh node (Zienkiewicz and Zhu, 1987). In order to recover a regular Hessian matrix, this approximation is elevated to the third order in *FEMS*, as suggested in the literature (Zhang and Naga, 2005). This recovered approximation is fitted in a least-squares sense to the values of $\mathbf{s}$ at neighboring mesh nodes, this neighborhood being called the patch.

Second, for adaption to a sensor variable that may evolve to a very heterogeneous field during the simulation, it is preferable to use the complexity as control parameter. Indeed, this would prevent the computational cost from blowing up during the simulation, for instance due to complex topological events or very localized phenomena. The goal is hence to estimate the approximation error on $\mathbf{s}$, and prescribe a metric tensor field to distribute this error uniformly on the domain, for a given complexity. The simulation is expected to capture only a certain level of detail that can be afforded with this prescribed complexity.

The continuous mesh framework for metric-driven mesh adaption can be used to achieve this goal for simplex-type Lagrange FEs of any order (Coulaud & Loseille, 2016; Loseille & Alauzet, 2011a,b; Shakoor & Park, 2021a). The definition of the metric tensor field is done in two steps. A geometric averaging operation is first used to define a single directional error tensor field $\mathbf{Q}$ for all components of $\mathbf{s}$. For linear (P1) FEs, this operation is defined $\forall \mathbf{x} \in \Omega$ as

$$\mathbf{Q}(\mathbf{x}) = \left(\exp\left(\frac{1}{\dim(\mathbf{s})}\sum_{i=1}^{\dim(\mathbf{s})} \log\left((\nabla\nabla s_i(\mathbf{x}))^{-\frac{1}{2}}\right)\right)\right)^{-2}. \tag{7}$$

Details on this geometric averaging operation can be found in the literature (Arsigny et al., 2006; Laug & Borouchaki, 2013). The extension of this operation for quadratic (P2) Lagrange FEs is given by (Shakoor and Park, 2021a)

$$\mathbf{Q}(\mathbf{x}) = \left(\exp\left(\frac{1}{\dim(s)\,d}\sum_{i=1}^{\dim(s)}\sum_{j=1}^{d}\log\left(\left(\nabla\nabla\frac{\partial s_i}{\partial x_j}(\mathbf{x})\right)^{-\frac{1}{2}}\right)\right)\right)^{-2}. \tag{8}$$

This first step includes a post-processing operation to control the element stretching and mesh size variations that will be induced by the error tensor field **Q**. This is done by computing the median eigenvalue Q_{med} of **Q** over the whole mesh, and then bounding all its eigenvalues so that none of them are higher than $Q_{med}h_{max}$, or lower than ${}^{Q_{med}}/_{h_{max}}$, where h_{max} is the prescribed ratio.

The second step is to convert **Q** into a metric tensor field **M** minimizing the total error while uniformly distributing local errors and controlling the complexity. The solution of this constrained minimization problem can be expressed as (Huang, 2005)

$$\mathbf{M}(\mathbf{x}) = \mathcal{N}_c^{\frac{2}{d}}\left(\int_\Omega (\det(\mathbf{Q}(\mathbf{x}))^{\frac{k+1}{2(k+1)+d}}\,\mathrm{d}\mathbf{x}\right)^{-\frac{2}{d}} (\det(\mathbf{Q}(\mathbf{x})))^{-\frac{1}{2(k+1)+d}}\,\mathbf{Q}(\mathbf{x}), \tag{9}$$

where $\mathcal{N}_c$ is the prescribed number of P1 nodes (in the P2 case, this excludes nodes at edge middles), and k is the order of the FE method (1 for P1, and 2 for P2).

Note again that Equation (7) requires to recover the second derivatives of **s** in the P1 case, and Equation (8) in the P2 case its third derivatives. The SPR operation can be used in both cases (Huang, 2005; Shakoor et al., 2015b; Shakoor & Park, 2021a).

2.2.1.2 Remeshing algorithm

Once the metric tensor field **M** is defined and computed at mesh nodes using any of the error estimators implemented in *FEMS*, mesh modifications can be operated to satisfy the mesh size and orientations prescribed by this field. As mentioned previously, this requires to define a compromise between maximizing elements volumes as per Equation (1) and bringing edge lengths as close as possible to 1 as per Equation (2). Two strategies are available in *FEMS* to combine these objective.

In the first strategy (Dobrzynski and Frey, 2008), the edge length criterion is first applied by looping over all edges, splitting those that have a length larger than $\sqrt{2}$, and collapsing those that have a length smaller than $\sqrt{2}^{-1}$. This is done using a Delaunay kernel to ensure the FE mesh remains valid. Second, an element quality criterion is applied by looping over all elements and performing local mesh modifications such as edge flips and node re-positioning when they improve element quality. More complex local mesh modifications are involved in 3D, as illustrated in Figure 3 of Reference (Dapogny et al., 2014). Element quality is defined as

$$\mathcal{Q}_{\mathbf{M}}(K) = \alpha_d \frac{h_{\mathbf{M}}(K)}{|K|_{\mathbf{M}}}, \tag{10}$$

$$h_{\mathbf{M}}(K) = \left(\sum_{m_K, n_K \in \mathcal{N}(K), m_K < n_K} ||\mathbf{A}_{\Pi_K(m_K)} \mathbf{A}_{\Pi_K(n_K)}||_{\mathbf{M}}^2 \right)^d, \tag{11}$$

where α_d is a normalization factor so that $\mathcal{Q}_{\mathbf{M}}(K) = 1$ for a regular simplex. Element quality $\mathcal{Q}_{\mathbf{M}}(K)$ must be minimized with this definition. With the edge length criterion based on $\sqrt{2}$, the authors of Reference (Dobrzynski and Frey, 2008) mention that the edge sizing and element improving steps can be performed again and again until no mesh improvement is found by the algorithm, with no risk of infinite loop. However, for efficiency purposes, they recommend to limit the overall number of iterations.

In the second strategy (Gruau & Coupez, 2005; Shakoor et al., 2017a), a single element quality measure combining both the element quality criterion and the edge length one is defined for each element as

$$\mathcal{Q}_{\mathbf{M}}(K) = \min \left(\frac{d!}{\sqrt{d+1}} 2^{\frac{d}{2}} \frac{|K|_{\mathbf{M}}}{h_{\mathbf{M}}(K)}, \; h_{\mathbf{M}}(K), \; \frac{1}{h_{\mathbf{M}}(K)} \right), \tag{12}$$

$$h_{\mathbf{M}}(K) = \left(\frac{2}{d(d+1)} \sum_{m_K, n_K \in \mathcal{N}(K), m_K < n_K} ||\mathbf{A}_{\Pi_K(m_K)} \mathbf{A}_{\Pi_K(n_K)}||_{\mathbf{M}}^2 \right)^{\frac{d}{2}}, \tag{13}$$

so that $\mathcal{Q}_{\mathbf{M}}(K) = 0$ for a degenerated simplex, and 1 for a regular simplex. Element quality $\mathcal{Q}_{\mathbf{M}}(K)$ must be maximized with this definition. This single element quality criterion is applied by looping over patches of elements neighboring all nodes and edges of the mesh, and applying local mesh modifications such as edge flips, node re-positioning, node removal, and node addition when they improve element quality. Examples of mesh modifications are shown in Fig. 5. This is only a subset, in fact a wider range of local mesh modifications is explored with this strategy as compared to the first strategy, even in 2D. This is illustrated in Figs. 1 and 2 of Reference (Shakoor et al., 2017a). This is performed again and again until no mesh improvement can be found by the algorithm. It is reported in Reference (Shakoor et al., 2017a) that this algorithm always converges in practice.

Due to their very general nature, the mesh modification operations used by both algorithms result in an unstructured simplex mesh, even if the initial mesh is structured. Additionally, both algorithms are restricted to linear simplex meshes both in 2D and 3D, and are fully compatible with either isotropic or anisotropic metric tensor fields.

Once the mesh adaption algorithm terminates, mechanical variables including the sensor variable must be transferred from the old mesh to the new (adapted) mesh. First, a space partitioning technique is used to locate efficiently the element of the old mesh containing each node of the new mesh. Variables values are then computed at each node of the new mesh using FE interpolation from the containing element of the old mesh.

For P2 FE meshes, since mesh adaption algorithms are restricted to linear simplexes, nodes at edge middles must be removed and then added back after the mesh has been optimized. This preserves the advantage of P2 FE interpolation during variables transfer from old to new mesh, both being P2 FE meshes, but forbids the use of isoparametric elements which could be interesting for body-fitted meshing of curved interfaces.

For variables defined at quadrature points, the space partitioning technique locates the element of the old mesh containing each quadrature point of the new mesh. Variables values are then directly copied from the

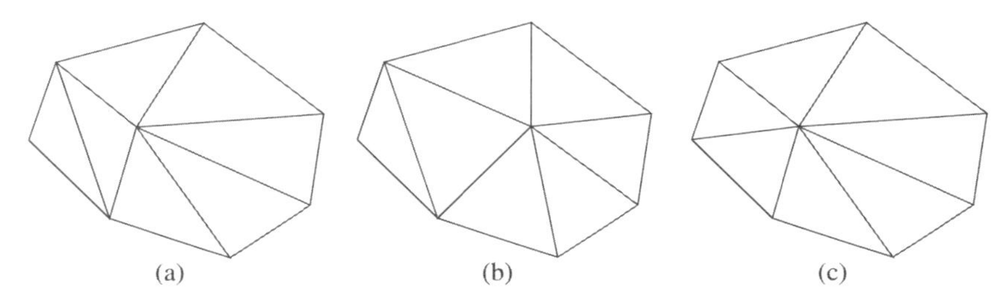

Fig. 5 Examples of mesh modifications in 2D: (A) initial patch of elements, (B) node re-positioning, (C) edge flip. *Reprinted from Shakoor, M. (2021). FEMS - A mechanics-oriented finite element modeling software. Computer Physics Communications, 260, 107729 with permission from Elsevier.*

closest quadrature point within that element of the old mesh to the quadrature point of the new mesh with no interpolation or smoothing.

As a conclusion, it can be seen that mesh adaption is not an easy task as it requires different mathematical theories to be understood and implemented. An error estimator is necessary to define a local mesh size criterion at each point of the domain, which in the case of *FEMS* can be an anisotropic metric tensor field. Then, this metric tensor field is used as input to a remeshing algorithm that will operate local modifications on the topology of the mesh and the position of its nodes to optimize a metric-based quality criterion. Finally, a variables transfer algorithm is necessary to transfer any variable defined on the old mesh to the new (adapted) mesh.

An interesting feature in *FEMS* is that the sensor variable can be defined as an LS function in order to adapt the mesh to an implicitly represented geometry. This representation can also automatically be made explicit, as presented in the sequel.

2.2.2 Mesh generation method

The idea of generating body-fitted FE meshes for geometries implicitly represented through signed distance functions (LS functions) has been proposed by various authors (Dapogny et al., 2014; Shakoor et al., 2015b). This is relevant for simple geometric entities such as circles, cylinders, ellipsoids, planes, squares and combinations (unions, intersections, complements), for which Equation (4) can be analytically computed. This is even more relevant for geometries segmented from 2D or 3D images which may be acquired using optical or electronic microscopy, or tomography (Shakoor et al., 2017b; Zhao et al., 2016).

Following Reference (Shakoor et al., 2017b), the methodology implemented in *FEMS* requires an LS function as input. It may be defined analytically, for instance from a sphere's center coordinates and radius, or voxel-wise on a background image. Note that an LS function can be computed directly on a segmented 2D or 3D image in linear complexity with respect to the number of voxels (Maurer et al., 2003), while such performance cannot be achieved on unstructured simplex meshes (Sethian & Vladimirsky, 2000; Shakoor et al., 2015a; Zhao et al., 2016).

This input LS function is used as sensor variable for mesh adaption using the isotropic curvature-based metric tensor field defined in Equations (3) and (5)). Depending on the initial mesh, which may be any mesh of the domain independently of the LS function, this mesh adaption process will generally have to be done in several iterations. Indeed, the LS function and hence the error estimator may not be well represented on the initial mesh, and the adaption process will improve this discretization up to convergence (usually in 5–7 iterations (Shakoor et al., 2017b).

At convergence, if the simulation requires a body-fitted FE mesh, a discretization of the LS function's zero iso-level can be reconstructed as a surface mesh using either a triangle and tetrahedron marching strategy (Dapogny et al., 2014) or a purely topological internal fitting strategy (Shakoor et al., 2015b). The former browses each triangle or tetrahedron of the mesh and splits it depending on how it is intersected with the zero iso-level. The latter browses each edge of the mesh and splits it if it has LS function values of opposing signs at its ends by inserting a new node at its intersection with the zero iso-level. Both strategies have a linear complexity with respect to the number of nodes, but the latter is simpler to implement as it only computes intersections at edges.

Once a body-fitted FE mesh has been constructed, the LS function is no longer required, except for a last mesh adaption step. Indeed, mesh quality as defined in Equation (12) or Equation (10) is likely to be deteriorated during the reconstruction of the zero iso-level's discretization. It must be restored using mesh adaption again, which must rely on a remeshing algorithm preserving the body-fitted mesh at internal interfaces, which is the case for both algorithms presented in Paragraph 2.2.1.2 (see References Dapogny et al., 2014; Shakoor et al., 2017a).

2.2.3 Applications

In the following, a compelling example from Reference (Shakoor, 2021) is first presented, and then applications where the objective is to compute effective linear properties for mechanical simulations. Applications mixing both mesh generation and adaption are voluntarily left out, as they will be detailed in Subsection 2.3.

2.2.3.1 Compelling example

This first set of simulations aims at showing the capabilities of *FEMS* regarding the generation of body-fitted FE meshes for geometries with internal interfaces initially represented through LS functions, as presented in Paragraph 2.2.2. The geometry for these simulations is based on the 2D *FEMS* image shown in Fig. 6A. The LS function to the surface of the letters is computed using the *Fiji* software (Schindelin et al., 2012) and is shown in Fig. 6B and C.

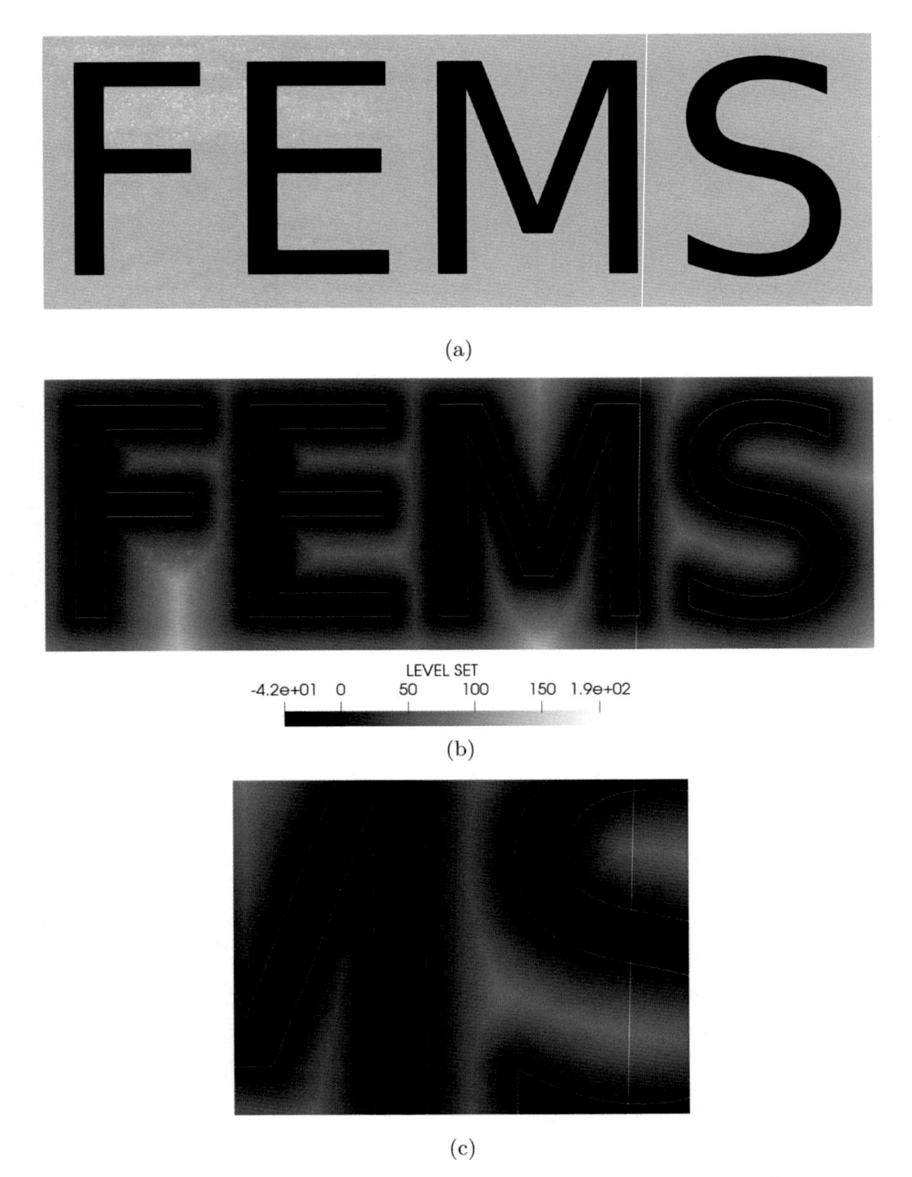

Fig. 6 *FEMS* image used for mesh generation simulations: (A) image of 1999 × 679 pixels, (B) LS function computed on the image with signed distances in pixels, (C) zoom on the LS function. *Reprinted from Shakoor, M. (2021). FEMS - A mechanics-oriented finite element modeling software. Computer Physics Communications, 260, 107729 with permission from Elsevier.*

As described in Paragraph 2.2.2, the metric tensor field used for mesh generation is that of Equations (3) and (5), with three metric parameters h_c, h_{min} and h_{max} to prescribe. In the sequel, body-fitted FE meshes are generated in 2D by projecting the image-based LS function shown in Fig. 6B and C to the initial FE mesh of the 2D domain. Parameters h_{min} and h_{max} are set respectively to 4 and 32 μm, while control parameter h_c is varied in order to show its influence.

The remeshing algorithm of the first strategy described in Paragraph 2.2.1.2 is used for these mesh generation simulations. To generate the body-fitted FE mesh, the strategy based on triangle and tetrahedron marching (Dapogny et al., 2014) is used.

As described in Paragraph 2.2.2, the image-based LS function shown in Fig. 6B and C is first interpolated to a structured FE mesh of the domain. The result is shown in Fig. 7A and B. Once the LS function is available at nodes of this FE mesh, the metric tensor field can be computed and mesh adaption can be performed. The result is shown in Fig. 7C and D for $h_c = 0.128$.

The LS function is then re-interpolated as it should be better captured and represented using the new adapted mesh, and the metric tensor is re-computed in order to re-adapt the mesh. The result after eight cycles is shown in Fig. 8A and B. These figures show how the mesh is automatically refined in regions with large local maximum principal curvature, which are mainly the regions of sharp angles in the letters.

Finally, this adapted mesh is modified through triangle marching and re-adapted in order to produce the mesh shown in Fig. 8C and D. This final body-fitted mesh accurately captures all features of the geometry, especially the *M* letter which has three regions with very sharp angles.

The accuracy is obviously guided by the choice of parameters h_{min} and h_{max}, which determine bounds on the prescribed mesh size. However, metric parameter h_c has a major influence on how the local maximum principal curvature influences mesh size. This is shown in Fig. 9A and B where $h_c = 0.256$ has been used, and Fig. 9C and D where $h_c = 0.512$ has been used.

On the one hand, if h_c is too large, the local principal curvatures do not have any influence and a uniform mesh size of h_{max} is prescribed everywhere. On the other hand, if h_c is too low, h_{min} is prescribed everywhere. It is thus necessary to choose an intermediary value so that fine features with large local maximum principal curvature are well described but a coarser mesh size is prescribed in regions with low local maximum principal curvature.

Fig. 7 (A) First steps of 2D mesh generation for the *FEMS* image using $h_c = 0.128$: (A,B) interpolation of the LS function from the image to the structured mesh, (C,D) mesh adapted once using the isotropic curvature-based metric tensor field. *Reprinted from Shakoor, M. (2021). FEMS - A mechanics-oriented finite element modeling software. Computer Physics Communications, 260, 107729 with permission from Elsevier.*

2.2.3.2 Permeability computation

It is proposed to compute the permeability of woven reinforcements from 3D images acquired by tomography. Details on this material can be found in Reference (Bodaghi et al., 2021).

Getting 3D images of a good resolution and properly segmented is not easy for woven reinforcements, especially in the out-of-plane direction. It is usual to find non-yarn voxels inside yarns, and to have a poor separation between *warp* and *weft* yarns.

Fig. 8 (A) Last steps of mesh generation for the *FEMS* image using $h_c = 0.128$: (A,B) mesh adapted eight times using the isotropic curvature-based metric tensor field, (C,D) mesh after internal fitting and body-fitted mesh adaption. *Reprinted from Shakoor, M. (2021). FEMS - A mechanics-oriented finite element modeling software. Computer Physics Communications, 260, 107729 with permission from Elsevier.*

This makes the generation of FE meshes with explicit interfaces and elements of a good quality particularly difficult for these images.

Meshes generated for the material of Reference (Bodaghi et al., 2021) are shown in Fig. 10. Mesh (A) corresponds to a little volume where a mesh size $h_{min} = 1\ \mu m$ can be used and mesh (B) to a larger volume where $h_{min} = 2\ \mu m$. There are nearly 500,000 elements for (A) and 1,000,000 for (B).

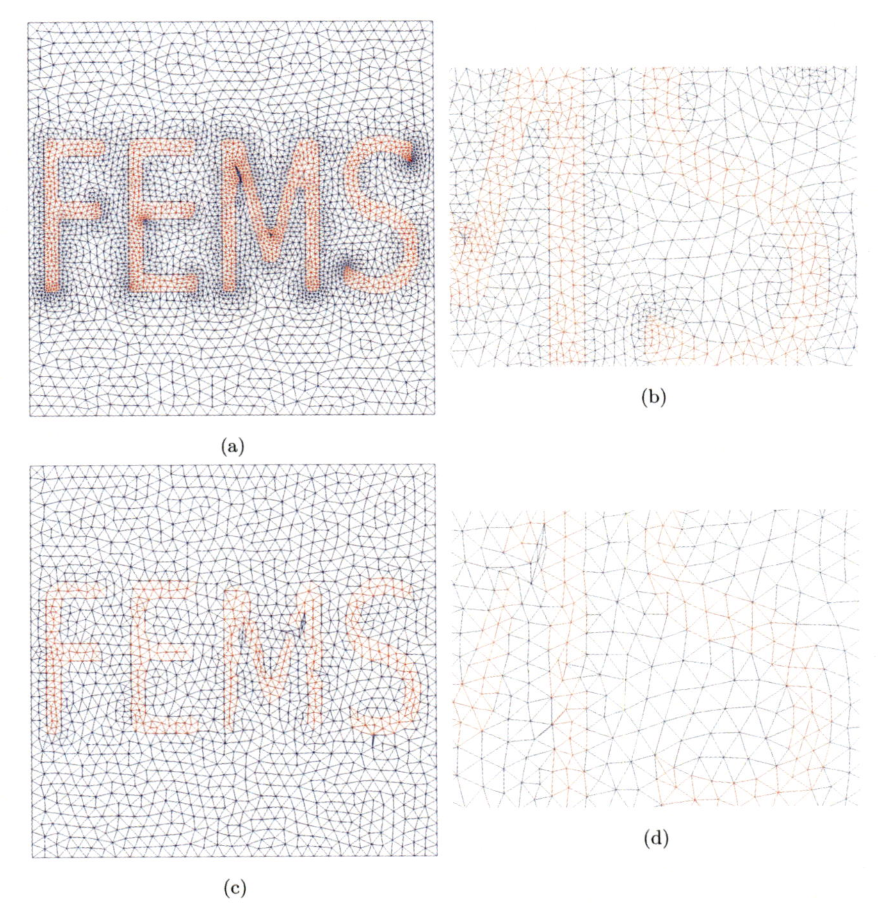

Fig. 9 Final meshes for the *FEMS* image with: (A,B) $h_c = 0.256$, (C,D) $h_c = 0.512$. *Reprinted from Shakoor, M. (2021). FEMS - A mechanics-oriented finite element modeling software. Computer Physics Communications, 260, 107729 with permission from Elsevier.*

For permeability computation, a Stokes-Brinkman solver has been developed in *FEMS*. A straightforward approach would be to suppress the yarns from the 3D mesh, and use a Stokes solver with a zero velocity imposed at the borders of the yarns. In practice, this kind of solver does not converge. A Brinkman term is hence introduced to regularize the transition between the yarns, which are assumed to be impermeable, and the rest of the domain. The boundary conditions used for these simulations are based on the multiscale approach with Lagrange multipliers

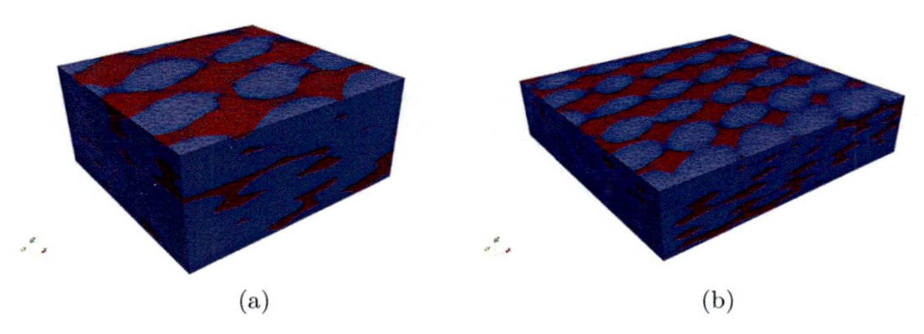

Fig. 10 Meshes used for permeability computation: (A) little volume, (B) large volume. The height of the volume (*z* direction) is about 1 mm for each volume.

presented in Subsection 5.3. FEs of Taylor-Hood P2/P1 type are used, which means a quadratic interpolation for the velocity and linear for the pressure.

Given the large size of the meshes, these simulations require the parallel implementation of *FEMS* using distributed memory, as detailed in Subsection 2.4. Permeability computation for 3D composites is, therefore, a good application for image meshing and also the development of efficient solvers and numerical methods.

An international team of researchers working on long fiber composites manufacturing processes has recently organized a benchmark on permeability characterization methods. After a successful work on experimental methods (Vernet et al., 2014), numerical methods have been considered. As detailed in References (May et al., 2021; Syerko et al., 2023), each participant had to numerically compute the permeability of a material from a 3D image acquired by tomography, which was provided both in raw and segmented form.

The benchmark objective was to assess the influence of boundary conditions, solver type (Stokes, Navier-Stokes, Stokes-Brinkman, *etc.*) and numerical method (finite volumes, FEs, *etc.*). The image was also provided in non-segmented form to assess the influence of the segmentation method. The benchmark focused on a unidirectional long fiber composite, for which 2D simulations along ten sections were possible.

Results for a section are shown in Fig. 11A. The zoom presented in Fig. 11C demonstrates that the circular form of the fibers is not accurately represented. This issue does not originate from the mesh generation method but from the poor resolution of the segmented image provided by the organizers. The segmentation methods for unidirectional long fiber composites are actually quite specific as it is possible to rely on the knowledge of the fiber shape. This is the case for the method used by the organizers, and

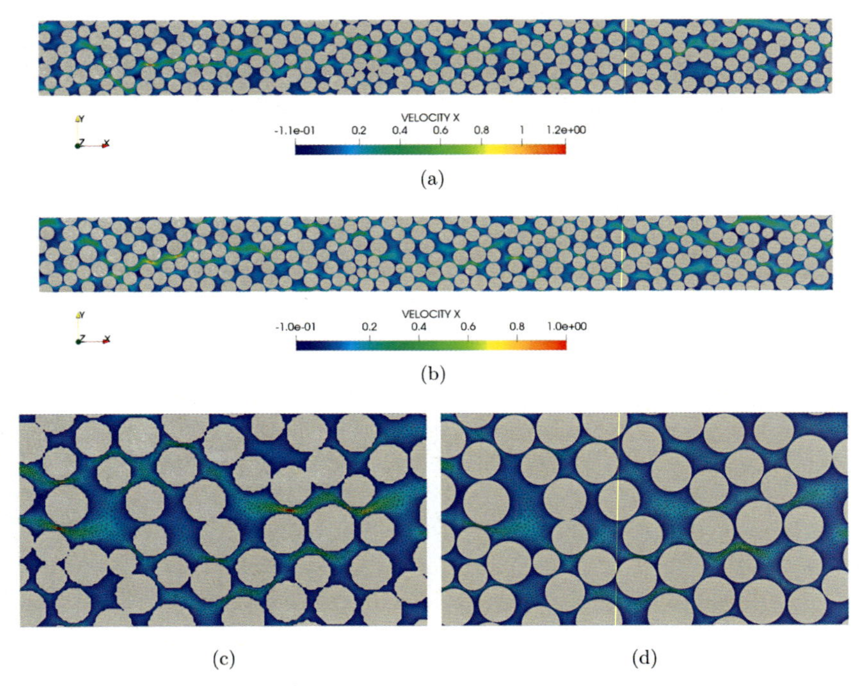

Fig. 11 Velocity field (mm s^{1}) in the horizontal direction for a section of the permeability benchmark image computed: (A) from the already segmented image, (B) from the raw image using the segmentation method proposed in Reference (Parvathaneni, 2020), (C) zoom on (A) showing the mesh around the fibers, (D) zoom on (B) showing the mesh around the fibers. Section height (*y* direction) is about 50 μm and the pressure gradient is applied in the *x* direction.

also for the one developed in Reference (Parvathaneni, 2020). The organizers nevertheless chose to re-project the circular forms on a 3D image. As an alternative, the raw image provided by the organizers has been segmented using the method proposed in Reference (Parvathaneni, 2020) and converted to a parametric form (centers and radii of the circles for each section). This is advantageous in order to generate meshes of a better quality, especially at fiber boundaries, as demonstrated in Fig. 11B and D. This illustrates another advantage of the mesh generation method: the geometry can be provided under various forms (image, parametric representation, *etc.*) as long as an LS function can be obtained.

This work on permeability computation demonstrates that the mesh generation method presented in Paragraph 2.2.2 is efficient. The generated meshes in 2D and 3D can indeed be exploited to compute useful flow properties.

2.3 Simulation of two-phase flows

Permeability computation as presented in Paragraph 2.2.3.2 is the first essential step towards the simulation of polymer composites manufacturing processes, especially regarding resin transfer molding processes. In those processes, the resin flows progressively through the mold, and is slowed down in its progression by the presence of the reinforcement. A poor impregnation can lead to the formation of bubbles which can remain trapped and deteriorate greatly the final properties of structural composites (Mehdikhani et al., 2019; Michaud, 2016; Park et al., 2011).

To predict the formation of bubbles and model their effect on final properties, simulating the interaction between the resin front and both the mesostructure and microstructure of reinforcements is essential. The second step, consequently, is the development of a numerical method for modeling two-phase flows in domains containing obstacles.

The equations governing these flows are summarized in Paragraph 2.3.1. The numerical method is detailed in Paragraph 2.3.2. Some results are presented in Paragraph 2.3.3.

2.3.1 Model

The simulation domain $\Omega \subset \mathbb{R}^d$, $d = 2, 3$ is split into two sub-domains Ω_1 and Ω_2, with $\Omega_1 \cup \Omega_2 = \Omega$ and $\Omega_1 \cap \Omega_2 = \Gamma_{1,2}$. These two sub-domains and the interface $\Gamma_{1,2}$ separating them evolve during time: $\Omega_1 = \Omega_1(t)$, $\Omega_2 = \Omega_2(t)$ and $\Gamma_{1,2} = \Gamma_{1,2}(t)$. The LS function ϕ is defined according to Equation (4) in order to implicitly represent the interface. Its evolution is described by the convection equation:

$$\frac{\partial \phi}{\partial t} + \mathbf{v}. \nabla \phi = 0 \tag{14}$$

where $\mathbf{v}$ is the flow velocity vector. This velocity is obtained by solving the Navier-Stokes equations modeling the incompressible and isotherm flow of two Newtonian fluids with surface tension:

$$\begin{cases} \rho \left(\frac{\partial \mathbf{v}}{\partial t} + \mathbf{v}. \nabla \mathbf{v} \right) - \nabla . \boldsymbol{\sigma}(\mathbf{v}, p) - \mathbf{f}_g + \mathbf{f}_s = \mathbf{0}, \\ \nabla . \mathbf{v} = 0. \end{cases} \tag{15}$$

These equations are completed by boundary conditions that depend on the problem to solve. In these equations, p is the pressure, ρ is the mass density and μ the dynamic viscosity. The Cauchy stress tensor $\boldsymbol{\sigma}$ is defined by

$$\boldsymbol{\sigma}(\mathbf{v}, p) = \mu (\nabla \mathbf{v} + \nabla^T \mathbf{v}) - p\mathbf{I}, \tag{16}$$

where $\mathbf{I}$ is the identity matrix. The force $\mathbf{f}_g = -\rho g \mathbf{e}_y$ is the gravity force, and g is the gravitational acceleration. The force $\mathbf{f}_s$ is the surface tension force and is modeled as a continuous surface force (Brackbill et al., 1992):

$$\mathbf{f}_s = \sigma_s \kappa_s \nabla H, \tag{17}$$

with σ_s the surface tension coefficient, κ_s the mean interface curvature and H the Heaviside function given by

$$H(\phi) = \begin{cases} 1, & \phi > 0, \\ \frac{1}{2}, & \phi = 0, \\ 0, & \phi < 0. \end{cases} \tag{18}$$

As proposed Reference (Sussman et al., 1998), this Heaviside function is also used to determine flow properties in Equation (15):

$$\begin{aligned} \rho(\mathbf{x}, t) &= \rho(\phi(\mathbf{x}, t)) = H(\phi(\mathbf{x}, t))\rho_1 \\ &\qquad + (1 - H(\phi(\mathbf{x}, t)))\rho_2 \\ \mu(\mathbf{x}, t) &= \mu(\phi(\mathbf{x}, t)) = H(\phi(\mathbf{x}, t))\mu_1 \\ &\qquad + (1 - H(\phi(\mathbf{x}, t)))\mu_2 \end{aligned} \tag{19}$$

where ρ_1, ρ_2 and μ_1, μ_2 are the mass densities and the viscosities of each fluid, respectively. Geometric properties of the interface such as the normal vector or the mean curvature can directly be computed from the LS function using the formulas

$$\begin{aligned} \mathbf{n}_s = \mathbf{n}_s(\phi) = \frac{\nabla\phi}{|\nabla\phi|}, \ \kappa_s &= \kappa_s(\phi) = -\nabla.\,\mathbf{n}_s(\phi) \\ &= \frac{\nabla\phi.\,\nabla\nabla\phi.\,\nabla\phi - |\nabla\phi|^2 \mathrm{tr}(\nabla\nabla\phi)}{|\nabla\phi|^3}, \end{aligned} \tag{20}$$

where $\nabla\nabla\phi$ is the Hessian matrix of ϕ and $|.|$ is the Euclidean norm. Due to the sign convention chosen for ϕ in Equation (4), the normal vector is directed towards fluid 1, which means that this fluid is the liquid and fluid 2 is the gas.

The coupled velocity-pressure-LS function problem is, moreover, nonlinear due to the auto-advection term, the surface tension force and the Heaviside function that is involved in the definition of flow properties. This adds up to the complexity of the numerical method used to solve these equations.

2.3.2 Numerical method

The first step for the numerical implementation of the equations detailed in Paragraph 2.3.1 is about the Heaviside function defined in Equation (18). It is possible to enrich the FE method to deal with this function's discontinuity (Rasthofer et al., 2011). The alternative approach proposed in the literature consists in regularizing this discontinuity and making up for the corollary loss of accuracy by refining the mesh at the interface. The advantage is that it is not necessary to modify FE shape functions. The regularized Heaviside function is given by

$$H_\epsilon(\phi) = \begin{cases} 1, & \phi > \epsilon, \\ \frac{1}{2}\left(1 + \frac{\phi}{\epsilon} + \frac{1}{\pi}\sin\left(\frac{\pi\phi}{\epsilon}\right)\right), & |\phi| \le \epsilon, \\ 0, & \phi < -\epsilon. \end{cases} \tag{21}$$

Since flow properties depend on this function (Equation (19)), parameter ϵ defines the half-thickness of the transition layer between the two fluids. This layer should be as thin as possible to converge to a discontinuous transition, but an accurate approximation of the regularized Heaviside function is only possible if this layer is large enough compared to the mesh size h. It is hence necessary to ensure that $\epsilon > h$ for linear FEs and $\epsilon > \frac{h}{2}$ for quadratic FEs (Pochet et al., 2013).

The second step is the choice of FE interpolation in the aim of approximating the surface tension force as defined by Equation (17). A quadratic interpolation for the LS function ϕ is particularly interesting for the computation of the mean curvature in Equation (17), which requires the second derivatives of the LS function. This is also advantageous for improving mass and energy conservation (Pochet et al., 2013).

Such a choice, nevertheless, imposes a significant restriction on the interpolation used for the pressure field p. It is essential, indeed, to ensure that the discretization of the surface tension force is balanced with respect to the one of the pressure (Francois et al., 2006). The discretization of the ∇H_ϵ term in Equation (17) should therefore be the same as the one of the ∇p term in Equation (15). The simplest solution if a P2 interpolation is used for the LS function is to choose the same interpolation for the pressure.

This choice, again, is not without consequences. The inf-sup condition, indeed, imposes a velocity interpolation order strictly greater than the pressure interpolation order. This restriction can be alleviated thanks to stabilization. The same quadratic interpolation can be used for the three fields by relying on a Residual-Based Variational MultiScale (RBVMS) stabilization (Bazilevs et al., 2007).

Details on the RBVMS formulation for the two-phase flow problem with stabilization terms can be found in Reference (Shakoor and Park, 2021a). This formulation is solved simultaneously for the three fields with an implicit Euler temporal scheme and hence a strong coupling between Equations (14) and (15). A Newton-Raphson algorithm with automatic differentiation is used to solve nonlinear terms, and the time step is automatically reduced when the nonlinear solver has difficulties to converge (Shakoor and Park, 2021a).

As mentioned previously, smoothing the properties transition between the two fluids and introducing the surface tension force imply a sufficiently fine discretization near the interface. To reach this goal while maintaining a reasonable computational cost during the simulation, the remeshing metric of Equation (9) is used with sensor variable $s_h = (H_\epsilon(\phi_h^n), H_\epsilon(2\phi_h^n - \phi_h^{n-1}))$, where ϕ_h^n is the LS function at the beginning of a time increment (before the Newton-Raphson solve) and $2\phi_h^n - \phi_h^{n-1}$ is an extrapolation of ϕ_h^{n+1}. Mesh adaption is triggered at the beginning of each increment.

The remeshing metric as well as the regularized Heaviside function defined in Equation (21) and the surface tension force defined in Equation (17) depend on a key property of the LS function, which is the distance property. During LS function convection, this property is generally lost and it should therefore be restored. This is achieved by an operation called re-initialization or re-distancing of the LS function. A direct or geometric approach consisting in projecting each node of the FE mesh onto a reconstruction of the interface has been implemented in this work (Pochet et al., 2013; Shakoor & Park, 2021a). This operation is triggered at each time increment, just after mesh adaption.

2.3.3 Results

An example of static simulation demonstrating that the surface tension force discretization is balanced with respect to the pressure field can be found in Reference (Shakoor and Park, 2021a). The accuracy of the obtained solution is of the same order as the tolerance used for the Newton-Raphson algorithm as long as the mean curvature is computed analytically. This demonstrates the interest of the balanced quadratic interpolation with RBVMS stabilization.

The results included in the following originate from dynamic simulations where all components of the numerical method are tested. The first examples presented in Paragraph 2.3.3.1 are based on a benchmark of numerical methods for modeling two-phase flows with surface tension (Hysing et al., 2009). The application of this method to resin transfer modeling processes has raised an issue that is specific to flows of low capillary number. This issue has been studied with an alternative approach based on a Lagrangian mesh. The results of this study are presented in Paragraph 2.3.3.2. An example of simulation combining mesh generation from an image and two-phase flow with an obstacle is finally presented in Paragraph 2.3.3.3.

2.3.3.1 Bubble rise

The benchmark conducted in Reference (Hysing et al., 2009) relies on the simulation of bubble rise in a 2D domain, as illustrated in Fig. 12.

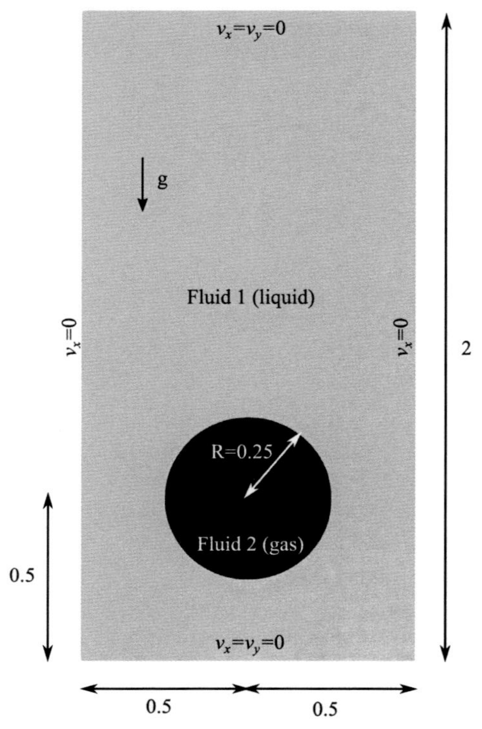

Fig. 12 Domain and boundary conditions for bubble rise simulations. *Reprinted from Shakoor, M., & Park, C. H. (2021a). A higher-order finite element method with unstructured anisotropic mesh adaption for two phase flows with surface tension. Computers & Fluids, 230, 105154 with permission from Elsevier.*

Two cases corresponding to two sets of non-dimensionalized fluid and flow properties (Hysing et al., 2009) are used:

- for case 1, properties are $\rho_1 = 1000$, $\rho_2 = 100$, $\mu_1 = 10$, $\mu_2 = 1$, $g = 0.98$ and $\sigma_s = 24.5$.
- and for case 2, $\rho_1 = 1000$, $\rho_2 = 1$, $\mu_1 = 10$, $\mu_2 = 0.1$, $g = 0.98$ and $\sigma_s = 1.96$.

Case 1 corresponds to the Reynolds number of Re = 35 and the Eötvös number of Eo = 10, while case 2 corresponds to Re = 35 and Eo = 125. For the definitions of these dimensionless numbers in this particular setup, the reader is referred to Reference (Hysing et al., 2009).

Simulations with and without mesh adaption and with different discretizations are conducted. The non-adaptive meshes are uniform and isotropic triangulations. The number of velocity-pressure-LS function degrees of freedom to solve is denoted NDOF, the number of FE mesh elements NEL and the number of time increments NTS. For simulations with an adaptive mesh, mean values are reported for NEL and NDOF. Since a finer mesh gives access to a smaller half-thickness ϵ for the transition layer, the value of ϵ is varied along with NDOF.

For each simulation, the quantities that are computed and monitored are the bubble area V_b, the height of its center of mass y_b and its rise velocity v_b. These quantities can be computed from the Heaviside function using:

$$
\begin{aligned}
V_b(t) &= \int_\Omega H(-\phi(\mathbf{x}, t))\,\mathrm{d}\Omega, \\
y_b(t) &= \frac{\int_\Omega H(-\phi(\mathbf{x}, t))\mathbf{x}\,\mathrm{d}\Omega}{V_b(t)} . \mathbf{e}_y, \\
v_b(t) &= \frac{\int_\Omega H(-\phi(\mathbf{x}, t))\mathbf{v}(t)\,\mathrm{d}\Omega}{V_b(t)} . \mathbf{e}_y.
\end{aligned}
$$

It is then possible to compute the relative error in L^2 norm integrated over the whole simulation duration:

$$
\mathrm{Error}(q) = \sqrt{\frac{\int_0^T (q(t) - q_{ref}(t))^2 \mathrm{d}t}{\int_0^T (q_{ref}(t))^2 \mathrm{d}t}},
$$

where $T = 3$ and $q = V_b$, y_b, ouv_b. The reference solution q_{ref} is πR^2 regarding bubble area (the flow is incompressible) and is a reference numerical simulation for the two other quantities. This reference is systematically the simulation giving the most accurate result for bubble area.

An example of result with anisotropic mesh adaption is shown in Fig. 13. The bubble becomes nearly elliptical in this case, with a lower side that becomes nearly flat. Fig. 13 demonstrates well the interest of anisotropic mesh adaption, which flattens and stretches elements when the third derivatives of the LS function are close to zero in order to prioritize regions where these derivatives have significant values.

Errors for case 1 are reported in Table 1. The number of time increments is quite greater than that initially prescribed for some simulations. This can be explained by the difficulty to converge for the Newton-Raphson algorithm and hence an automatic reduction of the time step to obtain convergence. The result with the 80×160 mesh and an NTS prescribed at 100 is, in fact, not reported in Table 1 because the final NTS was close to 200.

Very good bubble area conservation is observed for all simulations thanks to the quadratic interpolation used for the LS function. This

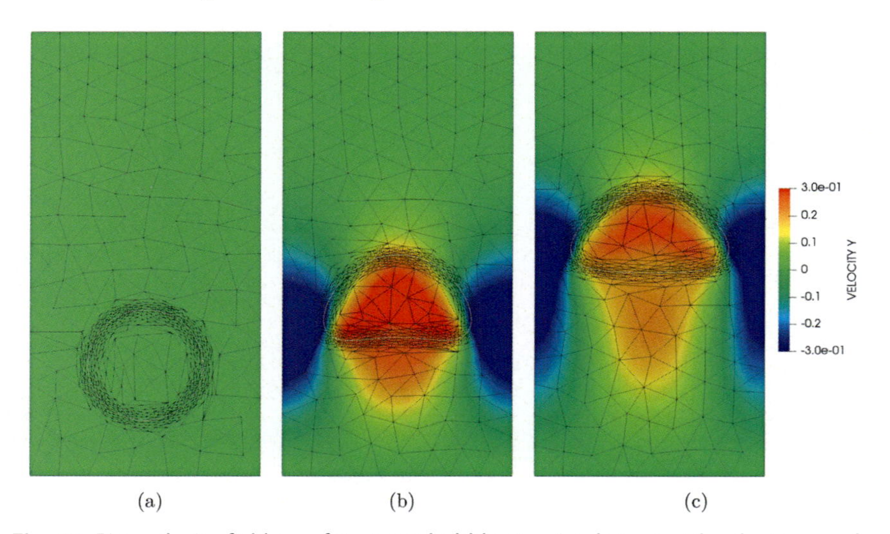

Fig. 13 Rise velocity field $\mathbf{v}.\mathbf{e}_y$ for case 1 bubble rise simulations with adaptive mesh, NDOF $\approx$ 11,500 and NTS = 407 at: (A) $t = 0$, (B) $t = 1.5$, (C) $t = 3.0$. *Reprinted from Shakoor, M., & Park, C. H. (2021a). A higher-order finite element method with unstructured anisotropic mesh adaption for two phase flows with surface tension. Computers & Fluids, 230, 105154 with permission from Elsevier.*

Table 1 Relative errors in L^2 norm for case 1 bubble rise simulations with different temporal and spatial discretizations, and with or without mesh adaption. *Reprinted from Shakoor, M., & Park, C. H. (2021a). A higher-order finite element method with unstructured anisotropic mesh adaption for two phase flows with surface tension. Computers & Fluids, 230, 105154 with permission from Elsevier.*

Mesh	20 × 40	40 × 80	80 × 160	Adaptive	Adaptive	Adaptive
ϵ	0.1	0.05	0.025	0.05	0.025	0.0125
NEL	1600	6400	25,600	≈740	≈1400	≈2800
NDOF	12,962	51,522	205,442	≈6000	≈11,500	≈23,000
NTS	102	100		100	104	165
Error(V_b)	5.45×10^{-2}	4.79×10^{-2}		4.89×10^{-2}	4.68×10^{-2}	4.32×10^{-2}
Error(y_b)	4.45×10^{-3}	2.46×10^{-3}		2.13×10^{-3}	1.81×10^{-3}	3.41×10^{-3}
Error(v_b)	8.05×10^{-2}	3.01×10^{-2}		2.74×10^{-2}	1.74×10^{-2}	1.62×10^{-2}
NTS	207	200	200	200	204	261
Error(V_b)	3.30×10^{-2}	2.40×10^{-2}	2.33×10^{-2}	2.05×10^{-2}	2.13×10^{-2}	2.14×10^{-2}
Error(y_b)	4.89×10^{-3}	2.51×10^{-3}	1.95×10^{-3}	2.99×10^{-3}	7.48×10^{-4}	1.32×10^{-3}
Error(v_b)	7.64×10^{-2}	2.44×10^{-2}	1.37×10^{-2}	2.38×10^{-2}	9.25×10^{-3}	9.59×10^{-3}

NTS	406	400	400	400	407	463
Error(V_b)	2.71×10^{-2}	1.06×10^{-2}	1.10×10^{-2}	2.25×10^{-3}	6.18×10^{-3}	8.22×10^{-3}
Error(y_b)	7.45×10^{-3}	2.24×10^{-3}	1.85×10^{-3}	1.52×10^{-3}	4.51×10^{-4}	1.86×10^{-3}
Error(ν_b)	8.39×10^{-2}	2.12×10^{-2}	1.05×10^{-2}	1.60×10^{-2}	5.07×10^{-3}	8.04×10^{-3}
NTS	802	800	800	800	804	880
Error(V_b)	3.10×10^{-2}	1.99×10^{-3}	3.92×10^{-3}	2.11×10^{-2}	7.17×10^{-3}	1.67×10^{-3}
Error(y_b)	1.03×10^{-2}	9.72×10^{-4}	2.05×10^{-3}	1.49×10^{-3}	9.10×10^{-4}	
Error(ν_b)	7.80×10^{-2}	1.71×10^{-2}	9.20×10^{-3}	1.48×10^{-2}	7.09×10^{-3}	

conservation is even better with an adaptive mesh. For instance, the simulation with NDOF ≈ 6000 and NTS = 400 leads to better results than all simulations with a non-adaptive mesh for the same NTS. The reduction ratio in terms of NDOF is nearly 35.

Results for case 1, overall, match well with those presented in Reference (Hysing et al., 2009), especially regarding the evolution γ_b and v_b, and the final shape of the bubble (Shakoor and Park, 2021a).

An example of result for case 2 is shown in Fig. 14. Thin gas filaments form in this case at the lateral extremities of the bubble. They then stretch progressively while the main part of the bubble remains similar to case 1. As detailed in Reference (Shakoor and Park, 2021a), simulations with a finer

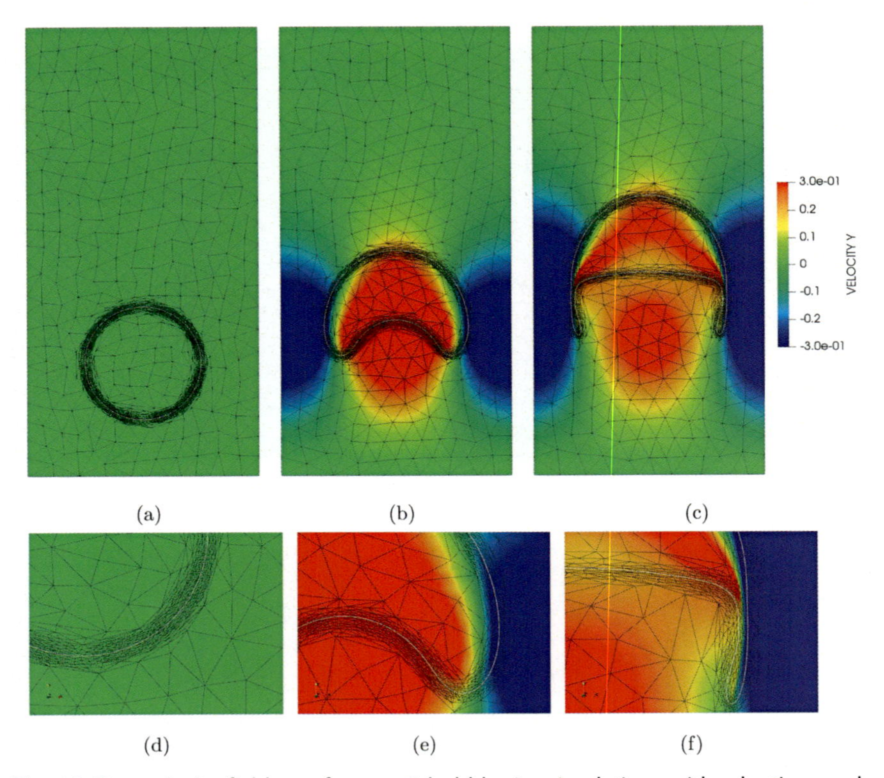

Fig. 14 Rise velocity field $\mathbf{v}.\mathbf{e}_y$ for case 2 bubble rise simulations with adaptive mesh, NDOF ≈ 23,000 and NTS = 530 at: (A) $t=0$, (B) $t=1.5$, (C) $t=3.0$. *(A) Zoom on the adapted mesh at: (D) t = 0, (E) t = 1.5, (F) t = 3.0. Reprinted from Shakoor, M., & Park, C.H. (2021a). A higher-order finite element method with unstructured anisotropic mesh adaption for two phase flows with surface tension. Computers & Fluids, 230, 105154 with permission from Elsevier.*

mesh and hence a smaller value of ϵ predict thinner filaments, but without breakup. The numerical methods compared in Reference (Hysing et al., 2009) did not all lead to the same result for case 2. Some methods predicted filaments breakup and some did not. The results obtained in this work match well with those obtained with the most efficient methods in terms of mass conservation.

Studying computational times is interesting for these case 2 simulations. As detailed in Reference (Shakoor and Park, 2021a), most of the computation time is spent solving the RBVMS formulation. Mesh adaption comes second, with a computation time including error estimation, SPR, metric computation, mesh adaption and variables transfer that varies between 1% and 15% of the total. This overhead of mesh adaption is small compared to the gain in terms of accuracy, especially regarding mass conservation. The quadratic interpolation used for the LS function largely contributes to this result as it enables to minimize diffusion during variables transfer whenever the mesh is adapted.

2.3.3.2 Bubble rise with a Lagrangian mesh

As demonstrated previously, the numerical method developed in this work has a number of advantages thanks to quadratic interpolation, anisotropic mesh adaption and strong coupling between velocity, pressure and LS function. This is particularly revealed by the very good mass conservation and the possibility to use large time steps. The first attempts to apply this method to resin transfer molding simulation at the microscopic scale have nevertheless not been successful. This has raised the interest for an alternative approach relying on an adaptive Lagrangian mesh and a weak coupling (Shakoor and Park, 2021c).

In this alternative approach, each node of the mesh is moved at the beginning of each time increment depending on the flow velocity. An iterative prediction-correction algorithm is used to preform this motion while triggering remeshing operations to avoid the formation of zero or negative volume elements (Shakoor et al., 2017a).

An isotropic remeshing metric is preferred to push the remeshing algorithm to form regular elements which facilitate mesh motion. This metric is the same as the one used in the Eulerian approach, but an additional operation is introduced after Equation (8) to obtain a scalar error. The latter is equal to the maximum eigenvalue of the error matrix defined by Equation (8).

Mesh motion implies, moreover, convection of the LS function, which makes the solution of Equation (14) superfluous. The auto-advection term is also eliminated from Equation (15). This is advantageous due to the computational cost spent on solving these equations in the Eulerian approach, but it implies a Courant-Friedrichs-Lewy (CFL) condition on the time step.

To summarize:

- The mesh is fixed with respect to the flow in the Eulerian approach, while it follows the flow in the Lagrangian approach. In the latter, mesh motion occurs at the beginning of the time increment and can require several remeshing operations.
- Both approaches are compatible with adaptive meshes, but anisotropic meshes can hardly be used in the Lagrangian approach. Mesh adaption occurs at the beginning of the time increment.
- There is an auto-advection term and a convection equation in the Eulerian approach, but not in the Lagrangian one.
- The Eulerian approach gives access to a strong coupling between the velocity-pressure system and the convection of the LS function, which is not possible in the Lagrangian approach. Consequently, there is a CFL condition in the Lagrangian approach and the Eulerian approach could give access to larger time steps. The time step, in addition, is automatically reduced when the CFL condition is violated in the Lagrangian approach, while it is also reduced in the Eulerian approach when the Newton-Raphson algorithm does not converge.

In a recent study (Denner et al., 2017), a relation has been established between parasitic currents in surface tension models and the type of coupling that is used. This study, in particular, concluded that surface tension modeling implies a stability condition on the time step which cannot be eliminated, even with a strong coupling. This condition depends, in fact, on the capillary number. It is interesting, consequently, to evaluate the interest of the Lagrangian approach for low capillary numbers.

As a reminder, the capillary number is defined as $\mathrm{Ca} = \frac{\mu v_c}{\sigma_s}$, where v_c is the characteristic velocity of the flow. Capillary numbers observed in polymer composites manufacturing processes are generally lower than 10^{-2}, and even lower than 10^{-5} in reservoir engineering applications (especially in the oil and gas industry).

Simulations of 3D bubble rise with two sets of fluid properties corresponding respectively to a capillary number $\mathrm{Ca} = 1.5$ and $\mathrm{Ca} = 0.015$ are conducted using both approaches. Boundary conditions for this extension to 3D are inspired from Reference (Safi et al., 2017) and consist in imposing a zero velocity at the whole domain boundary.

First, for a capillary number of Ca = 1.5, simulations using different spatial discretizations are conducted with the same prescribed time step for both approaches. Second, in order to compare fairly both approaches in terms of accuracy, additional Eulerian simulations are conducted with the same NTS as that obtained for each Lagrangian simulation.

An example of result with both approaches for nearly the same NTS and a capillary number of Ca = 1.5 is presented in Fig. 15. Both approaches seem to lead to a qualitatively similar result. The final shape of the bubble is similar to that observed for case 2 in 2D. Although the number of degrees of freedom is larger for the Eulerian approach, these two simulations are equivalent in terms of spatial discretization. The Eulerian approach, indeed,

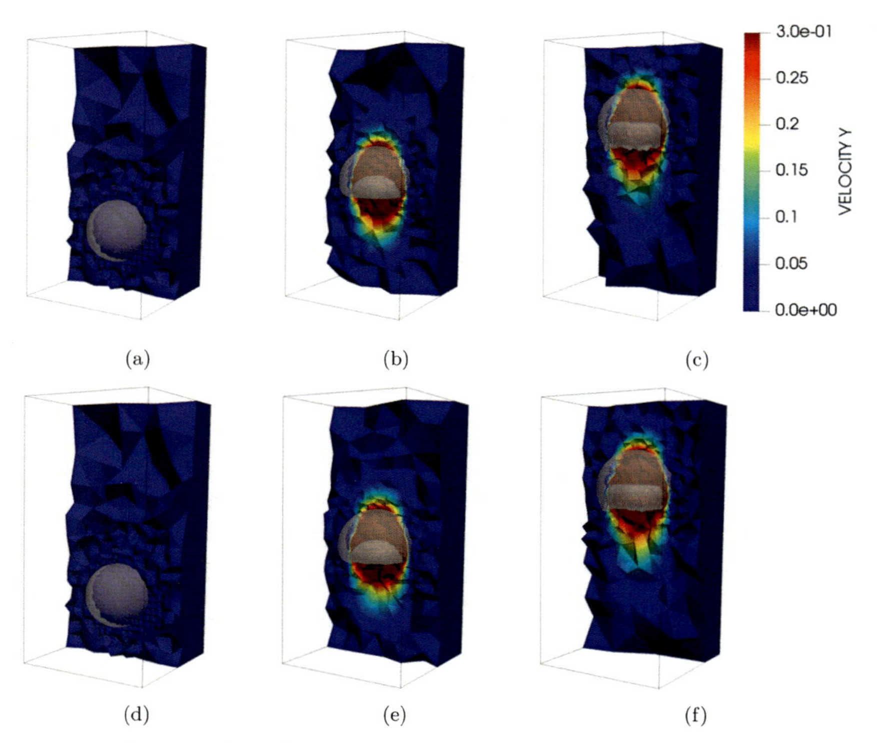

Fig. 15 Inside view of the domain showing the rise velocity field $\mathbf{v}.\mathbf{e}_y$ for bubble rise simulations with Ca = 1.5. Eulerian simulation with NDOF ≈ 205,000 and NTS = 328 at: (A) $t = 0$, (B) $t = 1.5$, (C) $t = 3.0$. Lagrangian simulation with NDOF ≈ 166,000 and NTS = 325 at: (D) $t = 0$, (E) $t = 1.5$, (F) $t = 3.0$.

implies solving a convection equation for the LS function and hence a larger number of degrees of freedom for the same number of elements.

Fig. 16 shows gas volume evolution for different discretizations. It is supposed to remain constant as the flow is incompressible. The volume appears to increase, which shows that for these simulations it is the time discretization error that dominates over the spatial discretization error. This is verified for both approaches.

The Eulerian approach, moreover, can run simulations with a number of time increments close to 100 whatever the spatial discretization, while the CFL condition prevents this even for the coarser mesh in the Lagrangian case. As a conclusion, the Eulerian approach is more interesting than the Lagrangian one for this capillary number, and it would be even more interesting with a more accurate temporal scheme. This approach, indeed, gives access to larger time steps and avoids any difficulty related to mesh motion.

Results with a capillary number Ca = 0.015 are presented in Fig. 17. The bubble maintains its spherical shape in this case. This can be explained by a more significant surface tension. The bubble also rises faster as compared to the Ca = 1.5 case. The Eulerian approach predicts a faster rise as compared to the Lagrangian one, which could be due to a larger number of time increments.

For this low capillary number, indeed, the Newton-Raphson algorithm of the Eulerian approach has a hard time converging, which leads to a significant increase of the number of time increments. The final number is five times the prescribed one. For the Lagrangian approach, on the

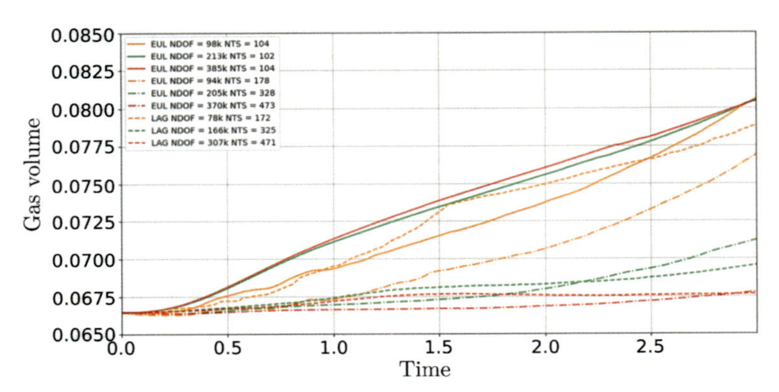

Fig. 16 Gas volume evolution with respect to time for the Lagrangian (LAG) and Eulerian (EUL) simulations with Ca = 1.5.

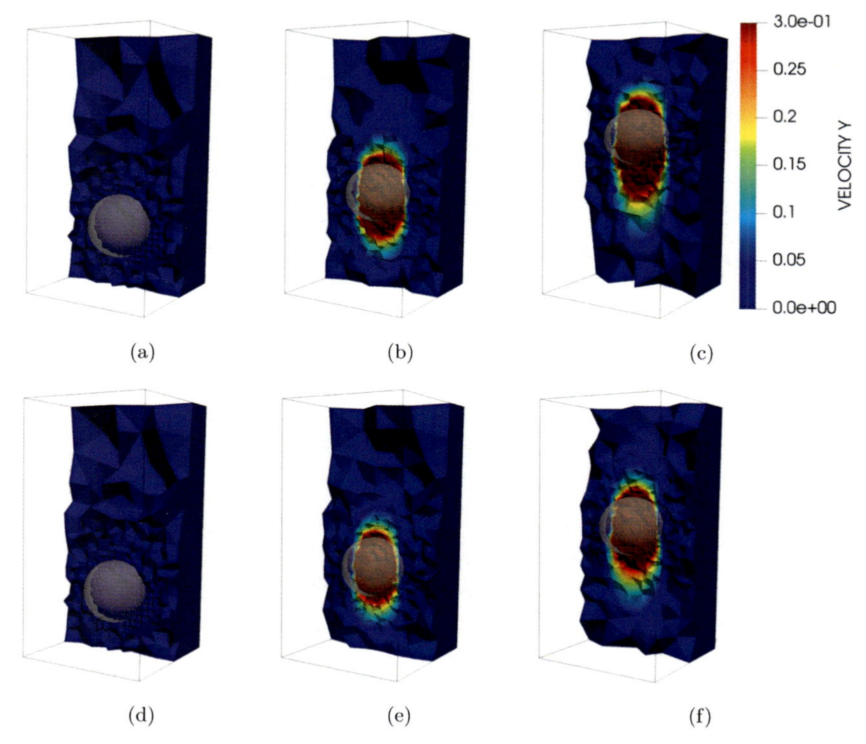

Fig. 17 Inside view of the domain showing the rise velocity field $\mathbf{v.e}_y$ for bubble rise simulations with Ca = 0.015. Eulerian simulation with NDOF $\approx$ 199,000 and NTS = 530 at: (A) $t = 0$, (B) $t = 0.75$, (C) $t = 1.5$. Lagrangian simulation with NDOF $\approx$160,000 and NTS = 218 at: (D) $t = 0$, (E) $t = 0.75$, (F) $t = 1.5$.

contrary, there is no convergence issue and the CFL condition can be satisfied for a number of time increments that is lower to that used in the Ca = 1.5 case.

As a conclusion, although the Eulerian approach developed in this work with a strong coupling may seem very promising and advantageous with a large capillary number, its interest can be questioned for a low one. The numerical developments conducted in this work around mesh adaption with a quadratic interpolation for the LS function and the balanced RBVMS formulation are still relevant, given that they are completed by a mesh motion algorithm to obtain a robust Lagrangian approach. It appears, in general, that for a low capillary number convergence difficulties as well as the CFL condition restrict the time step to a value so small that a weak coupling with a Lagrangian mesh become appealing.

Other examples exploiting the Lagrangian approach for applications related to mechanics of materials with a nonlinear behavior such as finite strain hyperelasticity or elasto-plasticity can be found in Reference (Shakoor, 2021).

2.3.3.3 Simulation including obstacles

Finally, a two-phase flow simulation with a domain containing obstacles is considered. Domain and boundary conditions are presented in Fig. 18A. A droplet is placed in the higher half of the domain, while the lower part contains obstacles.

The first step is to generate an FE mesh with an explicit discretization of obstacles. The procedure described in Subsection 2.2 is used and the obstacles geometry is the one considered in Paragraph 2.2.3.1. Parameters $h_{min} = 8\ \mu m$, $h_{max} = 32\ \mu m$ and $h_c = 0.128$ are chosen for the remeshing metric based on the maximum principal curvature and the obtained mesh is presented in Fig. 18B after eight adaption cycles.

The second step is to introduce the LS function for the liquid/gas interface in this FE mesh with explicit discretization of obstacles. In this aim, the remeshing metric is replaced by the one defined in Equation (9) with a sensor variable $\mathbf{s} = H_\epsilon(\phi^0)$ where ϕ^0 is the LS function of the liquid/gas interface in the initial configuration. Parameters $h_{max} = 1000$, $\mathcal{N}_c = 4096$ and $\epsilon = 8\ \mu m$ are chosen to obtain the mesh presented in Fig. 18C after five more adaption cycles. During these new cycles and also during the whole flow simulation, the mesh inside the obstacles is frozen. Fig. 18D clearly shows the combination of the explicit discretization of obstacles and the implicit discretization of the liquid/gas interface with an adaptive mesh.

Finally, the two-phase flow simulation can be conducted. A zero velocity is imposed inside and at obstacles boundaries, so that they could have been eliminated from the FE mesh. The liquid is of mass density $1\ kg\ mm^3$ and dynamic viscosity $10\ MPa\ ms$. The surface tension coefficient is of $1.96\ g\ ms^2$. The gas is of mass density $1\ g\ mm^3$ and dynamic viscosity $0.1\ MPa\ ms$. Gravitational acceleration is set to $0.98\ mm\ ms^2$.

The Eulerian approach with strong coupling described in Paragraph 2.3.2 is chosen, with a slight modification consisting in triggering mesh adaption only if the quality of at least one mesh element as defined in Equation (12) falls below 0.33. The prescribed time step is 0.1 ms. The simulation ends at $T = 10$ ms in 750 time increments, among which 454 include mesh adaption due to a drop of mesh quality. The liquid/gas interface as well as the adapted meshes are shown in Fig. 19 at different instants.

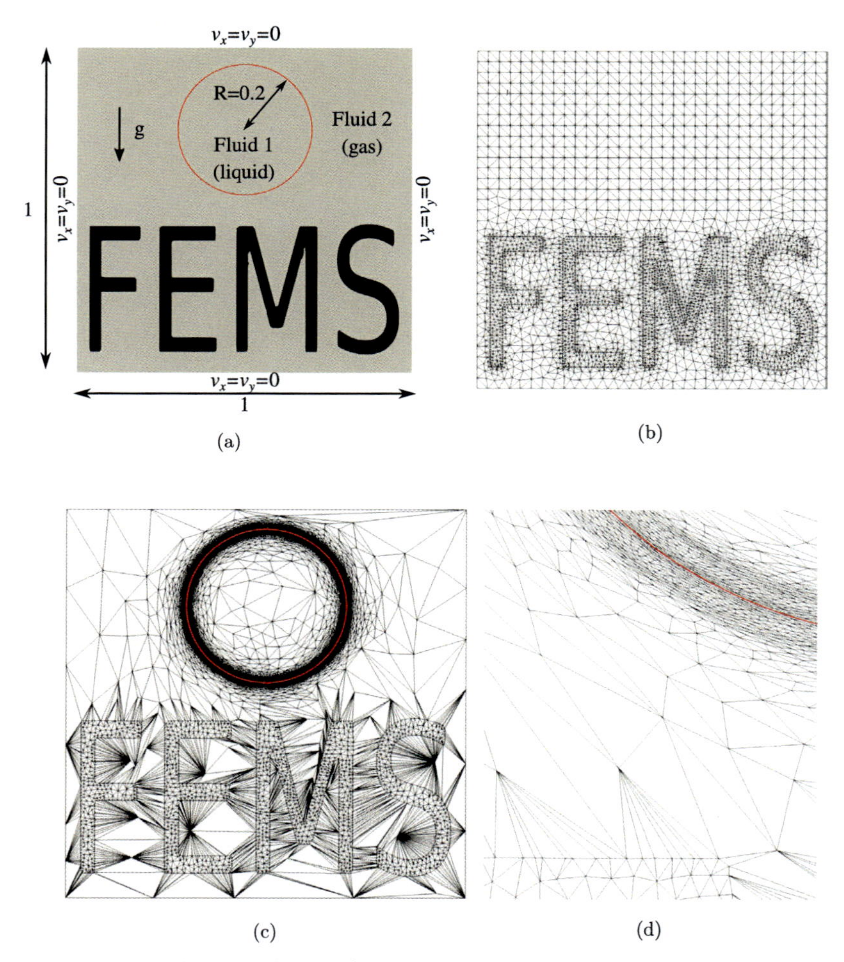

Fig. 18 (A) Boundary conditions for the two-phase flow problem with obstacles. Lengths are in millimeters and velocities in meters per second. (B) FE mesh generated with an explicit discretization of obstacles. *(C) Adapted FE mesh with explicit obstacles discretization and implicit liquid/gas interface discretization. (D) Zoom on the adapted mesh. Reprinted from Shakoor, M. (2021). FEMS - A mechanics-oriented finite element modeling software. Computer Physics Communications, 260, 107729 with permission from Elsevier.*

The droplet contacts with the obstacles in Fig. 19A. It then spreads due to inertia in Fig. 19B, before pouring down between obstacles until the final state shown in Fig. 19C. The zoom on the adapted mesh shown in Fig. 19D can be compared to the one in Fig. 18D. It appears that freezing the computational cost with parameter $\mathcal{N}_c$ enforces a pragmatic

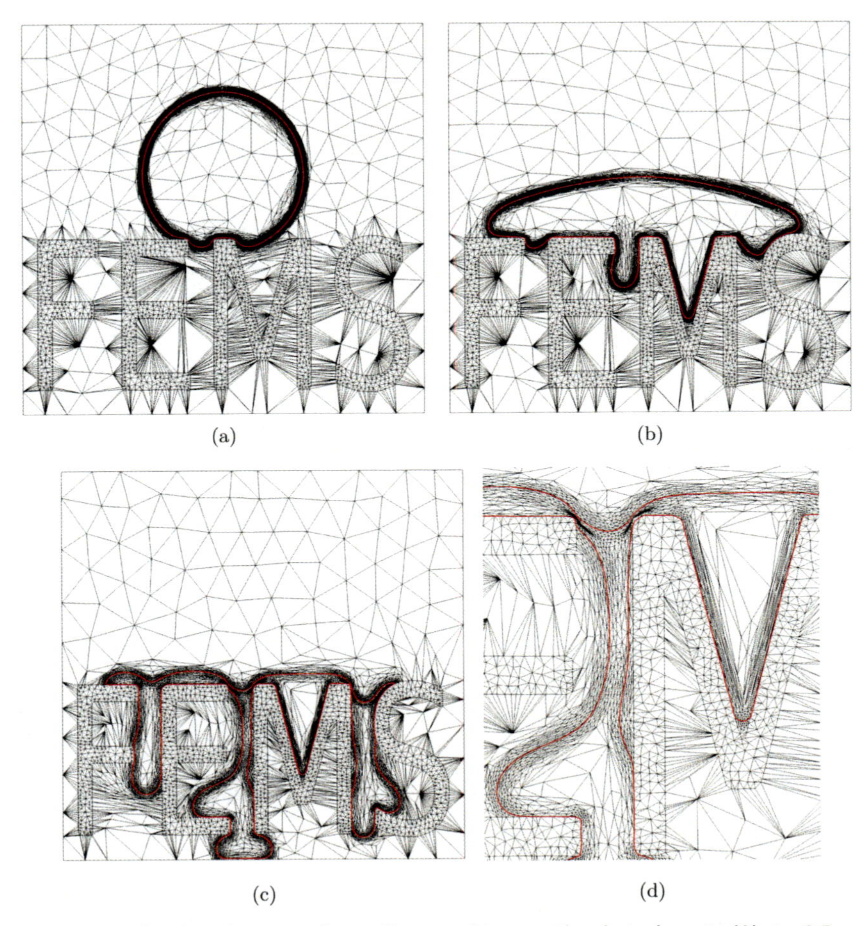

Fig. 19 Results for the two-phase flow problem with obstacles at: (A) $t = 0.5$ ms, (B) $t = 1$ ms, (C) 10 ms. *(D) Zoom on the adapted mesh at t = 10 ms. Reprinted from Shakoor, M. (2021). FEMS - A mechanics-oriented finite element modeling software. Computer Physics Communications, 260, 107729 with permission from Elsevier.*

management of the discretization. The morphology of the interface, indeed, becomes way more complex during the simulation than it was at the beginning. Stretching elements where the interface flattens out enables to maintain a good level of refinement where the curvature of the interface increases, and this for a constant computational cost.

This simulation with obstacles demonstrates quite well the interest of the numerical method developed in this work. The LS method with quadratic interpolation and mesh adaption is interesting to ensure a very good mass conservation. It can be completed with a balanced RBVMS formulation with strong or weak coupling depending on the capillary number.

These numerical developments have been exploited for the investigation of bubble migration at the microscopic scale in long fiber composites (Shakoor & Park, 2019,2021b).

2.4 Software engineering and parallel computing

Managing the scientific computing code *FEMS* where all developments of this section have been integrated has been a key part of this research work, and it should not be ignored. Multiple objectives were to be dealt with:

- Some collaborators who were not initiated to programming could be brought to use the code.
- Some collaborators who did not have a significant scientific computing experience could be brought to contribute to the code.
- The code had to be maintained with some homogeneity.
- It had to be efficient enough to exploit as much as possible the supercomputing capabilities of research centers, while remaining accessible for beginners in scientific computing.

In the first version of *FEMS*, detailed in Reference (Shakoor, 2021), the code has been divided into three blocks. Users can use the code without programming anything thanks to input files in *Extensible Markup Language* format, and meshes in *MSH* format. Meshes with this format can be created with the *Gmsh* software (Geuzaine and Remacle, 2009). These input files form the first block.

The second block is constituted of a whole part of the code that does not require managing parallel aspects. This is helpful to implement operations that are local to a node, an integration point or an element of the mesh without paying any particular attention to parallel computing.

The operations of the third block, finally, are those that require knowledge in parallel computing. This includes for instance remeshing, SPR, and re-initialization.

This first version of *FEMS* relied on shared memory parallel computing and various existing sequential linear algebra libraries. It has been published as free software (Shakoor, 2022) along with a technical documentation written with the *Doxygen* tool and tutorials published in Reference (Shakoor, 2021).

Research on two-phase flows has opened the opportunity to experiment parallel computing on graphical processing units. These units are, indeed, equipped with specific multi-processors that carry a great number of computing units. They are tailored for performing the same operation in parallel for a great number of inputs. The LS re-initialization operation typically falls in that category. Results published in Reference (Shakoor and Park, 2021a) clearly demonstrate that even with a brute force algorithm consisting in projecting every node of the FE mesh to each segment or triangle of the reconstructed surface mesh of the zero iso-level of the LS function, the computation time spent re-initializing is negligible when a graphical processing unit is involved.

Motivated by these remarkable results, an attempt to exploit the same technology for the remeshing algorithm has been investigated (Shakoor and Delbeke, 2021). This study has not been satisfactory as limitations of parallel computing on graphical processing units have been identified. Although remeshing, indeed, consists in performing the same operation for a great number of inputs, this operation requires a great amount of data on elements topology, for which memory accesses could not be optimized (Shakoor and Delbeke, 2021).

Finally, the parallel implementation of *FEMS* has recently been revisited to switch to distributed computing. This new version has played a key role in the research work on multiscale modeling that is presented in Section 5. Distributed memory parallel computing is, indeed, essential not only for tapping in the power of supercomputers and their hundreds of computing nodes, but also to spread a great amount of data on several nodes.

These efforts regarding the management of the free software *FEMS* and its parallel implementation are not at the core of the present work but they are a necessary contribution. It seemed, therefore, opportune to summarize them in this section.

2.5 Conclusions

This section has dealt with FE modeling of two-phase flows in domains containing obstacles. The first topic has been the discretization of obstacles with a geometry provided in a parametric form or in the form of a 3D image acquired by tomography. A mesh generation method based on LS functions has been developed. LS functions can, indeed, be computed directly from a parametric representation and also an image. The robustness of this mesh generation method with an explicit discretization of obstacles has been demonstrated for the computation of the permeability of reinforcements.

Regarding two-phase flows, the LS method with a quadratic FE interpolation has been presented in order to implicitly represent the liquid/gas interface. A real gain in terms of accuracy has been demonstrated. This is interesting as it makes up for some deficiencies of the LS method, especially regarding mass conservation. Mesh adaption has been revisited with a remeshing metric extended to quadratic interpolation. This gives additional benefits in terms of accuracy and computation time.

The presented surface tension model is balanced. The specificity of this approach lies in the intrinsic balance between the discretization used for the pressure field and that used for the LS function. The RBVMS stabilization plays a key role in this balance because it gives the possibility to use the same order of interpolation for the velocity, the pressure and the LS function.

The numerical method presented in this work is flexible and gives access to both a strong coupling with a simultaneous solution of the velocity-pressure system and the LS function, and a weak coupling where those equations are solved separately. In the second case, a Lagrangian approach with a mesh motion algorithm has been proposed and shown to be promising for problems where the Capillary number is small.

All numerical developments regarding the FE method have been integrated in a free software. The management and parallel implementation of this software has constituted a non-negligible complementary part of this research work. This supplementary effort has facilitated data generation in Section 4, and opened the possibility of simulations of a great complexity in Section 5.

3. Fracture modeling with a phase-field and the fast Fourier transform

3.1 Introduction

Mesh generation methods such as those presented in Subsection 2.2 have a great interest. Their numerical implementation, however, is technically difficult and has encouraged researchers to consider alternative routes because:

- The generated meshes can contain elements of a poor quality. These elements can locally alter numerical results, and even prevent any numerical solution.

- Mesh generation in itself is a supplementary step that has to be added after the construction of the geometry or the segmentation of the 3D image (if the geometry has been acquired through a tomography scan).
- The efficient implementation of these algorithms is not straightforward, especially if parallel computing is considered (see Subsection 2.4).

The numerical method based on the Fast Fourier Transform (FFT) has known an increasing popularity since its invention in 1994 (Moulinec and Suquet, 1994). This method is based on two fundamental ingredients: Green's functions and the FFT. It additionally relies on a voxel mesh, hence avoiding any mesh generation or adaption step. This is both an advantage and a limitation, as detailed in the following. This solution method, moreover, does not require the assembly of a global stiffness matrix as compared to the Finite Element (FE) method. All operations are local to a voxel, except for the FFT. This is very advantageous for parallel implementations, as very efficient open source libraries can easily be found for the parallel computation of the FFT on multi-processors as well as on graphical processing units and supercomputers (Frigo and Johnson, 2005).

The main objective of this section is to model the fracture of brittle and quasi-brittle materials such as polymer matrix composites. This section can be read as a direct sequel to Section 2. Given an accurate prediction of the morphology and distribution of voids in these composite materials, indeed, the next step would be to predict the final properties of these materials depending on this porosity.

Due to its advantages, the numerical method based on the FFT is chosen to reach that objective. This method nevertheless brings some drawbacks, among which the presence of parasitic oscillations in the computed mechanical fields. Various explanations can be found in the literature regarding the sources of these oscillations:

- In Reference (Moulinec and Suquet, 1998), it is suggested that the approximation of the Fourier transform by a discrete one entails a significant truncation error that comes out as oscillations. Shannon's theorem requires that the number of discretization points per unit length should be larger than twice the maximum local gradient of the represented fields. For strains and stresses in a domain composed of multiple materials, this maximum gradient is infinite at interfaces between materials and Shannon's theorem is impossible to satisfy. The truncation error is hence significant.

- In Reference (Eloh et al., 2019), Gibbs' phenomenon is mentioned as a possible explanation. This phenomenon is characterized by the apparition of oscillations in a discrete field close to discontinuities.
- It is well known that an FE discretization with hexahedra and a single integration point per element brings parasitic deformation modes known as hourglass modes. The numerical method based on the FFT relies on a voxel discretization, which are hexahedra, and on a voxel-wise constant strain field. Hourglass modes have been observed in the literature for this method (Leuschner and Fritzen, 2018).
- Interface discretization with a voxel mesh is obviously a source of oscillations. It has also been identified in the literature (Doitrand et al., 2015).

Given the number of possible explanations and the great number solutions proposed to attenuate them, it seems necessary to analyze and rank them.

Numerical modeling of damage and fracture is a broad research field Shakoor et al. (2019c). Phase-field modeling is an interesting route that has been popularized quite recently (Bourdin et al., 2008; Miehe et al., 2010a,b). It is inspired from continuous damage models, which circumvent numerical difficulties incidental to the explicit representation of a changing discontinuity in a discretization, as well as those related to the prediction of crack initiation and branching phenomena (Shakoor et al., 2019c). These models, indeed, represent cracks implicitly through a continuous damage variable, which can also be used to model the initiation and growth of micro-cracks and micro-voids in addition to (macro-)cracks (Besson, 2010). These models have a well known drawback which is the dependence of the predictions on the discretization (Bažant & Jirásek, 2002; Peerlings et al., 1996). This dependence can be eliminated using a regularization which consists in introducing a non-local damage variable through a local smoothing or the solution of a reaction-diffusion equation (Bažant & Jirásek, 2002; Peerlings et al., 1996).

Phase-field models can be seen as a variational approach to non-local continuous damage models. They are based on a unified formulation featuring a unique damage variable with built-in regularization (Miehe et al., 2010a,b).

These phase-field models are particularly interesting for solution by the FFT-based numerical method, as it seems difficult to explicitly represent a crack in a voxel mesh. The implementation of a phase-field model in the FFT-based numerical method has not been addressed in the literature, with the exception of Reference (Chen et al., 2019).

The work in Reference (Chen et al., 2019) relies on a simplification of the reaction-diffusion equation for the phase-field. The potential influence of this simplification on results has been reported in the literature (Jeulin, 2020), but has not been analyzed numerically.

The FFT-based numerical method and an example of result are presented in Subsection 3.2.

An analysis of the causes of the parasitic oscillations in results computed with this method is summarized in Subsection 3.3. A focus on oscillations due to the irregular discretization of interfaces by a voxel mesh is presented as well.

The extension of the FFT-based numerical method to fracture modeling with a phase-field approach is presented in Subsection 3.4.

3.2 Basics

The FFT-based numerical method is presented in Paragraph 3.2.1. An application establishing clearly the interest of this approach is presented in Paragraph 3.2.2.

3.2.1 *Method*

The simulation domain is the box $\Omega = [0,\ L_1] \times [0,\ L_2] \times [0,\ L_3]$, with $(L_1,\ L_2,\ L_3) \in \mathbb{R}^{3+*}$. The objective is to solve the small strain equilibrium problem

$$\begin{cases} -\nabla.\,\boldsymbol{\sigma}(\boldsymbol{x}) = \mathbf{0},\ \boldsymbol{x} \in \Omega, \\ \boldsymbol{\sigma}(\boldsymbol{x}) = \mathbb{C}(\boldsymbol{x}) : \boldsymbol{\varepsilon}(\boldsymbol{x}),\ \boldsymbol{\varepsilon}(\boldsymbol{x}) = \nabla^{sym}\boldsymbol{u}(\boldsymbol{x}) \\ = \frac{1}{2}\left(\nabla\boldsymbol{u}(\boldsymbol{x}) + \nabla^T\boldsymbol{u}(\boldsymbol{x})\right), \\ \boldsymbol{u}(\boldsymbol{x}) = \boldsymbol{\varepsilon}^M.\,\boldsymbol{x} + \boldsymbol{u}^{\#}(\boldsymbol{x}), \\ \boldsymbol{u}^{\#}\begin{pmatrix} 0 \\ x_2 \\ x_3 \end{pmatrix} = \boldsymbol{u}^{\#}\begin{pmatrix} L_1 \\ x_2 \\ x_3 \end{pmatrix},\ \boldsymbol{u}^{\#}\begin{pmatrix} x_1 \\ 0 \\ x_3 \end{pmatrix} = \boldsymbol{u}^{\#}\begin{pmatrix} x_1 \\ L_2 \\ x_3 \end{pmatrix}, \\ \boldsymbol{u}^{\#}\begin{pmatrix} x_1 \\ x_2 \\ 0 \end{pmatrix} = \boldsymbol{u}^{\#}\begin{pmatrix} x_1 \\ x_2 \\ L_3 \end{pmatrix}. \end{cases} \tag{22}$$

The unknown to find is the displacement field $\boldsymbol{u}$. This problem involves the strain field $\boldsymbol{\varepsilon}$, the Cauchy stress field $\boldsymbol{\sigma}$ and the stiffness tensor $\mathbb{C}$. The displacement field is decomposed into a linear part and a periodic part. The linear part depends on a macroscopic strain $\boldsymbol{\varepsilon}^M$ which is an input to the problem, as well as the stiffness tensor. For the sake of simplicity, Hooke's law is used herein, which corresponds to a linear elastic behavior. Numerous studies have shown that the FFT-based numerical method can adapt to nonlinear material behavior laws both in small and finite strains (Kabel et al., 2014).

The equilibrium problem in Equation (22) is actually a homogenization problem, as detailed in Section 5. It involves heterogeneous coefficients as the stiffness tensor $\mathbb{C}$ depends on $\boldsymbol{x}$. This is due to the fact that the domain is assumed to be composed of multiple materials.

The first step consists in introducing a stiffness tensor $\mathbb{C}^0$ associated to a purely fictitious reference material:

$$\begin{aligned} -\nabla.\ \boldsymbol{\sigma}(\boldsymbol{x}) = \boldsymbol{0} &\Leftrightarrow -\nabla.\ (\mathbb{C}(\boldsymbol{x}) : \boldsymbol{\varepsilon}(\boldsymbol{x})) = \boldsymbol{0}, \\ &\Leftrightarrow -\nabla.\ ((\mathbb{C}(\boldsymbol{x}) - \mathbb{C}^0) : \boldsymbol{\varepsilon}(\boldsymbol{x})) = \nabla.\ (\mathbb{C}^0 : \boldsymbol{\varepsilon}(\boldsymbol{x})), \\ &\Leftrightarrow \nabla.\ (\mathbb{C}^0 : \boldsymbol{\varepsilon}(\boldsymbol{x})) = -\nabla.\ ((\mathbb{C}(\boldsymbol{x}) - \mathbb{C}^0) : \boldsymbol{\varepsilon}(\boldsymbol{x})). \end{aligned}$$

By introducing the polarization field $\boldsymbol{\tau}$, a linear problem with homogeneous coefficients is obtained:

$$\begin{cases} \nabla.\ (\mathbb{C}^0 : \boldsymbol{\varepsilon}(\boldsymbol{x})) = -\nabla.\ \tau(\boldsymbol{x}),\ \boldsymbol{x} \in \Omega, \\ \tau(\boldsymbol{x}) = (\mathbb{C}(\boldsymbol{x}) - \mathbb{C}^0) : \boldsymbol{\varepsilon}(\boldsymbol{x}),\ \varepsilon(\boldsymbol{x}) = \nabla^{sym}\boldsymbol{u}(\boldsymbol{x}), \\ \boldsymbol{u}(\boldsymbol{x}) = \boldsymbol{\varepsilon}^M .\ \boldsymbol{x} + \boldsymbol{u}^{\#}(\boldsymbol{x}), \\ \boldsymbol{u}^{\#}\begin{pmatrix} 0 \\ x_2 \\ x_3 \end{pmatrix} = \boldsymbol{u}^{\#}\begin{pmatrix} L_1 \\ x_2 \\ x_3 \end{pmatrix},\ \boldsymbol{u}^{\#}\begin{pmatrix} x_1 \\ 0 \\ x_3 \end{pmatrix} = \boldsymbol{u}^{\#}\begin{pmatrix} x_1 \\ L_2 \\ x_3 \end{pmatrix}, \\ \boldsymbol{u}^{\#}\begin{pmatrix} x_1 \\ x_2 \\ 0 \end{pmatrix} = \boldsymbol{u}^{\#}\begin{pmatrix} x_1 \\ x_2 \\ L_3 \end{pmatrix}. \end{cases} \tag{23}$$

The solution by a Green's function relies on the simplified problem:

$$\begin{cases} \nabla.\ (\mathbb{C}^0 : \mathbf{G}^0(\boldsymbol{x})) = \begin{pmatrix} 1 \\ 1 \\ 1 \end{pmatrix} \delta(\boldsymbol{x}),\ \boldsymbol{x} \in \Omega, \\ \int_\Omega \mathbf{G}^0 \mathrm{d}\Omega = 0, \end{cases} \tag{24}$$

with $\delta(\boldsymbol{x}) = \begin{cases} 1,\ \boldsymbol{x} = \boldsymbol{0}, \\ 0,\ \boldsymbol{x} \neq \boldsymbol{0}, \end{cases}$ being Dirac's δ function. It is easy to prove that if $\mathbf{G}^0$ is a solution to problem (24), then the solution to problem (23) can be written as

$$\boldsymbol{\varepsilon}(\boldsymbol{x}) = \boldsymbol{\varepsilon}^M - (\mathbf{G}^0 * \nabla.\ \boldsymbol{\tau})(\boldsymbol{x}), \tag{25}$$

with * the convolution operator. It is timely to remind that a differential operator with homogeneous coefficients can be integrated in a convolution indifferently on the left or right side of the operands, and that Dirac's

δ function is the neutral element of the convolution. It can hence be verified that

$$\begin{aligned}\nabla.\left(\mathbb{C}^0:\boldsymbol{\varepsilon}(\boldsymbol{x})\right) &= \nabla.\left(\mathbb{C}^0:\left(\boldsymbol{\varepsilon}^M-\left(\mathbf{G}^0*\nabla.\,\boldsymbol{\tau}\right)(\boldsymbol{x})\right)\right)\\ &= -\left(\nabla.\left(\mathbb{C}^0:\mathbf{G}^0\right)*\nabla.\,\boldsymbol{\tau}\right)(\boldsymbol{x})\\ &= -\left(\begin{pmatrix}1\\1\\1\end{pmatrix}\delta(\boldsymbol{x})*\nabla.\,\boldsymbol{\tau}\right)(\boldsymbol{x})\\ &= -\nabla.\,\boldsymbol{\tau}(\boldsymbol{x}).\end{aligned}$$

This result shows that the formula in Equation (25) is truly a solution to problem (23). To use this formula, a solution to problem (24) is needed. This solution $\mathbf{G}^0$ is the Green's function associated to problem (23). It can be computed easily by applying the Fourier transform $\mathcal{F}$:

$$\begin{aligned}\mathcal{F}\left(\nabla.\left(\mathbb{C}^0:\mathbf{G}^0\right)\right)(\boldsymbol{\xi}) &= J\boldsymbol{\xi}.\,\mathcal{F}\left(\mathbb{C}^0:\mathbf{G}^0\right)\\ &= J\boldsymbol{\xi}.\left(\mathbb{C}^0:\mathcal{F}\left(\mathbf{G}^0\right)\right),\\ &= J\boldsymbol{\xi}.\left(\mathbb{C}^0:\widehat{\mathbf{G}}^0\right),\end{aligned}$$

$$\mathcal{F}\left(\begin{pmatrix}1\\1\\1\end{pmatrix}\delta\right)(\boldsymbol{\xi})=\begin{pmatrix}1\\1\\1\end{pmatrix},$$

with $J=\sqrt{-1}$ and $\widehat{\mathbf{G}}^0=\mathcal{F}(\mathbf{G}^0)$. Problem (24) is therefore equivalent in Fourier space to:

$$\begin{cases}J\boldsymbol{\xi}.\left(\mathbb{C}^0:\widehat{\mathbf{G}}^0\right)=\begin{pmatrix}1\\1\\1\end{pmatrix}, & \boldsymbol{\xi}^2\neq 0,\\ \widehat{\mathbf{G}}^0=\mathbf{0}, & \boldsymbol{\xi}^2=0.\end{cases}$$

This problem can seem difficult to inverse but with a suitable choice of reference material, it becomes easy. For instance, for $\mathbb{C}^0_{ijkl}=\delta_{ik}\delta_{jl}$ the result is

$$\widehat{\mathbf{G}}^0_{ij}=\begin{cases}-\frac{\xi_j J}{\xi^2}, & \boldsymbol{\xi}^2\neq 0,\\ 0, & \boldsymbol{\xi}^2=0.\end{cases}$$

It is preferable to rewrite solution (25) as

$$\begin{aligned}\boldsymbol{\varepsilon}(\boldsymbol{x}) &= \boldsymbol{\varepsilon}^M-\left(\mathbf{G}^0*\nabla.\,\boldsymbol{\tau}\right)(\boldsymbol{x})=\boldsymbol{\varepsilon}^M-\left(\nabla.\,\mathbf{G}^0*\boldsymbol{\tau}\right)(\boldsymbol{x})\\ &= \boldsymbol{\varepsilon}^M-\left(\boldsymbol{\Gamma}^0*\boldsymbol{\tau}\right)(\boldsymbol{x}),\end{aligned}\tag{26}$$

where $\widehat{\boldsymbol{\Gamma}}^0 = \mathbf{I} \otimes J\boldsymbol{\xi}.\ \widehat{\mathbf{G}}^0$ is coined Green operator, with $\mathbf{I}$ being the identity matrix. In Reference (Moulinec and Suquet, 1994), the reference material was chosen as $\mathbb{C}^0_{ijkl} = \mu^0(\delta_{ik}\delta_{jl} + \delta_{il}\delta_{jk}) + \lambda^0\delta_{ij}\delta_{kl}$, which leads to the Green operator

$$\widehat{\boldsymbol{\Gamma}}^0_{ijkl} = \frac{(\delta_{ki}\xi_l\xi_j + \delta_{li}\xi_k\xi_j + \delta_{kj}\xi_l\xi_i + \delta_{lj}\xi_k\xi_i)}{4\mu_0\xi^2} - \frac{\lambda_0 + \mu_0}{\mu_0(\lambda_0 + 2\mu_0)}\frac{\xi_i\xi_j\xi_k\xi_l}{\xi^4}. \tag{27}$$

This approach based on a Green operator requires an initial trick which is the introduction of the polarization field $\boldsymbol{\tau}$ as right member, although this field actually depends on the solution. The price to pay for this trick is that an iterative solution algorithm is necessary to solve the initial problem (22).

Different algorithms have been proposed in the literature for this solution, including Newton-Krylov methods (Kabel et al., 2014) which are particularly interesting for problems featuring nonlinear material behavior (*i.e.*, where $\mathbb{C}$ depends on $\boldsymbol{\varepsilon}$). In Reference (Moulinec and Suquet, 1994), the following fixed-point algorithm was proposed:

$$\begin{array}{ll}
\text{Initialization:} & \\
(a_0) & \boldsymbol{\varepsilon}^1(\boldsymbol{x}) = \boldsymbol{\varepsilon}^M \\
(b_0) & \boldsymbol{\sigma}^1(\boldsymbol{x}) = \mathbb{C}(\boldsymbol{x}) : \boldsymbol{\varepsilon}^1(\boldsymbol{x}) \\
\text{Iteration}\,(i+1): & \boldsymbol{\varepsilon}^i \text{ and } \boldsymbol{\sigma}^i \text{ known} \\
(\text{a}) & \boldsymbol{\tau}^i(\boldsymbol{x}) = \boldsymbol{\sigma}^i(\boldsymbol{x}) - \mathbb{C}^0 : \boldsymbol{\varepsilon}^i(\boldsymbol{x}) \\
(\text{b}) & \hat{\boldsymbol{\tau}}^i = \mathcal{F}(\boldsymbol{\tau}^i) \\
(\text{c}) & \hat{\boldsymbol{\varepsilon}}^{i+1}(\boldsymbol{\xi}) = \begin{cases} -\widehat{\Gamma}^0(\boldsymbol{\xi}) : \hat{\boldsymbol{\tau}}^i(\boldsymbol{\xi}), & \boldsymbol{\xi}^2 \neq 0, \\ \boldsymbol{\varepsilon}^M, & \boldsymbol{\xi}^2 = 0, \end{cases} \\
(\text{d}) & \boldsymbol{\varepsilon}^{i+1} = Re\{\mathcal{F}^{-1}(\hat{\boldsymbol{\varepsilon}}^{i+1})\} \\
(\text{e}) & \boldsymbol{\sigma}^{i+1}(\boldsymbol{x}) = \mathbb{C}(\boldsymbol{x}) : \boldsymbol{\varepsilon}^{i+1}(\boldsymbol{x}) \\
(\text{f}) & \text{Convergence test}
\end{array} \tag{28}$$

This algorithm involves a forward Fourier transform and a backward Fourier transform (of which only the imaginary part is ignored). The goal is to convert the convolution to a simple multiplication and to use the formula in Equation (27) which provides the Green operator directly in Fourier space. This transform is relevant only if it can be computed rapidly.

This is the reason why the FFT is used, although it restricts the method to regular computational grids, *i.e.*, voxel meshes.

It is important to note that the only global operation is the FFT, all other operations being local to a voxel. The numerical solution, moreover, is conducted directly for the strain field, and naturally satisfies periodic boundary conditions.

3.2.2 Compelling example

In the following example based on Reference (Gao et al., 2019), the FFT-based numerical method is used to compute the effective stiffness of the nano-composite shown in Fig. 20. This nano-composite is composed of a rubber matrix and silica particles.

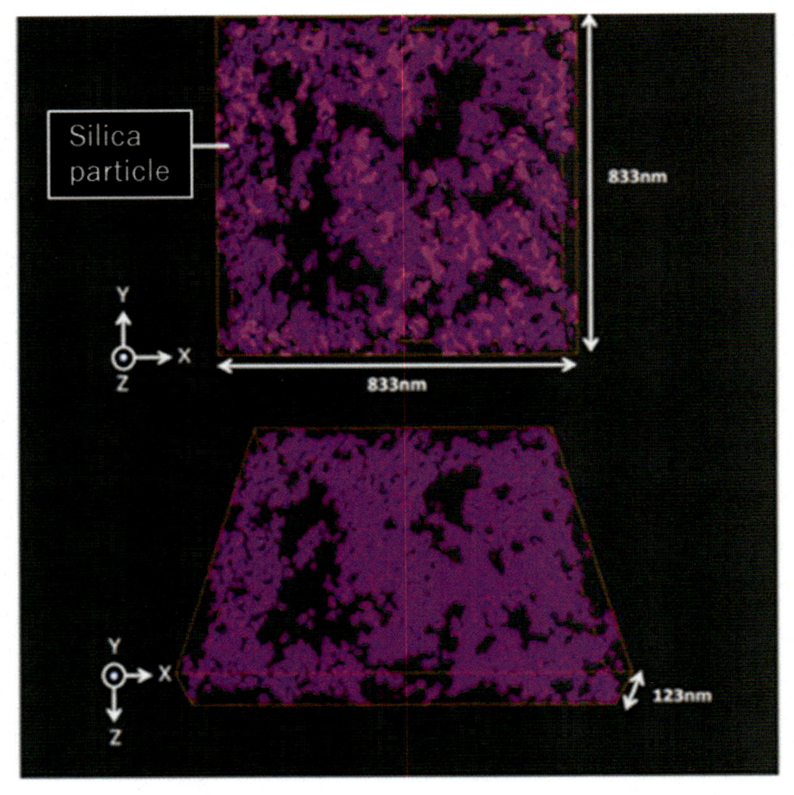

Fig. 20 Image of a nano-composite acquired by 3D high resolution transmission electron micro-tomography. *Reprinted from Gao, J., Shakoor, M., Jinnai, H., Kadowaki, H., Seta, E., & Liu, W. K. (2019). An inverse modeling approach for predicting filled rubber performance. Computer Methods in Applied Mechanics and Engineering, 357, 112567 with permission from Elsevier.*

The behavior of the particles is assumed to be linear isotropic elastic. Rubber, however, has a visco-elastic behavior for small strains, which means that its mechanical response depends on the frequency at which it is loaded. This response can be characterized using Dynamic Mechanical Analysis (DMA). This experimental method consists in applying a sinusoidal stress at a prescribed frequency. The resulting strain is delayed due to the visco-elastic behavior of the material. The phase angle δ is characteristic of this delay and is generally presented as a curve showing $\tan(\delta)$ as a function of frequency. The elastic properties of the material can be obtained from this curve for each frequency.

For the following calculations, instead of using a visco-elastic material law for rubber, a simple linear isotropic elastic law is used, but a separate simulation is conducted for each frequency. The numerical method presented in Algorithm (28) is used.

Particles are assumed to have a very large Young's modulus of 300 MPa and a Poisson's coefficient of 0.19. For the rubber matrix, Poisson's coefficient is set to 0.499 to obtain a nearly incompressible behavior and the stiffness is directly read from a DMA conducted on pure rubber specimens.

A simulation domain of $513 \times 513 \times 75$ voxels is chosen, which corresponds to an eighth of the domain shown in Fig. 20. Voxel size is 1.62 nm. Simulations are conducted for 17 frequencies. Each simulation took between 100 and 200 h.

The $\tan(\delta)$ curve computed by the FFT-based numerical method is compared to experimental measurements obtained by a DMA on nanocomposite specimens in Fig. 21. A slight difference can be observed between experimental and numerical results. In the literature, this difference is often attributed to the presence of a third material, coined interphase, in a small layer around particles. This interphase's behavior is neither that of the matrix nor the particles. In Reference (Gao et al., 2019), a numerical framework is proposed to identify by inverse analysis the elastic properties of this material for each frequency.

This compelling example shows quite well the advantages of the FFT-based numerical method:

- once segmented, a 3D image acquired by tomography can directly serve as computation grid, without any mesh generation or adaption step,
- it is easy to crop the initial image to conduct simulations on a sub-domain,
- the method can be implemented easily in any programming language by relying on existing FFT computation libraries,
- computation times are reasonable, even for large volumes.

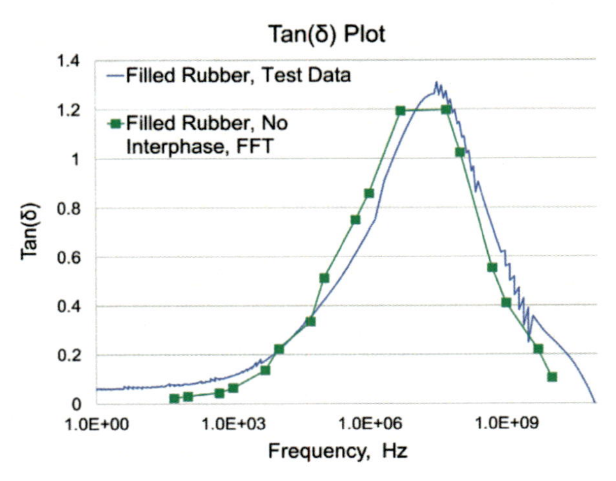

Fig. 21 Evolution of *tan*(δ) with respect to frequency for the nano-composite. The blue curve represents experimental measurements obtained by DMA. The green curve represents simulation results computed using the FFT-based numerical method. *Reprinted from Gao, J., Shakoor, M., Jinnai, H., Kadowaki, H., Seta, E., & Liu, W. K. (2019). An inverse modeling approach for predicting filled rubber performance. Computer Methods in Applied Mechanics and Engineering, 357, 112567 with permission from Elsevier.*

3.3 Sources of parasitic oscillations and solutions

An example of parasitic oscillations in a solution computed using the FFT-based numerical method is shown in Fig. 22. Such oscillations are a major issue for the fracture mechanics problems which are targeted in the present section.

In the following analysis based on Reference (Ma et al., 2021), the three setups shown in Fig. 23 are used to analyze the causes of oscillations. Young's modulus and Poisson's coefficient correspond to glass fibers in the red material and epoxy in the blue one.

Uniaxial tension loading is applied in the x direction shown in Fig. 23. In the FFT-based numerical method, this loading is imposed by setting the macroscopic strain component $\boldsymbol{\varepsilon}_{xx}^{M}$ to the chosen value and all other components to zero for the macroscopic stress field $\boldsymbol{\sigma}$. Controlling components of the macroscopic stress requires an extension of Algorithm (28) that can be found in the literature (Kabel et al., 2016; Moulinec & Suquet, 1998).

The stress field is expected to be homogeneous and the strain field discontinuous for setup A, as it is uni-dimensional. Setup B introduces four stress concentration zones. Setup C adds the irregular *zig-zag* discretization of the interface due to the voxel mesh.

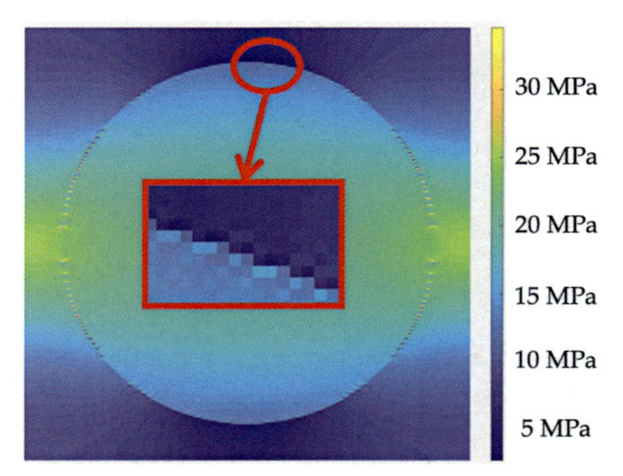

Fig. 22 Example of computation using the FFT-based numerical method for a heterogeneous material featuring a circular inclusion and subjected to tension in the horizontal direction. Parasitic oscillations are clearly visible in the stress field although the grid resolution is 255 × 255.

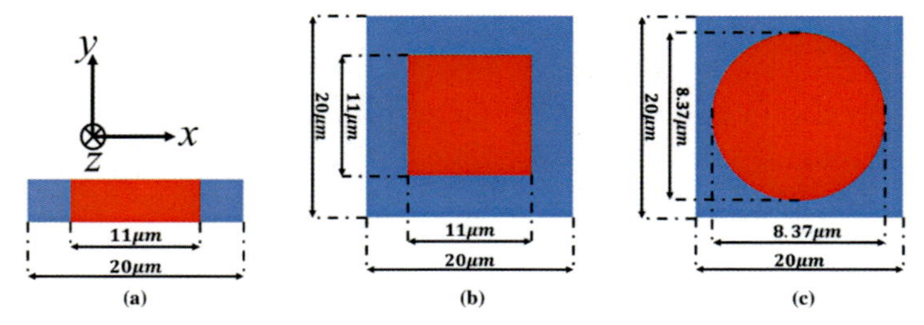

Fig. 23 Illustration of the three setups used for analyzing oscillations: (A) setup A, (B) setup B and (C) setup C. The three setups are discretized using voxel meshes with an identical resolution in directions *x* and *y*, and a unique voxel in direction *z*. *Reprinted from Ma, X., Shakoor, M., Vasiukov, D., Lomov, S. V., & Park, C. H. (2021). Numerical artifacts of Fast Fourier Transform solvers for elastic problems of multi-phase materials: their causes and reduction methods. Computational Mechanics, 67, 1667–1683 with permission from Springer Nature.*

Oscillations are analyzed for the three setups using different variants of the FFT-based numerical method in Paragraph 3.3.1, while Paragraph 3.3.2 focuses on oscillations due to the *zig-zag* discretization of the interface.

3.3.1 Analysis and general solutions

The analysis is conducted with the original FFT-based numerical method presented in Paragraph 3.2.1, and also with two variants proposed in the literature. The first one, called Willot's method (Willot, 2015), is based on a modification of the Green operator. The second one, called Schneider's method (Schneider et al., 2017), is based on both a modification of the Green operator and the integration scheme. This method, indeed, introduces eight integration points per voxel, which increases the computational cost significantly as all operations of Algorithm (28) have to be done eight times. All methods have been implemented in a *Python* code.

In order to compare the accuracy of these different variants, the jump of the first component of the strain field $D_j = \varepsilon_{j+1} - \varepsilon_j$ along a given line is reported. For different variables α, the absolute error $MD_j = |D_j^{\alpha} - D_j^{REF}|$ with respect to a reference solution *REF* computed by the FE method on the same voxel mesh is also reported. Finally, the mean error $\overline{MD}$ is computed by averaging MD over a complete line.

The reference FE solution is obtained for setups A and B using full integration, while for setup C reduced integration is used with a non-voxel mesh adapted to the geometry but of equivalent resolution.

Fig. 24 shows that for setup A all variants lead to a nearly exact result, whatever the resolution. Even though the strain field is discontinuous and Shannon's theorem is not satisfied, no oscillation is observed. This potential cause of oscillation can hence be eliminated, as well as Gibbs' phenomenon.

Mean absolute errors along two different lines for setup B are shown in Fig. 25. Errors for Willot's method are close to those of the reduced integration FE method, while errors for Schneider's method are close to those of the full integration FE method. Both methods are more accurate than the original FFT-based numerical method. Schneider's method, however, is more accurate than Willot's method. This can be attributed to the presence of hourglass modes in the fields computed with reduced integration.

Fig. 26 shows mean absolute errors for setup C. Again, Willot's method corresponds well to the reduced integration FE method, while Schneider's method corresponds well to the full integration FE method. For this setup, however, it is the original method that seems to give the best result. This can be attributed to a dominant role of the *zig-zag* discretization of the interface as cause of oscillations. None of the three variants, in fact, is designed to deal with this source of oscillations.

Errors in Fig. 26, moreover, are larger than those in Fig. 25. The irregular discretization of the interface, consequently, is more problematic than the hourglass modes due to the reduced integration scheme.

To summarize, this analysis has evidenced the presence of oscillations in the fields computed by the FFT-based numerical method. Two causes have been identified. Hourglass modes are one of them but they can be eliminated by using Schneider's method. This method seems nevertheless too

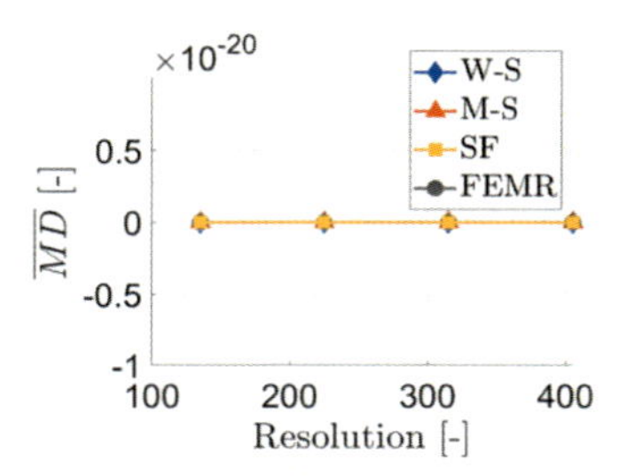

Fig. 24 Mean absolute error $\overline{MD}$ along a horizontal line for setup A, using different resolutions. The variants are denoted *M-S* for the original method based on Equation (27), *W-S* for Willot's method and *SF* for Schneider's method. Solutions obtained by the FE method with reduced integration are denoted *FEMR*. *Reprinted from Ma, X., Shakoor, M., Vasiukov, D., Lomov, S. V., & Park, C. H. (2021). Numerical artifacts of Fast Fourier Transform solvers for elastic problems of multi-phase materials: their causes and reduction methods. Computational Mechanics, 67, 1667–1683 with permission from Springer Nature.*

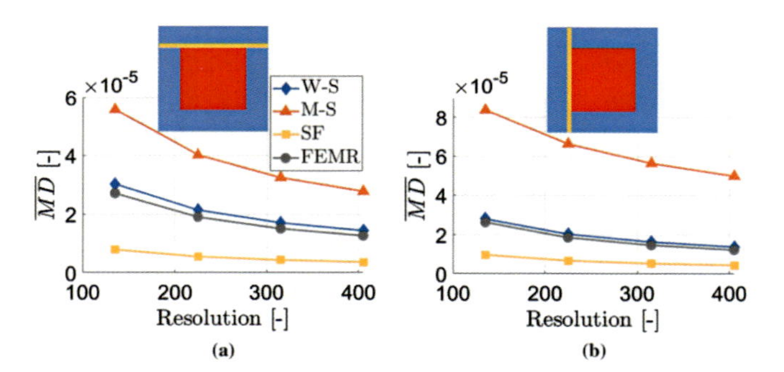

Fig. 25 Mean absolute error $\overline{MD}$ for setup B, using different resolutions: (a) along a horizontal line and (b) along a vertical line. The variants are denoted *M-S* for the original method based on Equation (27), *W-S* for Willot's method and *SF* for Schneider's method. Solutions obtained by the FE method with reduced integration are denoted *FEMR*. *Reprinted from Ma, X., Shakoor, M., Vasiukov, D., Lomov, S. V., & Park, C. H. (2021). Numerical artifacts of Fast Fourier Transform solvers for elastic problems of multi-phase materials: their causes and reduction methods. Computational Mechanics, 67, 1667–1683 with permission from Springer Nature.*

expensive computationally as oscillations due to hourglass modes are negligible compared to those caused by the irregular discretization of the interface.

In the following, a focus is made on the oscillations due to the *zig-zag* discretization of the interface and numerical methods to attenuate them.

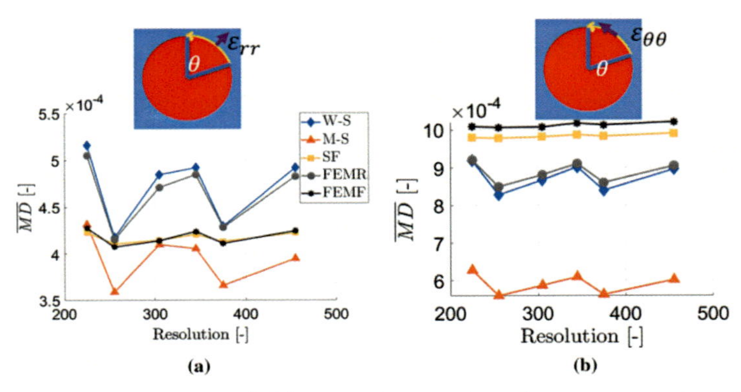

Fig. 26 Mean absolute error $\overline{MD}$ along the interface for setup C, using different resolutions: (A) radial component of the strain field and (B) tangential component of the strain field. The variants are denoted *M-S* for the original method based on Equation (27), *W-S* for Willot's method and *SF* for Schneider's method. Solutions obtained by the FE method with reduced integration are denoted *FEMR. Reprinted from Ma, X., Shakoor, M., Vasiukov, D., Lomov, S. V., & Park, C. H. (2021). Numerical artifacts of Fast Fourier Transform solvers for elastic problems of multi-phase materials: their causes and reduction methods. Computational Mechanics, 67, 1667–1683 with permission from Springer Nature.*

3.3.2 Solutions for interfaces

The method of composite voxels has been proposed in Reference (Kabel et al., 2015) to specifically reduce oscillations due to the *zig-zag* discretization of the interface. In this method, the voxels that are crossed by the interface for geometries like in setup C are called composite voxels. Their mechanical response is defined using the mixture law or any other analytical approach that can model the presence of two different materials in the same voxel.

The classical theory of laminates is shown to give the best results in Reference (Kabel et al., 2015). From a numerical point of view, this method requires to access the local volume fraction of each material within the composite voxel and the normal vector to the interface.

Both quantities were computed in Reference (Kabel et al., 2015) by sub-dividing each composite voxel into a finer sub-grid. A *zig-zag* discretization of the interface was then used at the sub-grid level, but both the local volume fraction and the normal vector were recovered at the composite voxel level with averaging operations. When the interface is provided as a parametric representation, this approach is easy to implement. For instance, for a circle with known center and radius as in setup C, it is easy to color each sub-voxel. For non-parametric representations such as geometries provided as 3D images acquired by tomography, the implementation of this method is not straightforward.

In Reference (Ma et al., 2021), an alternative approach based on LS functions (see Equation (4)) and regularized Heaviside functions (see Equation (21)) has been proposed. The voxel-wise regularized Heaviside function is directly used as local volume fraction, while the gradient of this function is used as local normal vector.

As mentioned in Paragraph 2.2.2, the advantage of LS functions is that they can easily be computed for any geometry, even if it originates from an image. Computing regularized Heaviside functions is also quite easy, although the thickness of the transition layer where composite voxels will be introduced should be defined. A thickness of two voxels is chosen in order to ensure the presence of at least one composite voxel all along the interface, and at most one on each side of the interface. This is different from the original composite voxels method where only voxels crossed by the interface are defined as composite.

Fig. 27 shows an example of result for setup C. Both the intensity of oscillations due to the irregular discretization of the interface and their attenuation with composite voxels are clearly visible.

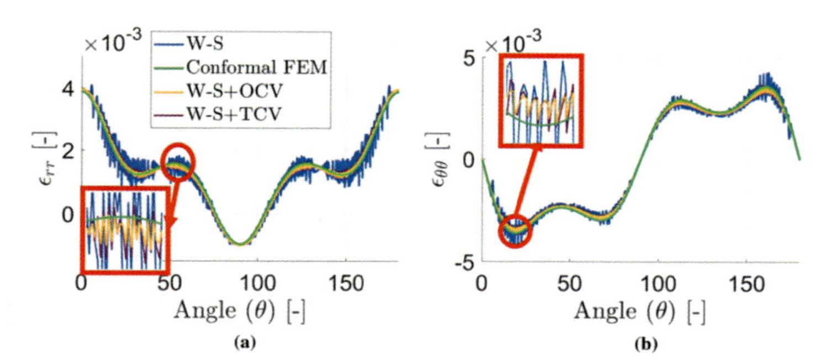

Fig. 27 Strain field along the interface for setup C using a resolution of $375 \times 375 \times 1$ voxels: (A) radial component of the strain field and (B) tangential component of the strain field. Willot's method is systematically used for the Green operator. The variants are denoted *W-S* for Willot's method without composite voxels, *W-S+TCV* for the original composite voxels method and *W-S+OCV* for the LS-based composite voxels method. *The solution obtained by the FE method with reduced integration and a non-voxel mesh adapted to the geometry but of equivalent resolution is denoted Conformal FEM. Reprinted from Ma, X., Shakoor, M., Vasiukov, D., Lomov, S. V., & Park, C. H. (2021). Numerical artifacts of Fast Fourier Transform solvers for elastic problems of multi-phase materials: their causes and reduction methods. Computational Mechanics, 67, 1667–1683 with permission from Springer Nature.*

A more detailed analysis can be found in Reference (Ma et al., 2021). It is found that the LS-based approach compares well to the original approach in terms of accuracy. This method is hence interesting given the advantageous simplicity of implementation.

3.4 Phase-field approach to fracture

Algorithm (28) assumes linear material behavior. This algorithm, nevertheless, can be applied effortlessly to nonlinear material behavior, although this can increase significantly the number of fixed-point iterations. Phase-field models introduce an additional variable, the phase-field d, and an additional equation to solve for this variable. Solving this equation by the FFT-based numerical method is not straightforward but it has already been addressed in the literature Chen et al. (2019).

The solution method in Reference (Chen et al., 2019) is based on a simplified phase-field evolution equation. A complete formulation has been proposed in Reference (Ma, 2022), and compared to the simplified one. This work is summarized in the following.

The model implemented in Reference (Chen et al., 2019) is Miehe's model (Miehe et al., 2010a,b). It is briefly presented in Paragraph 3.4.1. Details on the variational approach leading to the construction of this model can be found in the literature (Miehe et al., 2010a,b).

The FFT-based numerical method for solving this model's phase-field evolution equation is presented in Paragraph 3.4.1, both for the simplified and complete formulations.

Simulation results comparing the two formulations are presented in Paragraph 3.4.3.

3.4.1 Model

Miehe's model relies on an additive split of the strain tensor into its tension and compression parts:

$$\boldsymbol{\varepsilon} = \sum_{n=1}^{3} \varepsilon_n \boldsymbol{p}_n \otimes \boldsymbol{p}_n = \boldsymbol{\varepsilon}^+ + \boldsymbol{\varepsilon}^-,$$

where ε_n and $\boldsymbol{p}_n$ denote respectively the eigenvalues and eigenvectors of the strain field. The tension $\boldsymbol{\varepsilon}^+$ and compression $\boldsymbol{\varepsilon}^-$ parts can be obtained by relations

$$\boldsymbol{\varepsilon}^+ = \sum_{n=1}^{3} \langle \varepsilon_n \rangle^+ \boldsymbol{p}_n \otimes \boldsymbol{p}_n, \text{ and } \boldsymbol{\varepsilon}^- = \sum_{n=1}^{3} \langle \varepsilon_n \rangle^- \boldsymbol{p}_n \otimes \boldsymbol{p}_n,$$

where $\langle w \rangle^{+} = \frac{w+|w|}{2}$ is the positive part operator and $\langle w \rangle^{-} = \frac{w-|w|}{2}$ is the negative part one. This additive split of the strain tensor is tailored to attenuate material stiffness only in regions of the domain where tension loading prevails. The Cauchy stress tensor is then given by

$$\begin{aligned}\boldsymbol{\sigma}(\boldsymbol{x}, t) = g(d(\boldsymbol{x}, t))(\lambda(\boldsymbol{x})\langle \mathrm{tr}(\boldsymbol{\varepsilon}(\boldsymbol{x}, t))\rangle^{+}\mathbf{I} + 2\mu(\boldsymbol{x})\boldsymbol{\varepsilon}^{+}(\boldsymbol{x}, t)) \\ + (\lambda(\boldsymbol{x})\langle \mathrm{tr}(\boldsymbol{\varepsilon}(\boldsymbol{x}, t))\rangle^{-}\mathbf{I} + 2\mu(\boldsymbol{x})\boldsymbol{\varepsilon}^{-}(\boldsymbol{x}, t)),\end{aligned} \tag{29}$$

where g is the energetic degradation function defined as $g(d) = (1-d)^2$, with d the phase-field. Lam's parameters λ and μ vary spatially due to the material's heterogeneity.

The additive split is also present in the expression of the initial strain energy:

$$\phi_0^{+}(\boldsymbol{x}, t) = \frac{1}{2}\lambda(\boldsymbol{x})(\langle \mathrm{tr}(\boldsymbol{\varepsilon}(\boldsymbol{x}, t))\rangle^{+})^2 + \mu(\boldsymbol{x})\boldsymbol{\varepsilon}^{+}(\boldsymbol{x}, t) : \boldsymbol{\varepsilon}^{+}(\boldsymbol{x}, t),$$

which is used to define the history variable for the phase-field model:

$$\mathcal{H}(\boldsymbol{x}, t) = \max_{s \le t} \phi_0^{+}(\boldsymbol{x}, s).$$

This history variable guarantees that damage is irreversible $\dot{d} \ge 0$ and restricts the values of d in the [0,1] interval. This variable plays a role in the phase-field evolution equation:

$$\begin{aligned}g'(d(\boldsymbol{x}, t))\mathcal{H}(\boldsymbol{x}, t) + \frac{G_c(\boldsymbol{x})}{l_c(\boldsymbol{x})}d(\boldsymbol{x}, t) - \nabla. \\ (G_c(\boldsymbol{x})l_c(\boldsymbol{x})\nabla d(\boldsymbol{x}, t)) = 0, \ \boldsymbol{x} \in \Omega.\end{aligned} \tag{30}$$

This law introduces Griffith's critical energy release rate G_c and the regularization length l_c, which are both material properties. They vary spatially due to the material's heterogeneity.

Compared to continuous damage models, the phase-field model has the advantage of combining multiple operations into a single one: the evolution of the damage variable d, its regularization with the diffusion term and its irreversibility with the history variable $\mathcal{H}$.

In Reference (Chen et al., 2019), Equation (30) has been simplified as

$$\begin{aligned}g'(d(\boldsymbol{x}, t))\mathcal{H}(\boldsymbol{x}, t) + \frac{G_c(\boldsymbol{x})}{l_c(\boldsymbol{x})}d(\boldsymbol{x}, t) - G_c(\boldsymbol{x})l_c(\boldsymbol{x})\Delta d(\boldsymbol{x}, t) \\ = 0, \ \boldsymbol{x} \in \Omega,\end{aligned} \tag{31}$$

which means neglecting the $\nabla(G_c(\boldsymbol{x})l_c(\boldsymbol{x})).\ \nabla d(\boldsymbol{x},\ t)$ term. This is without consequence for a homogeneous material but it might have a significant influence for a heterogeneous one.

3.4.2 Numerical method

Although the following simulations will be performed in a static setting, they are conducted over several time increments corresponding actually to load increments (time plays a fictitious role). In Equation (29), moreover, it appears that the equilibrium problem to solve for $\boldsymbol{\varepsilon}$ depends on d through the degradation function g. In Equation (30), as well, the evolution equation to solve for d depends on $\boldsymbol{\varepsilon}$ through the history variable $\mathcal{H}$. To solve this problem, a weak coupling is chosen in the following, which implies using a small time step. The equilibrium problem is first solved with the phase-field computed at the previous time increment, and then the obtained strain is used to solve the phase-field evolution equation.

With this weak coupling, Algorithm (28) can be used to solve the equilbrium problem. Only the computation of the stress should be changed, as it should rely on Equation (29). This is not an issue even though Equation (29) is nonlinear due to the additive split. In order to reduce the number of iterations, the fixed-point scheme is accelerated with Anderson's algorithm. The latter is already implemented in the *AMITEX* software, which is used in the following. The Green operator of Equation (27) is chosen.

The only obstacle to overcome is the numerical solution of Equation (30). In this aim, the approach based on Green's functions and the FFT presented in Subsection 3.2 can be a source of inspiration. The first step is to replace $g'(d) = -2(1 - d)$ by its expression in Equation (30) to obtain.

$$\left(2\mathcal{H}(\boldsymbol{x},\ t) + \frac{G_c(\boldsymbol{x})}{l_c(\boldsymbol{x})}\right)d(\boldsymbol{x},\ t) - \nabla.\ (G_c(\boldsymbol{x})l_c(\boldsymbol{x})\nabla d(\boldsymbol{x},\ t))$$
$$= 2\mathcal{H}(\boldsymbol{x},\ t).$$

Two operators can be identified in the left member of this equation. They are associated to reaction and diffusion, and they both feature heterogeneous coefficients. To eliminate these heterogeneous coefficients, the functions $A(\boldsymbol{x}) = 2\mathcal{H}(\boldsymbol{x},\ t) + \frac{G_c(\boldsymbol{x})}{l_c(\boldsymbol{x})}$ and $Q(x) = G_c(\boldsymbol{x})l_c(\boldsymbol{x})$ and the reference constants

$$A^0 = \frac{\max\limits_{\boldsymbol{x}\in\Omega} A(\boldsymbol{x}) + \min\limits_{\boldsymbol{x}\in\Omega} A(\boldsymbol{x})}{2} \quad \text{and} \quad Q^0 = \frac{\max\limits_{\boldsymbol{x}\in\Omega} Q(\boldsymbol{x}) + \min\limits_{\boldsymbol{x}\in\Omega} Q(\boldsymbol{x})}{2}$$

are introduced. Then, operators with homogeneous coefficients are obtained in the left member:

$$A^0 d(\boldsymbol{x}, t) - Q^0 \Delta d(\boldsymbol{x}, t)$$
$$= 2\mathcal{H}(\boldsymbol{x}, t) - (A(\boldsymbol{x}) - A^0) d(\boldsymbol{x}, t) + \nabla.\left((Q(\boldsymbol{x}) - Q^0)\nabla d(\boldsymbol{x}, t)\right).$$

The polarization field

$$\eta(\boldsymbol{x}) = 2\mathcal{H}(\boldsymbol{x}, t) - (A(\boldsymbol{x}) - A^0) d(\boldsymbol{x}, t) + \nabla.\left((Q(\boldsymbol{x}) - Q^0)\nabla d(\boldsymbol{x}, t)\right)$$

can be identified in this equation. In Fourier space, the solution

$$\hat{d} = \frac{\hat{\eta}}{A^0 + Q^0 \boldsymbol{\xi}^2}$$

can be derived. Similarly as in Subsection 3.2, the dependence of the polarization field on the solution requires the introduction of a fixed-point algorithm:

$$
\begin{array}{lll}
\text{Initialization:} & & \\
(a_0) & & d^1(\boldsymbol{x}) = 0 \\
(b_0) & & A(\boldsymbol{x}) = 2\mathcal{H}(\boldsymbol{x}, t) + \frac{G_c(\boldsymbol{x})}{l_c(\boldsymbol{x})} \text{ and } Q(x) \\
& & = G_c(\boldsymbol{x}) l_c(\boldsymbol{x}) \\
(c_0) & & A^0 = \frac{\max_{x\in\Omega} A(\boldsymbol{x}) + \min_{x\in\Omega} A(\boldsymbol{x})}{2} \\
& & \text{and } Q^0 = \frac{\max_{x\in\Omega} Q(\boldsymbol{x}) + \min_{x\in\Omega} Q(\boldsymbol{x})}{2} \\
\text{Iteration } (i+1): & & d^i \text{ known} \\
(a) & & \eta^i(\boldsymbol{x}) = 2\mathcal{H}(\boldsymbol{x}, t) \\
& & - (A(\boldsymbol{x}) - A^0) d^i(\boldsymbol{x}, t) \\
& & + \nabla.\left((Q(\boldsymbol{x}) - Q^0)\nabla d^i(\boldsymbol{x}, t)\right) \\
(b) & & \hat{\eta}^i = \mathcal{F}(\eta^i) \\
(c) & & \hat{d}^{i+1} = \frac{\hat{\eta}^i}{A^0 + Q^0\boldsymbol{\xi}^2} \\
(d) & & d^{i+1} = Re\{\mathcal{F}^{-1}\}(\hat{d}^{i+1}) \\
(e) & & \text{Convergence test}
\end{array}
\tag{32}
$$

The simplification between Equations (30) and (31) has no influence on the algorithm because the left member originates from reference terms that rely systematically on constant coefficients. It is in the computation of the polarization field that the simplification plays a role. With the simplification, indeed, step (a) of Algorithm (32) requires computing

$$\begin{aligned} \eta^i &= 2\mathcal{H} - (A - A^0)\, d^i + (Q - Q^0)\,\Delta d^i \\ &= 2\mathcal{H} - (A - A^0)\, d^i - (Q - Q^0)\,\mathcal{F}^{-1}(\boldsymbol{\xi}^2 \mathcal{F}(d^i)), \end{aligned}$$

while without simplification it becomes

$$\begin{aligned} \eta^i &= 2\mathcal{H} - (A - A^0)\, d^i + \nabla.\, ((Q - Q^0)\,\nabla d^i) \\ &= 2\mathcal{H} - (A - A^0)\, d^i \\ &\quad + \mathcal{F}^{-1}(J\boldsymbol{\xi}.\, \mathcal{F}((Q - Q^0)\,\mathcal{F}^{-1}(J\boldsymbol{\xi}\mathcal{F}(d^i)))). \end{aligned}$$

This computation requires going back and forth to Fourier space twice in the complete formulation instead of once.

3.4.3 Results

Simulations presented in this part are based on a test case with two half-fibers separated from some distance. Studying the influence of the inter-fiber distance on the material's mechanical response has been fundamental to understand the limitations of the simplified formulation.

The setup with an inter-fiber distance of 3.29 μm is shown in Fig. 28. It is discretized with a grid of 225 × 225 × 1 voxels, which results in a voxel size of 88.9 nm. Glass fibers with a Young's modulus of 74 GPa, a Poisson's coefficient of 0.2 and a critical energy release rate of 60 N mm^{1} are considered, as well as an epoxy matrix with a Young's modulus of 4650 MPa, a Poisson's coefficient of 0.35 and a critical energy release rate of 0.96 N m^{1}.

The length l_c is assumed homogeneous and is varied between 0.54 and 3.30 μm. Even for the smallest value, Miehe's criterion is satisfied as l_c is at least twice larger than voxel size.

Uniaxial tension loading is applied in the x direction shown in Fig. 28, with a strain increment of 5×10^{-7}.

With the values of G_c used for the matrix and fibers, fiber cracking is nearly impossible for this test case, whatever the value of l_c. With the simplified method, however, fibers are clearly damaged. The maximum value of the phase-field d is non-zero and even increases with a larger length l_c, as shown in Table 2.

Fig. 29 shows clearly that with an increasing value of l_c, there is damage in fibers. It seems that the phase-field is uni-dimensional, as it does not seem to depend on y, despite the presence of fibers.

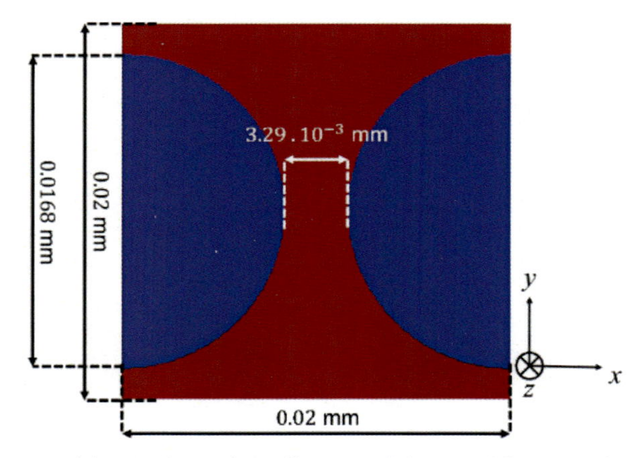

Fig. 28 Setup used for studying the influence of the simplification of the phase-field evolution equation. Fibers are in blue and the matrix in red. *Reprinted from Ma, X., Chen, Y., Shakoor, M., Vasiukov, D., Lomov, S. V., & Park, C. H. (2023). Simplified and complete phase-field fracture formulations for heterogeneous materials and their solution using a Fast Fourier Transform based numerical method. Engineering Fracture Mechanics, 279, 109049 with permission from Elsevier.*

Table 2 Maximum phase-field value computed inside fibers after failure with the simplified formulation and different values of the regularization length.

l_c	0.54 μm	0.81 μm	1.30 μm	2.20 μm	3.30 μm
$\max_{x \in \text{fibers}} d(\boldsymbol{x})$	0.15	0.24	0.38	0.56	0.68

Fibers damage is even more visible in Fig. 30, which shows phase-field evolution along a horizontal line. Clearly, fibers damage can be observed, even with the smallest value used for l_c. This damage of fibers increases with a larger value of l_c. This phenomenon seems to be due to diffusion.

To verify this assumption, G_c is increased in fibers. Results are presented in Fig. 31. Again, a diffusion of damage from matrix to fibers can be observed, even with an absurdly great critical energy release rate in fibers. This property, in fact, does not seem to influence the results, which indicates that the value of $60\,\text{N}\,\text{mm}^1$ used in the first simulations was already large enough.

The complete formulation is considered in the following simulations. The first result is that the maximum value of the phase-field d in fibers remains zero whatever the value of l_c, and this up to machine precision.

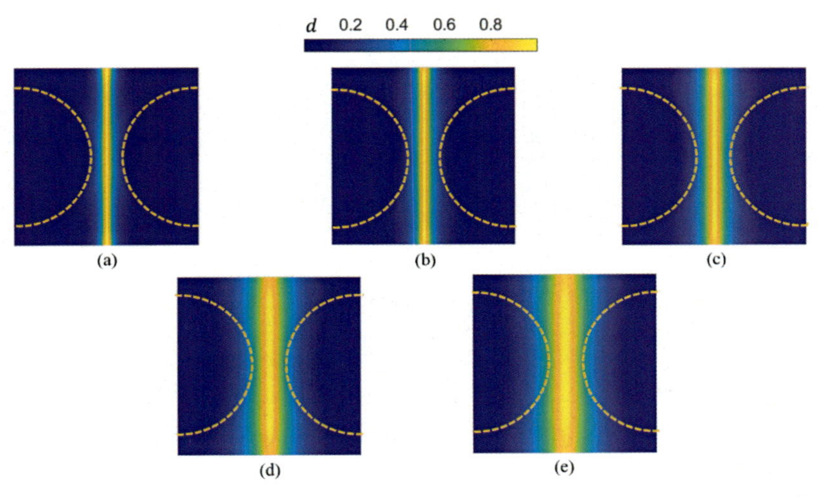

Fig. 29 Phase-field after failure using the simplified formulation and a length of: (A) 0.54 μm, (B) 0.81 μm, (C) 1.30 μm, (D) 2.20 μm and (E) 3.30 μm. The fibers/matrix interface is delimited in dashed lines. *Reprinted from Ma, X., Chen, Y., Shakoor, M., Vasiukov, D., Lomov, S. V., & Park, C. H. (2023). Simplified and complete phase-field fracture formulations for heterogeneous materials and their solution using a Fast Fourier Transform based numerical method. Engineering Fracture Mechanics, 279, 109049 with permission from Elsevier.*

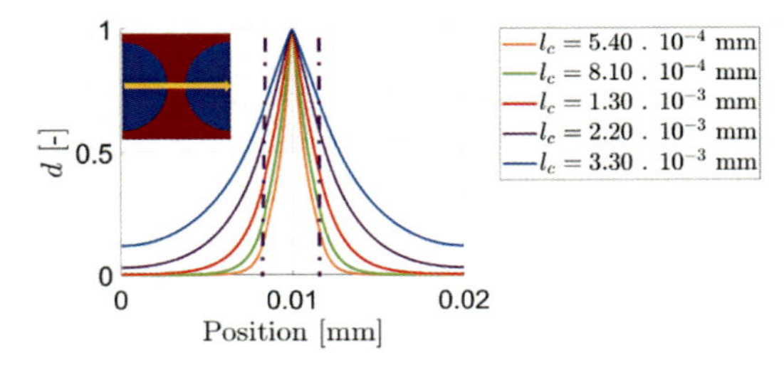

Fig. 30 Evolution of the phase-field along a horizontal line after failure for simulations using the simplified formulation. The fibers/matrix interface is delimited in dashed lines. *Reprinted from Ma, X., Chen, Y., Shakoor, M., Vasiukov, D., Lomov, S. V., & Park, C. H. (2023). Simplified and complete phase-field fracture formulations for heterogeneous materials and their solution using a Fast Fourier Transform based numerical method. Engineering Fracture Mechanics, 279, 109049 with permission from Elsevier.*

The phase-fields after failure using the complete formulation are shown in Fig. 32. The comparison with Fig. 29 demonstrates quite well the influence of the simplification. With the complete formulation, the phase-field varies in both directions. The fibers, in fact, seem to form a bottleneck that restricts damage in the horizontal direction at the center of the domain.

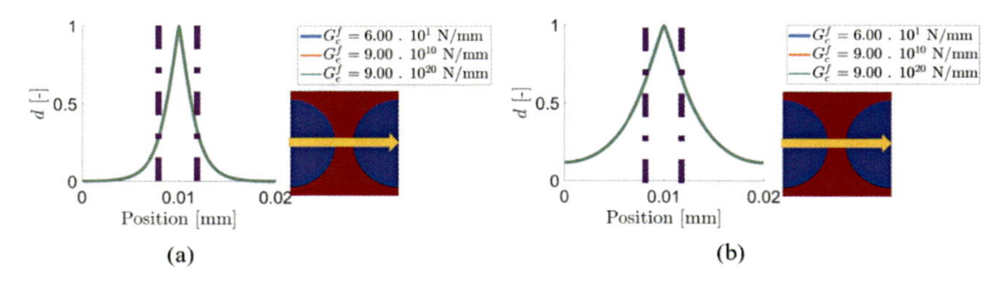

Fig. 31 Evolution of the phase-field along a horizontal line after failure for simulations using the simplified formulation, a greater critical energy release rate in the fibers, and a length l_c of: (A) 1.30 μm and (B) 3.30 μm. The fibers/matrix interface is delimited in dashed lines. *Reprinted from Ma, X., Chen, Y., Shakoor, M., Vasiukov, D., Lomov, S. V., & Park, C. H. (2023). Simplified and complete phase-field fracture formulations for heterogeneous materials and their solution using a Fast Fourier Transform based numerical method. Engineering Fracture Mechanics, 279, 109049 with permission from Elsevier.*

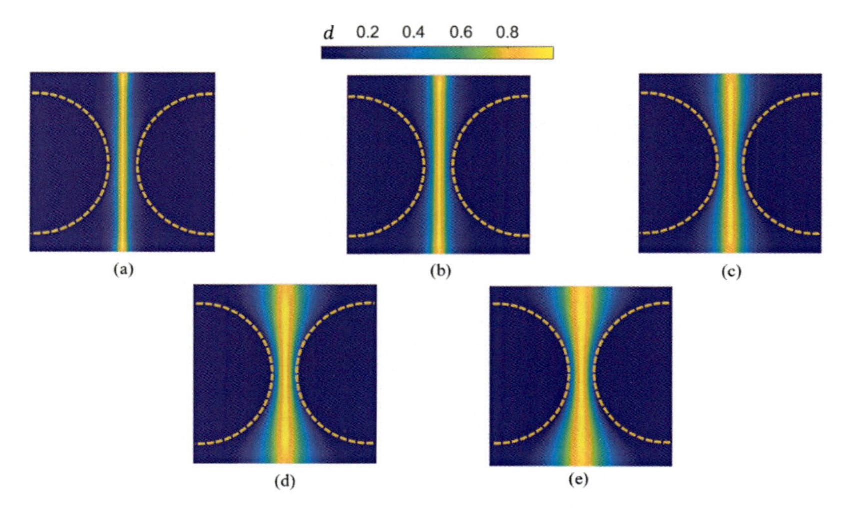

Fig. 32 Phase-field after failure using the complete formulation and a length l_c of: (A) 0.54 μm, (B) 0.81 μm, (C) 1.30 μm, (D) 2.20 μm and (E) 3.30 μm. The fibers/matrix interface is delimited in dashed lines. *Reprinted from Ma, X., Chen, Y., Shakoor, M., Vasiukov, D., Lomov, S. V., & Park, C. H. (2023). Simplified and complete phase-field fracture formulations for heterogeneous materials and their solution using a Fast Fourier Transform based numerical method. Engineering Fracture Mechanics, 279, 109049 with permission from Elsevier.*

Fig. 33 demonstrates clearly that a sharp transition of the phase-field between matrix and fibers is retrieved with the complete formulation. The influence of l_c on this transition is completely eliminated.

It is also important to consider the macroscopic mechanical response. Stress-strain curves for the two formulations and different values of l_c are presented in Fig. 34. These curves demonstrate the significant influence of damage diffusion across interfaces using the simplified formulation, from both local and global perspectives. Other examples can be found in Reference (Ma et al., 2023).

As a conclusion, phase-field modeling of fracture with the FFT-based numerical method has been revisited in this work, both regarding the model and its implementation. A simplification has been lifted and its influence on results has been analyzed both on the local and global level.

3.5 Conclusions

The FFT-based numerical method has been presented in this section with the objective of modeling fracture phenomena in (quasi-)brittle heterogeneous materials such as polymer matrix composites. Some properties that are intrinsic to this method have been exploited. On the one hand, this method only relies on the FFT and operations that are local to a voxel, and does not require the assembly of a global matrix. On the other hand, a simulation on a 3D image acquired by tomography has been conducted directly on the voxel mesh of the image, and this without any mesh generation or adaption step.

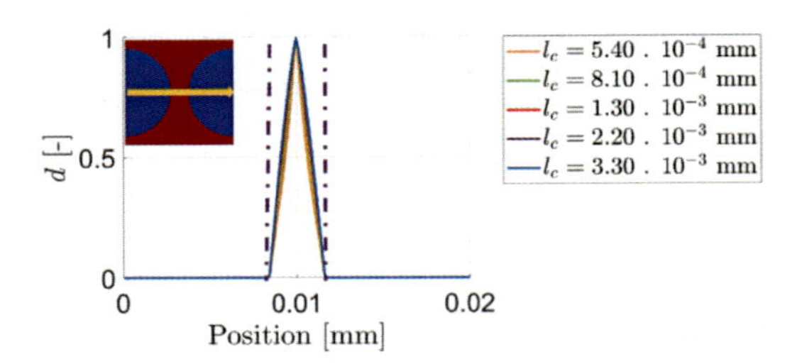

Fig. 33 Evolution of the phase-field along a horizontal line after failure using the complete formulation. The fibers/matrix interface is delimited in dashed lines. *Reprinted from Ma, X., Chen, Y., Shakoor, M., Vasiukov, D., Lomov, S. V., & Park, C. H. (2023). Simplified and complete phase-field fracture formulations for heterogeneous materials and their solution using a Fast Fourier Transform based numerical method. Engineering Fracture Mechanics, 279, 109049 with permission from Elsevier.*

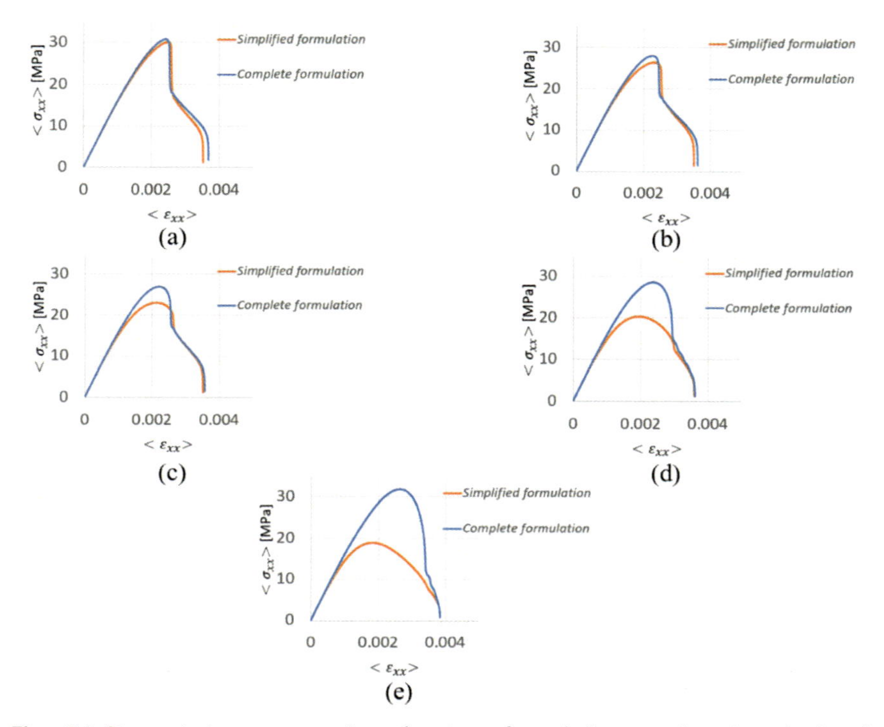

Fig. 34 Stress-strain curves using the two formulations and a length l_c of: (A) 0.54 µm, (B) 0.81 µm, (C) 1.30 µm, (D) 2.20 µm and (E) 3.30 µm. *Reprinted from Ma, X., Chen, Y., Shakoor, M., Vasiukov, D., Lomov, S. V., & Park, C. H. (2023). Simplified and complete phase-field fracture formulations for heterogeneous materials and their solution using a Fast Fourier Transform based numerical method. Engineering Fracture Mechanics, 279, 109049 with permission from Elsevier.*

Some problematic drawbacks of the FFT-based numerical methods have been addressed, especially regarding the presence of parasitic oscillations in solutions. Those can indeed be particularly inconvenient for applications to fracture mechanics. An analysis based on three well chosen setups has enabled to evaluate the oscillations causes and to rank them in terms of importance. The major cause that has been identified is the irregular discretization of the interfaces due to the voxel mesh.

The method of composite voxels has been introduced to attenuate these oscillations, and its efficiency has been proved on the three setups. An alternative composite voxels method based on LS functions has been proposed. This original approach gives results that are comparable to the initial method, but with a facilitated implementation thanks to LS functions. This improved method, indeed, eases the application to non-parametric representations of the geometry such as 3D images acquired by tomography.

To model fracture phenomena, an existing implementation of the phase-field approach with the FFT-based numerical method has been revisited. A particular attention has been given to the numerical solution of the phase-field evolution equation. The influence of a simplification assumed in the literature has been evaluated and a complete implementation (*i.e.*, without simplification) has been developed. Results show that the simplification leads to an abnormal diffusion of damage between the different components of the material and hence justifies the interest of the complete implementation.

4. Model order reduction

4.1 Introduction

As demonstrated in Sections 2 and 3, numerical modeling of highly nonlinear mechanical phenomena in heterogeneous materials requires significant computational resources. This is challenging when those computations are to be repeated several times, for instance in the context of a sensitivity analysis, a design optimization, an uncertainty analysis or a multiscale simulation (see Section 5). Parallel computing as presented in Subsection 2.4 is not sufficient to deal with such situations and alternatives are worth considering. The key aspect of the model order reduction methods presented in this section is that they exploit the partial redundancy of the computations.

The approaches considered in this section consist in:

- Generating data using a conventional approach such as the Finite Element (FE) method or the Fast Fourier Transform (FFT)-based numerical method. In this context, the conventional approach is coined Full Order Model (FOM). The generated data can consist in a large collection of displacement, velocity or stress fields obtained under different boundary conditions.
- Compressing the data with approaches borrowed from data science and more generally artificial intelligence.
- Replacing the FOM by a Reduced Order Model (ROM) where the number of unknowns to solve in order to find a result is drastically reduced thanks to compression. Predictions can be computed using an

ROM purely based on artificial intelligence, such as a deep neural network, or a hybrid ROM integrating some fundamental principles of mechanics.

The final objective of this section is to design a model order reduction approach in three steps (generation, compression, prediction). The FOMs targetted in this section can rely on the FE method or the FFT-based numerical method. The common aspect is the high nonlinearity of the mechanical phenomena to predict.

Database generation can be conducted using random sampling (Goury et al., 2016), Gaussian processes (Goury et al., 2016), Sobel sequences (Bessa et al., 2017) or other sampling methods. With these stochastic methods, the number of samples necessary to cover sufficiently the space might be quite large, so deterministic methods have also been considered by some authors (Liu et al., 2016; Yvonnet & He, 2007). For instance, instead of using a great number of random loading conditions to generate the database, it has been shown in Reference (Liu et al., 2016) that reliable results could be obtained for a small strain elasto-plastic analysis using only six orthogonal loadings of a small amplitude.

Among compression methods, Proper Orthogonal Decomposition (POD) (Liang et al., 2002) has been widely used in computational mechanics. This compression method consists in replacing any field of the database by a linear combination of eigenvectors. The data is hence projected onto the reduced space formed by the coefficients of this linear combination. Eigenvectors, moreover, are global fields, as opposed to FE shape functions which are local. It is also possible to project all linear systems to solve during the FE simulation onto the reduced space (Lieu et al., 2006; Ryckelynck, 2005; Yvonnet & He, 2007). If the compression error can be sufficiently minimized with a reasonable number of eigenvectors, then the obtained ROM can drastically reduce the computation time.

POD has two major drawbacks. Firstly, it is a linear compression method that does not perform very well for nonlinear problems. The solution of a fracture mechanics problem can for instance hardly be written as a linear combination of eigenvectors obtained using solutions associated to other loading conditions. This is the reason why approaches have been proposed to automatically update eigenvectors during the simulation (Ryckelynck, 2005). Secondly, when POD is coupled to the FE method, only the size of the linear problems to solve is reduced. It is necessary to add another ingredient to the ROM in order to also reduce the computational cost related to integration and assembly of the linear systems. Consequently, some authors have investigated hyper-reduction techniques (Ryckelynck, 2005) as well as the discrete empirical interpolation method (Chaturantabut and Sorensen, 2010).

Attempts using other linear compression methods can also be found in the literature. Among the most widely spread ones in computational mechanics, there is the proper generalized decomposition (Hitchcock, 1927). This method is quite similar to POD and can also be coupled to the FE method. The main difference is that each dimension of the problem is separated in a multiplicative expansion. This is quite useful for the introduction of additional parameters in the ROM, such as control or optimization variables (Chinesta et al., 2013). Proper generalized decomposition nevertheless shares the same drawbacks as POD regarding the application to highly nonlinear problems and the need for hyper-reduction.

It is worth mentioning that different compression methods have also been coupled to conventional numerical methods other than the FE method. The non-uniform transformation analysis method, for instance, couples POD to the FFT-based numerical method (Michel and Suquet, 2003). Although this combination has the same drawbacks as the combination with the FE method, it is interesting as it can be related to mean-field models.

This relation is exploited by the Self-consistent Clustering Analysis (SCA) method, where POD is replaced by K-means clustering (Liu et al., 2016). This approach relies on a voxel mesh, where each voxel is associated to a cluster. Like in the FFT-based numerical method, the strain field is the unknown to solve, but it is discretized cluster-wise instead of voxel-wise. The strain is, indeed, assumed to be constant per cluster. This has the interesting consequence of reducing both the number of unknowns and the cost of integration and linear system assembly in the ROM.

Nonlinear compression methods based on deep Artificial Neural Networks (ANNs) have known an increasing popularity in computational mechanics. Among these ANNs, autoencoders are particularly interesting as they rely on the same three-step process as POD-based approaches (Hu & Song, 2009; Wang et al., 2016). An autoencoder is a forward-propagation neural network that associates a low-dimensional code to a given input field, and then reconstructs the full field from the code. It can be viewed as a generalization of POD, as an autoencoder with a single hidden layer and linear activation functions is equivalent to a POD (Plaut, 2018). The advantage of autoencoders over POD is that they can be deep and rely on nonlinear activation functions in order to perform nonlinear compression.

Autoencoders have recently been used in nonlinear mechanics, especially in fluid mechanics (Fukami et al., 2021; Murata et al., 2020). These studies focus mainly on compressing complex flow data. An ROM based on an autoencoder and an additional ANN has been developed to predict unsteady flows in Reference (Pant et al., 2021). A similar approach relying on a combination of POD and an autoencoder has been developed in Reference (Ahmed et al., 2021) to model turbulent flows. Both these ROMs do not mimic the iterative process of the FOM, where the solution is predicted time increment by time increment. This aspect is fundamental for replacing FOMs by ROMs relying on deep learning in engineering. Recent works on autoencoders, in addition, focus mainly on fluid mechanics, and highly nonlinear mechanical problems such as fracture mechanics problems have not been addressed yet.

The SCA approach and some contributions to this method are presented in Subsection 4.2. Given that it is based on a linear compression method, an alternative approach based on autoencoders is proposed in Subsection 4.3.

4.2 Self-consistent clustering analysis

The SCA method has initially been proposed in Reference (Liu et al., 2016) under the assumption of small strains and for elastic and elasto-plastic behavior. The present work has mainly focused on its extension to finite strains with plastic strain localization and ductile damage. The integration of this ROM to a multiscale method has also been addressed, as detailed in Section 5.

Based on Reference (Shakoor et al., 2019a), the finite strain version of the SCA method is summarized in Paragraph 4.2.1, and some results are presented in Paragraph 4.2.2.

4.2.1 Method

The SCA method is a model order reduction approach that relies on a background conventional method which is the FFT-based numerical method (Liu et al., 2016). Some elements of Subsection 3.2 can, therefore, be borrowed, especially the problem to solve which is the extention to finite strains of Equation (22) in a total Lagrangian setting:

$$\begin{cases} -\nabla_{\mathbf{X}}.\ \mathbf{P}(\boldsymbol{X}) = \mathbf{0},\ \boldsymbol{X} \in \Omega_0, \\ \mathbf{P}(\boldsymbol{X}) = \mathbf{P}(\mathbf{F}(\boldsymbol{X})),\ \mathbf{F}(\boldsymbol{X}) = \mathbf{I} + \nabla_{\mathbf{X}}\,\boldsymbol{u}(\boldsymbol{X}), \\ \boldsymbol{u}(\mathbf{X}) = (\mathbf{F}^M - \mathbf{I}).\ \boldsymbol{X} + \boldsymbol{u}^{\#}(\boldsymbol{X}), \\ \boldsymbol{u}^{\#}\begin{pmatrix} 0 \\ X_2 \\ X_3 \end{pmatrix} = \boldsymbol{u}^{\#}\begin{pmatrix} L_1 \\ X_2 \\ X_3 \end{pmatrix},\ \boldsymbol{u}^{\#}\begin{pmatrix} X_1 \\ 0 \\ X_3 \end{pmatrix} = \boldsymbol{u}^{\#}\begin{pmatrix} X_1 \\ L_2 \\ X_3 \end{pmatrix}, \\ \boldsymbol{u}^{\#}\begin{pmatrix} X_1 \\ X_2 \\ 0 \end{pmatrix} = \boldsymbol{u}^{\#}\begin{pmatrix} X_1 \\ X_2 \\ L_3 \end{pmatrix}. \end{cases} \tag{33}$$

It is reminded that $\mathbf{F}(\boldsymbol{X})$ is the deformation gradient, $\mathbf{P}(\boldsymbol{X})$ is the first Piola-Kirchhoff stress tensor, Ω_0 is the domain in the initial configuration (*i.e.*, non deformed), and $\nabla_{\mathbf{X}}$ denotes a gradient with respect to coordinates in the initial configuration.

The SCA method relies on two major changes with respect to the FFT-based numerical method: clustering and a self-consistent scheme.

4.2.1.1 Clustering

The first leg of the SCA method is clustering. It has to be well-elaborated as all mechanical fields (strains and stresses) will be discretized as constant per cluster. It is hence suitable to place in the same cluster voxels which will have similar strains and stresses whatever the loading conditions. This is the reason why the database used for clustering should be as representative as possible in terms of strains and stresses distribution.

This is a non-supervised learning problem, where the objective is to minimize the similarity of mechanical fields in the database within each cluster or, equivalently, to maximize the difference between two distinct clusters. This can be achieved using various methods. Good results have been obtained using K-means clustering (Liu et al., 2016; Shakoor et al., 2019a).

Examples of clustering results are shown in Fig. 35. It can be noted that for domains composed of different components (matrix, inclusion, void, *etc.*), it is preferred to cluster each component independently. This avoids using mixture laws inside clusters. It can also be noted that a cluster may be disconnected, which is not problematic at all when computing predictions with the self-consistent scheme.

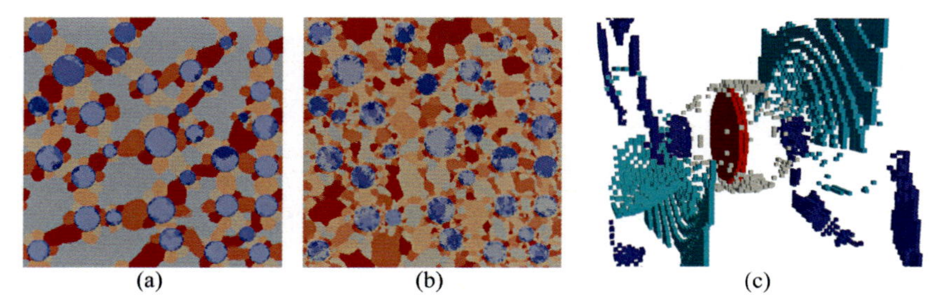

Fig. 35 Examples of simulation domains discretized with pixel/voxel meshes and then clustered by K-means: (A) 2D example with eight clusters, (B) 2D example with 65 clusters, (C) 3D example with 217 clusters showing only two clusters in blue for the matrix, one cluster in red for the inclusion and one cluster in gray for the void. *Reprinted from Shakoor, M., Kafka, O. L., Yu, C., & Liu, W. K. (2019a). Data science for finite strain mechanical science of ductile materials. Computational Mechanics, 64(1), 33–45 with permission from Springer Nature.*

4.2.1.2 Self-consistent scheme

Voxels clustering and discretization with a cluster-wise constant deformation gradient are introduced:

$$\mathbf{F}(\boldsymbol{X}) \approx \sum_{k=1}^{K} \mathbf{F}^k \chi^k(\boldsymbol{X}),\ \boldsymbol{X} \in \Omega_0,$$

where K is the number of clusters, $\mathbf{F}^k$ is the value of the deformation gradient within cluster k and χ^k is the characteristic function of this cluster:

$$\chi^k(\boldsymbol{X}) = \begin{cases} 1, & \boldsymbol{X} \in \Omega_0^k, \\ 0, & \boldsymbol{X} \notin \Omega_0^k, \end{cases}$$

with $\boldsymbol{\Omega}_0^k$ the part of domain Ω_0 occupied by the voxels of cluster k. With clustering, the polarization field becomes $\boldsymbol{\tau}_0^k = \mathbf{P}^k - \mathbb{C}^0 : \mathbf{F}^k$. Strains being cluster-wise constant, indeed, the stresses are as well, which clearly proves that the SCA method reduces simultaneously the number of unknowns and the cost of integration.

The reference material, moreover, is slightly changed in order to eliminate the symmetry, as proposed in Reference (Kabel et al., 2014). The Green operator used in the following is hence slightly different from the one in Equation (27). Details can, again, be found in Reference (Kabel et al., 2014).

Switching to finite strains and adding clustering, Equation (26) becomes

$$
\begin{aligned}
\mathbf{F}(\boldsymbol{X}) &= \mathbf{F}^{M} - (\boldsymbol{\Gamma}^{0} * \boldsymbol{\tau})(\boldsymbol{X}),\ \boldsymbol{X} \in \Omega_{0}, \\
\Leftrightarrow \mathbf{F}(\boldsymbol{X}) &= \mathbf{F}^{M} - \int_{\Omega_{0}} \boldsymbol{\Gamma}^{0}(\boldsymbol{X}, \boldsymbol{X}'): \boldsymbol{\tau}(\boldsymbol{X}')\,\mathrm{d}\boldsymbol{X}',\ \boldsymbol{X} \in \Omega_{0}, \\
\Leftrightarrow \mathbf{F}(\boldsymbol{X}) &= \mathbf{F}^{M} - \int_{\Omega_{0}} \boldsymbol{\Gamma}^{0}(\boldsymbol{X}, \boldsymbol{X}'): \left(\sum_{k'=1}^{K} \boldsymbol{\tau}^{k'} \chi^{k'}(\boldsymbol{X}') \right) \mathrm{d}\boldsymbol{X}',\ \boldsymbol{X} \in \Omega_{0}, \\
\Leftrightarrow \mathbf{F}(\boldsymbol{X}) &= \mathbf{F}^{M} - \sum_{k'=1}^{K} \left(\int_{\Omega_{0}} \boldsymbol{\Gamma}^{0}(\boldsymbol{X}, \boldsymbol{X}') \chi^{k'}(\boldsymbol{X}')\,\mathrm{d}\boldsymbol{X}' \right): \boldsymbol{\tau}^{k'},\ \boldsymbol{X} \in \Omega_{0}, \\
\Leftrightarrow \mathbf{F}(\boldsymbol{X}) &= \mathbf{F}^{M} - \sum_{k'=1}^{K} (\boldsymbol{\Gamma}^{0} * \chi^{k'})(\boldsymbol{X}): \boldsymbol{\tau}^{k'},\ \boldsymbol{X} \in \Omega_{0}.
\end{aligned}
$$

Multiplying by χ^{k} and averaging over Ω_0 leads to

$$
\begin{aligned}
\frac{1}{|\Omega_{0}|} \int_{\Omega_{0}} \mathbf{F}(\mathbf{X}) \chi^{k}(\mathbf{X})\,\mathrm{d}\mathbf{X} &= \frac{|\Omega_{0}^{k}|}{|\Omega_{0}|} \mathbf{F}^{k} \\
&= \frac{|\Omega_{0}^{k}|}{|\Omega_{0}|} \mathbf{F}^{M} - \sum_{k'=1}^{K} \left(\frac{1}{|\Omega_{0}|} \int_{\Omega_{0}} \chi^{k}(\mathbf{X}) (\boldsymbol{\Gamma}^{0} * \chi^{k'})(\mathbf{X}) \right. \\
&\qquad \left. \mathrm{d}\mathbf{X} \right): \boldsymbol{\tau}^{k'}, \\
\Leftrightarrow \mathbf{F}^{k} &= \mathbf{F}^{M} - \sum_{k'=1}^{K} \left(\frac{1}{|\Omega_{0}^{k}|} \int_{\Omega_{0}} \chi^{k}(\mathbf{X}) (\boldsymbol{\Gamma}^{0} \star \chi^{k'})(\mathbf{X})\,\mathrm{d}\mathbf{X} \right): \boldsymbol{\tau}^{k'}, \\
\Leftrightarrow \mathbf{F}^{k} &= \mathbf{F}^{M} - \sum_{k'=1}^{K} \mathbb{D}^{0}_{kk'}: \boldsymbol{\tau}^{k'},\ \mathbb{D}^{0} \\
&= \left(\frac{1}{|\Omega_{0}^{k}|} \int_{\Omega_{0}} \chi^{k}(\mathbf{X}) (\boldsymbol{\Gamma}^{0} * \chi^{k'})(\mathbf{X})\,\mathrm{d}\mathbf{X} \right)_{k,k'=1\ldots K},
\end{aligned}
\tag{34}
$$

where $\mathbb{D}^0$ is called interaction tensor. It is computed with the FFT:

$$
\mathbb{D}^{0} = \left(\frac{1}{|\Omega_{0}^{k}|} \int_{\Omega_{0}} \chi^{k}(\mathbf{X}) \mathcal{F}^{-1}(\widehat{\boldsymbol{\Gamma}}^{0} \mathcal{F}(\chi^{k'}))(\mathbf{X})\,\mathrm{d}\mathbf{X} \right)_{k,k'=1\ldots K},
$$

which has a significant computational cost in terms of time and memory, because it is a dense tensor. It can, however, be pre-computed and memory consumption can be minimized if the number of clusters is reasonable.

It can be noted that when the number of clusters K is equal to the number of voxels, Equation (34) is the one that appears in the finite strain version of the FFT-based numerical method (Kabel et al., 2014,2016). This is actually an alternative version of the method where a matrix (*i.e.*, the interaction tensor) is assembled by pre-computing all FFTs instead of applying them at each fixed-point iteration.

In practice, the number of clusters being minimized, Equation (34) leads to an over-constrained problem. It is generally not possible to satisfy simultaneously compatibility, boundary conditions and equilibrium. It is reminded that, according to Equation (33), compatibility imposes that the deformation gradient field should derive from a continuous displacement field, that boundary conditions impose that this displacement field should be composed of a linear part and a periodic part, and that the linear part should be driven by $\mathbf{F}^M$. It is possible to rewrite this last condition as

$$\frac{1}{|\Omega_0|}\sum_{k=1}^{K}|\Omega_0^k|\mathbf{F}^k = \mathbf{F}^M.$$

It can be introduced as a constraint with a Lagrange multiplier $\mathbf{F}^0$ so that Equation (34) can be transformed into

$$\begin{cases}\mathbf{F}^k + \sum_{k'=1}^{K} \mathbb{D}^0_{kk'}:(\mathbf{P}(\mathbf{F}^{k'}) - \mathbb{C}^0:\mathbf{F}^{k'}) - \mathbf{F}^0 = \mathbf{0},\ k = 1...K,\\ \frac{1}{|\Omega_0|}\sum_{k=1}^{K}|\Omega_0^k|\mathbf{F}^k = \mathbf{F}^M,\end{cases}$$

where the polarization field has been replaced by its expression. This nonlinear system can be solved using a Newton-Raphson algorithm (Shakoor et al., 2019a).

It has been shown in Reference (Liu et al., 2016) that using a self-consistent scheme enables to satisfy boundary conditions and equilibrium as much as possible, while partially sacrificing compatibility (the displacement is not continuous at inter-cluster boundaries). This self-consistent scheme consists in updating Lam's parameters of the reference material λ^0 and μ^0 during the solve. This does not require to reconstruct completely the interaction tensor. It is indeed possible to decompose it as $\mathbb{D}^0 = f^1(\lambda^0, \mu^0)\,\mathbb{D}^1 + f^2(\lambda^0, \mu^0)\,\mathbb{D}^2$, where only nonlinear functions $f^1(\lambda^0, \mu^0)$ and $f^2(\lambda^0, \mu^0)$ should be recomputed at each update of reference material, and where tensors $\mathbb{D}^1$ and $\mathbb{D}^2$ can be pre-computed with FFTs. Details on this decomposition can be found in Reference (Shakoor et al., 2019a), where the following finite strains SCA algorithm has been proposed:

Initialization:

(a_0)	Compute $\mathbb{D}^1$ and $\mathbb{D}^2$
(b_0)	Initialize λ^0 and μ^0

Iteration $(i + 1)$:

(a) $\mathbb{D}^0 = f^1(\lambda^0, \mu^0)\,\mathbb{D}^1 + f^2(\lambda^0, \mu^0)\,\mathbb{D}^2$

(b) Solve using a Newton-Raphson algorithm:

$$\begin{cases} \mathbf{F}^k + \sum_{k'=1}^{K} \mathbb{D}^0_{kk'} : (\mathbf{P}(\mathbf{F}^{k'}) - \mathbb{C}^0 : \mathbf{F}^{k'}) - \mathbf{F}^0 = \mathbf{0}, \; k = 1...K, \\ \frac{1}{|\Omega_0|} \sum_{k=1}^{K} |\Omega_0^k| \mathbf{F}^k = \mathbf{F}^M, \end{cases}$$

(d) $\mathbf{P}^M = \frac{1}{|\Omega_0|} \sum_{k=1}^{K} |\Omega_0^k| \mathbf{P}^k$

(e) $(\lambda^0, \mu^0) = \arg \min_{(\lambda^*, \mu^*)} ||\mathbf{P}^M - \mathbb{C}^* : (\mathbf{F}^0 - \mathbf{I})||_2$

(f) Convergence test

(35)

Three loops are embedded. Each iteration of the self-consistent scheme requires the solution of a nonlinear problem using a Newton-Raphson algorithm, and each Newton-Raphson algorithm requires solving a linear system. Computational cost is minimized by using a small number of clusters K compared to the initial number of voxels, and also by pre-computing all FFTs.

Step (e) is key to the self-consistent scheme, as it consists in finding the optimal reference material depending on the computed macroscopic behavior. This is not obvious as the reference material is assumed isotropic linear elastic, and the current implementation could be improved. Using an isotropic linear elastic reference material is nevertheless advantageous as it simplifies the update of the interaction tensor at each iteration of the self-consistent scheme.

4.2.2 Results

In Algorithm (35), it is possible to use any behavior law for the relation $\mathbf{P}^k = \mathbf{P}(\mathbf{F}^k)$ within each cluster k. As explained in Paragraph 4.2.1.1, if the domain is composed of multiple materials of different behaviors, the part of the domain occupied by a cluster can only be composed of a unique material. This facilitates the implementation and the use of new material laws. Different hyperelastic material laws have hence been integrated in the SCA method, as well as a multiplicative von Mises plasticity model and a fracture criterion to model ductile damage. Results presented in the following are restricted to ductile damage modeling.

4.2.2.1 Plasticity

In this part, a domain composed of matrix and inclusions is considered. A multiplicative von Mises plasticity model (Simo, 1992) with linear isotropic hardening is used for the matrix. Inclusions are assumed brittle, although their fracture is not modeled. The domain geometry with inclusions is presented in Fig. 36(a) while material properties are given in Table 3. Fig. 36A clearly shows the significant size of the 100 × 100 × 100 voxels mesh that is used to obtain accurate results for this model with the FFT-based numerical method.

Reference results are obtained for unidirectional tension loading up to 25% logarithmic strain with strain increments of 0.001. The database for K-means clustering is simply composed of deformation gradient tensors computed for the first load increment of this simulation. For such a small loading magnitude, material response is linear elastic. The

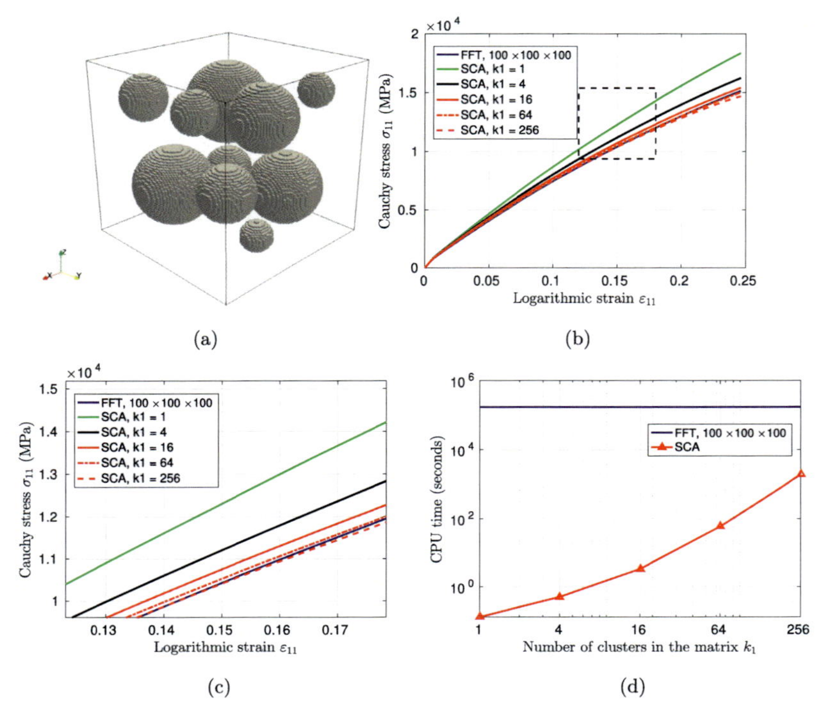

Fig. 36 Simulations with inclusions: (A) Domain geometry with an inclusions volume fraction of 20%; (B) Stress-strain curves computed with the SCA method using different numbers of clusters in the matrix (k1) and with the FFT-based numerical method with a mesh of 100 × 100 × 100 voxels; (C) Zoom on the rectangle delimited by dashes in (B); (D) Computation times for the different results shown in (B). *Reprinted from Shakoor, M., Kafka, O. L., Yu, C., & Liu, W. K. (2019a). Data science for finite strain mechanical science of ductile materials. Computational Mechanics, 64(1), 33–45 with permission from Springer Nature.*

Table 3 Material properties for simulations with elasto-plastic behavior. *Reprinted from Shakoor, M., Kafka, O.L., Yu, C., & Liu, W.K. (2019a). Data science for finite strain mechanical science of ductile materials. Computational Mechanics, 64(1), 33–45 with permission from Springer Nature.*

Property	Matrix	Inclusions	Voids	Unit
Young's modulus	70.0	400.0	70.0e-3	GPa
Poisson's coefficient	0.33	0.2	0.0	—
Yield stress	400	—	—	MPa
Hardening modulus	1333	—	—	MPa

computation time spent generating the database and clustering it, moreover, is negligible.

The number of clusters is varied between $k_1 = 1, 4, 16, 64, 256$ in the matrix and $k_2 = 1, 1, 4, 13, 26$ in inclusions. Stress-strain curves computed by the FFT-based numerical method and the SCA method using different numbers of clusters are presented in Fig. 36B and C. With at least 16 clusters in the matrix, results obtained with the SCA method are close enough to the reference solution.

The influence of the number of clusters on computation time for prediction using Algorithm (35) is analyzed in Fig. 36D. It can be estimated from this curve that with only thousands of clusters, the SCA method would become more computationally demanding than the FFT-based numerical method. This stems from the fact that the interaction tensor is a dense matrix where FFTs are pre-computed, as opposed to the FFT-based numerical method which does not assemble any global matrix. This is not an issue as with only 16 clusters in the matrix, predictions of the SCA method are already quite reliable, while the computation time is nearly 1000 times smaller as compared to the FFT-based numerical method.

A second set of simulations where inclusions are replaced by voids is considered. Both the FFT-based numerical method and the SCA method require discretizing voids in order to maintain the integrity of the voxel mesh (*i.e.*, avoid holes in the mesh). Voids behavior, therefore, has to be modeled. The same hyperelastic model as for inclusions is used but with a stiffness 1000 times lower as compared to the matrix.

Loading is also changed to switch to uniaxial tension and maintain a constant stress state during the simulation. This requires a slight modification of

Algorithm (35) in order to enable imposing a macroscopic deformation for component $\mathbf{F}_{11}^{M}$, and a zero macroscopic stress for all other components. This modification does not raise any particular difficulty and is detailed in Reference (Shakoor et al., 2019a).

Reference stress-strain curves are compared with those computed by the SCA method in Fig. 37A and B. Convergence appears to be slower when increasing the number of clusters. This can be explained by more intense local plastic strains as compared to the case with inclusions.

As shown in Fig. 37C, despite this slight inaccuracy of the SCA method, computation times remain very advantageous as compared to the FFT-based numerical method.

These first simulations with the SCA method in finite strains and with a multiplicative plasticity model demonstrate quite well the potential of the

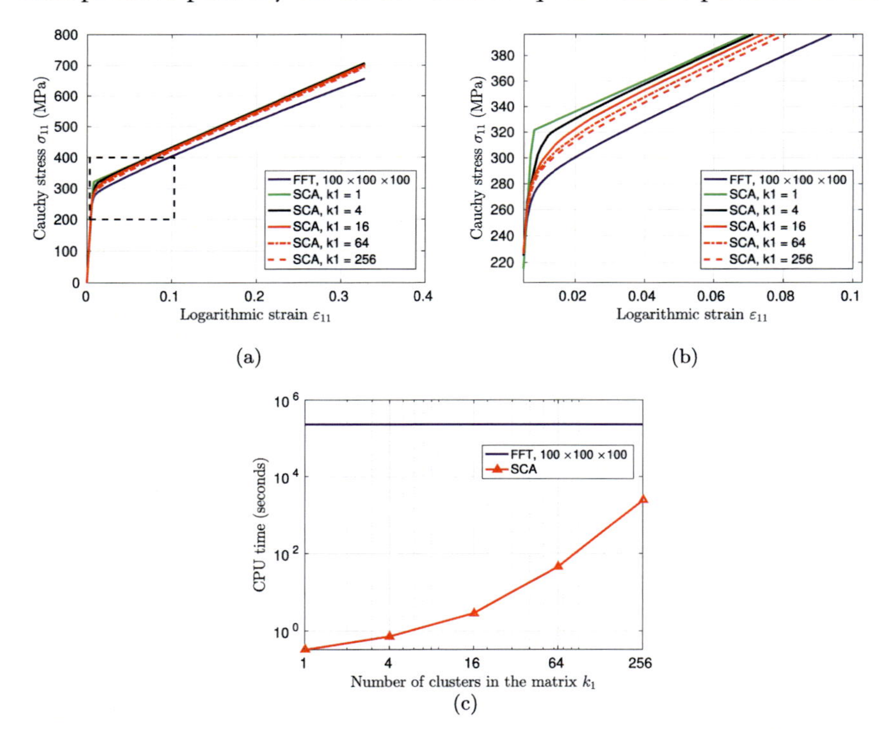

Fig. 37 Simulations with voids: (A) Stress-strain curves computed with the SCA method using different numbers of clusters in the matrix (k1) with the FFT-based numerical method with a mesh of $100 \times 100 \times 100$ voxels; (B) Zoom on the rectangle delimited by dashes in (A); (C) Computation times for the different results shown in (A). *Reprinted from Shakoor, M., Kafka, O. L., Yu, C., & Liu, W. K. (2019a). Data science for finite strain mechanical science of ductile materials. Computational Mechanics, 64(1), 33–45 with permission from Springer Nature.*

SCA method to reduce computation times. The database used for these simulations is of a very small size, as it contains only a single deformation gradient field. These simulations nevertheless show some limits, especially regarding the generalization to arbitrary loading conditions. A larger database could obviously be considered. Automatic database and compression update approaches developed for POD (Ryckelynck, 2005) could also be developed for the SCA method.

4.2.2.2 Ductile damage

Simulations conducted in the following are related to research on artificial prostheses and heart valves. These structures are manufactured from Nickel-Titanium tubes that are shaped using various thermo-mechanical processes. A focus is made in the following on cold drawing and the influence of this process on a tube's microstructure and its fatigue life.

Biaxial compression loading is applied to a microstructure composed of an inclusion that is already debonded from the matrix in the initial state. To predict the evolution of this microstructure during manufacturing, an inclusion fragmentation crterion is integrated in the SCA method. The post-manufacturing microstructure predicted by the SCA method is then extracted at different steps of the drawing process and used as input for a fatigue life prediction model (Kafka et al., 2018). Details and results regarding fatigue life prediction are not mentioned in the following as it is chosen to focus on processing. They can be found Reference (Shakoor et al., 2019a).

Material properties are kept identical to those in Table 3. Regarding the inclusion, a regularized Tresca failure criterion is introduced and coupled to a size criterion in order to model fragmentation. The Tresca criterion defines the shear stress σ_{Tresca} as

$$\sigma_{Tresca} = \frac{1}{2} \max |\sigma_1 - \sigma_2|, |\sigma_2 - \sigma_3|, |\sigma_3 - \sigma_1|$$

where σ_1, σ_2, σ_3 are principal stresses computed within each inclusion cluster. The shear stress σ_{Tresca} is then averaged over the whole inclusion, or inclusion fragment if failure has already occurred, and this average value $\overline{\sigma}_{Tresca}$ is compared to the fracture strength σ^c_{Tresca}.

The equivalent radius of this inclusion or inclusion fragment r is also compared to a critical size r^c. If the fracture strength has been reached and $r \geq r_c$, then the behavior of the cluster containing the σ_{Tresca}-weighted barycenter of this inclusion or fragment is changed to void. Fracture properties are given in Table 4.

Inclusion-to-void change is illustrated in Fig. 38. In practice, this change consists in decreasing by a factor of 1000 the Young's modulus of the considered cluster, and this over several load increments. This procedure is triggered at the end of each load increment, after the prediction of mechanical fields by the SCA method. As shown in Fig. 38, the fragmentation crack orientation is predefined as orthogonal to the drawing direction in the initial configuration. This requires to constrain clustering in the inclusion. K-means clustering is hence used in matrix and voids, while the inclusion is clustered in the horizontal direction into disks of a thickness of three voxels as shown in Fig. 38.

An example of result is presented in Fig. 39. The inclusion is initially ellipsoidal and has been fragmented into two parts after 45% of compression

Table 4 Fracture properties for cold drawing modeling. Reprinted from Shakoor, M., Kafka, O.L., Yu, C., & Liu, W.K. (2019a). Data science for finite strain mechanical science of ductile materials. *Computational Mechanics, 64*(1), 33–45 with permission from Springer Nature.

Property	Value	Unit
Fracture strength	3000	MPa
Critical size	0.17	mm

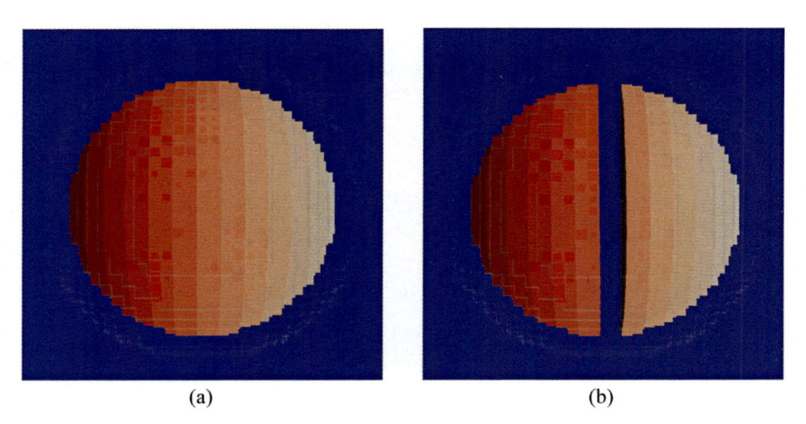

Fig. 38 Clusters (shades of red) in the pre-debonded inclusion: (A) before fragmentation, (B) after fragmentation with one cluster changed to void. Drawing direction is horizontal. *Reprinted from Shakoor, M., Kafka, O. L., Yu, C., & Liu, W. K. (2019a). Data science for finite strain mechanical science of ductile materials. Computational Mechanics, 64(1), 33–45 with permission from Springer Nature.*

(orthogonal to the drawing direction). Plastic strain localization is observed in the matrix around the inclusion. Other results with different compression levels and different inclusion shapes can be found in Reference (Shakoor et al., 2019a), where fatigue life prediction is also proposed.

These examples demonstrate the applicability of the SCA method to real problems. Some questions nevertheless remain regarding database construction and the method's capability to generalize to different loading conditions under great strains implying plastic localization and fracture phenomena. The fields' accuracy is also limited, as they are considered constant per cluster. The fragmentation model, moreover, relies on a specific clustering that assumes a prior knowledge of the crack's shape.

After significant research, it appears that the SCA method is mostly tailored for the prediction of global quantities, such as stress-strain curves or a material's macroscopic response. The fields' accuracy at the local level is limited and it does not seem suitable for modeling localized phenomena. The simultaneous reduction of both the number of unknowns and the cost of integration thanks to clustering, as opposed to POD-based approaches that require hyper-reduction, is hence both a strength and a weakness of the SCA method.

The SCA method can, in short, be placed in the category of mean-field approaches, where it is distinguishable thanks to its data-driven clustering. It is hence possible to model quite complex phenomena as long as only the macroscopic response is of interest, as presented in Subsection 5.2 (Gao et al., 2020). Macroscopic responses computed by the SCA method have also been exploited for inverse analysis based on the work presented in Paragraph 3.2.2

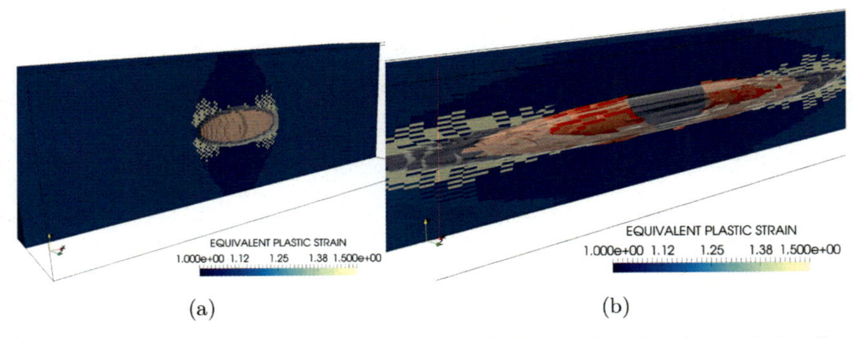

Fig. 39 Result of a drawing simulation at 45% of compression showing inclusion fragments in red, voids in grey, and equivalent plastic strain in the matrix: (A) in the initial configuration, (B) in the deformed configuration. *Reprinted from Shakoor, M., Kafka, O. L., Yu, C., & Liu, W. K. (2019a). Data science for finite strain mechanical science of ductile materials. Computational Mechanics, 64(1), 33–45 with permission from Springer Nature.*

(Gao et al., 2019). For those problems, the SCA method reduces computation times drastically as compared to the FFT-based numerical method.

To go beyond the strengths and weaknesses of the SCA method and also POD, nonlinear compression methods based on deep learning are worth considering.

4.3 Deep learning

It can be learned from research on the SCA method that a method relying on clustering cannot predict full fields and their local variations accurately. From the extensive literature on POD and other linear compression methods, a number of difficulties can be identified regarding generalization to highly nonlinear problems.

This has motivated research on nonlinear compression methods based on deep ANNs and in particular autoencoders. This work, which has been published in Reference (Shinde et al., 2023), is summarized in the following. The principle of autoencoders is presented in Paragraph 4.3.1. Paragraph 4.3.2 describes the nonlinear ROM developed for solving highly nonlinear brittle fracture problems. In Paragraph 4.3.3, the autoencoder's efficiency for compression and the ROM's efficiency for prediction are analyzed for an application related to fracture mechanics.

4.3.1 Compression

In the following, autoencoders are presented, and in particular convolutional autoencoders for nonlinear dimensionality reduction. A quantitative criterion is finally proposed to assess the accuracy of dimensionality reduction methods.

4.3.1.1 Autoencoders

Let $X_1, X_2, ..., X_M \in \mathbb{R}^N$ be M data points and X be a data matrix such that $X_1, X_2, \ldots, X_M$ are the rows of this matrix. It is assumed to be normalized, *i.e.*, column means are equal to zero and column standard deviations are equal to one.

The simplest autoencoder that can be designed is a feed-forward neural network which encodes its input $X_m \in \mathbb{R}^N$ (where N is the number of input neurons) into a hidden representation H_m as

$$H_m = g\,(X_m W + b),\ n = 1...N, \tag{36}$$

where $H_m \in \mathbb{R}^K$. Here K is the number of hidden layer neurons, which in the special case of an autoencoder is called encoding dimension, $W \in \mathbb{R}^{N \times K}$ is a weight matrix, g is a so-called activation function that has to be chosen appropriately, and $b \in \mathbb{R}^K$ is a bias vector. The hidden layer represents the compressed input data (see Fig. 40), often called encoded, latent, or low dimensional space.

The autoencoder decodes a reconstruction $\tilde{X}_m$ from the hidden layer representation as

$$\tilde{X}_m = f(H_m V + c), \tag{37}$$

where $V \in \mathbb{R}^{K \times N}$ is another weight matrix, f is another activation function, and $c \in \mathbb{R}^N$ is another bias vector. The autoencoder model is trained to minimize a certain loss function such as the mean squared error function:

$$\min_{W,V,b,c} \frac{1}{M} \sum_{m=1}^{M} \sum_{n=1}^{N} (\tilde{X}_{mn} - X_{mn})^2, \tag{38}$$

which ensures that $\tilde{X}_m$ is close to X_m, $m = 1 \ldots M$. The autoencoder can be trained just like a regular feed-forward network using back propagation. As shown in Fig. 40, an autoencoder neural network can be split into two parts: the encoder, which maps the input $X_m \in \mathbb{R}^N$ to its reduced representation $H_m \in \mathbb{R}^K$, $K \ll N$, and the decoder, which maps the reduced state to the reconstructed output $\tilde{X}_m$.

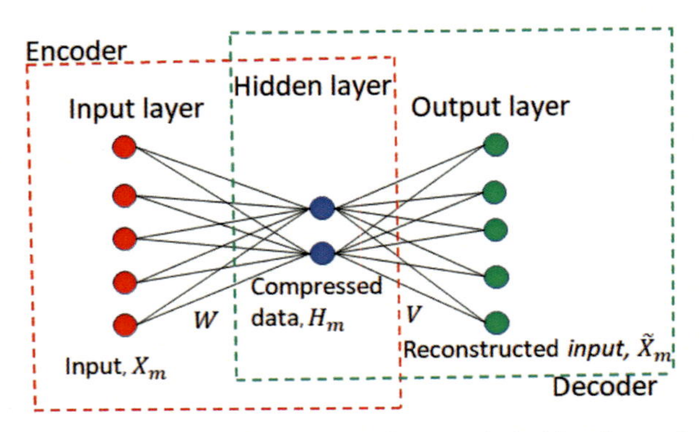

Fig. 40 Example of a single autoencoder with a single hidden layer. *Reprinted from Shinde, K., Itier, V., Mennesson, J., Vasiukov, D., & Shakoor, M. (2023). Dimensionality reduction through convolutional autoencoders for fracture patterns prediction. Applied Mathematical Modelling, 114, 94–113 with permission from Elsevier.*

4.3.1.2 Deep convolutional autoencoders

In the neural network shown in Fig. 40, each neuron of layer L is connected to all neurons of layer $L-1$. This kind of layer is called fully-connected or dense layer. A Convolutional Neural Network (CNN) contains layers where connections are more efficient for input data that is structured as in time series, 2D pictures, videos, *etc.* Fully-connected or dense layer connections can create problems in memory and computations if a neural network is large, as compared to convolutional layers.

In convolutional layers, each neuron in layer L depends on a subset of the neurons in the previous layer $L-1$. Kernels or filters are helpful for this purpose. For each neuron in layer L, the same convolutional filter is applied to the corresponding subset of neurons in layer $L-1$. Hence CNNs benefit in terms of sparse interaction between layers and parameter sharing between neurons of hidden layers, which makes them suitable for high-dimensional data. The use of convolutional layers makes sense for feature extraction from spatially distributed data, mainly in the form of pictures.

In the following, an autoencoder of the type shown in Fig. 41 is used for dimensionality reduction. Convolutional layers in the autoencoder can efficiently capture nonlinear patterns in the data, which helps finding the nonlinear low dimensional space. Both the encoder and decoder make use of convolutional and dense layers.

In Fig. 41, the encoder part of the autoencoder takes images as input tensors with a shape: (input height) × (input width) × (input channels). Then the

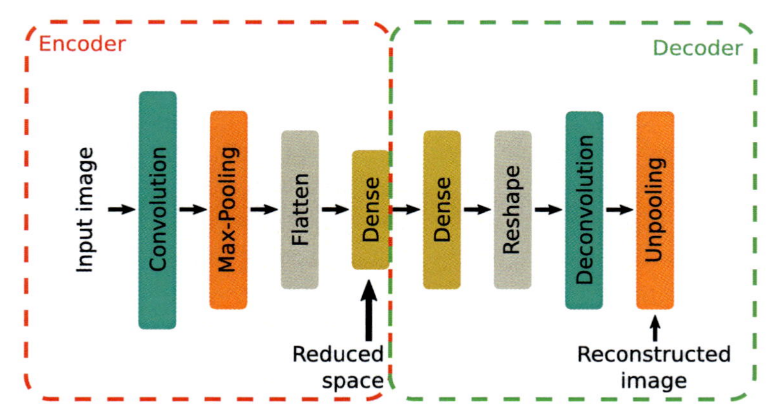

Fig. 41 Example of convolutional autoencoder. *Reprinted from Shinde, K., Itier, V., Mennesson, J., Vasiukov, D., & Shakoor, M. (2023). Dimensionality reduction through convolutional autoencoders for fracture patterns prediction. Applied Mathematical Modelling, 114, 94–113 with permission from Elsevier.*

convolutional layer performs the operation of convolution using filters ((feature map height) × (feature map width) × (feature map channels)) along with an activation function to capture the local features of the input image. Commonly used activation functions in convolutional layers are *sigmoid*, *tanh*, *elu*, and *relu*.

The convolutional layer in Fig. 41 is followed by pooling layers (max-pooling) to reduce the features' spatial size, decreasing the required computations and weights. The output of the max-pooling layer is then flattened and a dense layer is used to compress the input image to the encoding dimension.

The decoder part takes a compressed input (output of the encoder) with the shape of the encoding dimension (dimension of latent space). After that, a dense layer decompresses the encoded output to a larger size, which is then reshaped into image form. The image is passed through a deconvolutional layer to reconstruct the original image back to its input shape. The deconvolution operation can be performed by a convolutional layer followed by an unpooling layer or alternatively by a single transposed convolutional layer (Conv2DTranspose).

In this way, the encoder and decoder part of the autoencoder can be used for image compression and decompression.

4.3.1.3 Compression error

Autoencoders can reduce the dimensionality of the data and then reconstruct it with some compression error. In the present work, the aim is to investigate this method's performance in terms of dimensionality reduction and reconstruction of nonlinear data (both on training and test data). To do this, three criteria are considered.

The first one concerns the dimension of the reduced space required by the method to produce a low dimensional or encoded space. Here, the dimension of the reduced space means the dimension of the latent or encoded space.

The second criterion concerns the reconstruction error on test data, which is new unseen data (data that is not present in the training data set used to train the autoencoder). The Root Mean Square Error (RMSE) is chosen for measuring this reconstruction error:

$$\mathrm{RMSE} = \sqrt{\frac{1}{N}\sum_{n=1}^{N}(X_{mn} - \tilde{X}_{mn})^2} \tag{39}$$

where $X_m \in \mathbb{R}^N$ is the normalized sample and $\tilde{X}_m \in \mathbb{R}^N$ is the normalized reconstructed sample. Due to normalization, this RMSE is adimensional.

Lastly, the computation time required to train the models and reconstruct unseen test data is reported.

4.3.2 Prediction

This part describes the development of a so-called ROM based on deep learning for a highly nonlinear brittle fracture problem. Crack initiation and propagation in a 2D plate $\Omega \subset \mathbb{R}^2$ containing a heterogeneity to influence the crack path is considered. For instance, the plate may be notched as shown in Fig. 42A and the crack may propagate differently depending on loading conditions as shown in Fig. 42B.

To vary loading conditions easily, displacements are imposed at all boundaries:

$$\begin{bmatrix} u_1(x, y, t) \\ u_2(x, y, t) \end{bmatrix} = G(t)\begin{bmatrix} x \\ y \end{bmatrix}, \ (x, y) \in \partial\Omega \tag{40}$$

where

$$G(t) = \begin{bmatrix} G_{xx}(t) & G_{xy}(t) \\ G_{yx}(t) & G_{yy}(t) \end{bmatrix} \tag{41}$$

with $t \in [0, T]$, $T > 0$. In the particular case where G has a linear relationship with respect to time, the loading path is said to be proportional. It is coined non-proportional in all other cases.

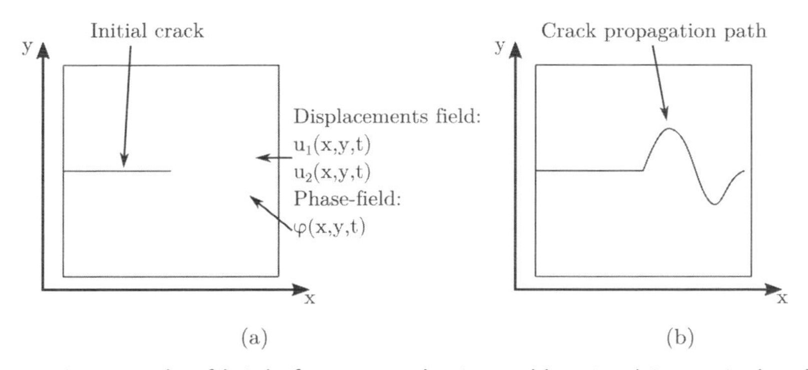

Fig. 42 An example of brittle fracture mechanics problem involving a single edge notch specimen: (A) initial setup, (B) example of crack propagation path. *Reprinted from Shinde, K., Itier, V., Mennesson, J., Vasiukov, D., & Shakoor, M. (2023). Dimensionality reduction through convolutional autoencoders for fracture patterns prediction. Applied Mathematical Modelling, 114, 94–113 with permission from Elsevier.*

In the FOM, loading is applied incrementally by solving balance equations at each discrete time step t_n with displacements imposed at all boundaries. To model crack initiation and propagation, the so-called phase-field cohesive zone model is used(Wu and Huang, 2020). There is hence an additional equation to solve at each time step, *i.e.*, the phase-field evolution equation. This damage model, moreover, introduces a history-dependence as the evolution of the phase-field variable at time step t_{n+1} depends on the fields at t_n. As a result, for each time step, displacements fields $u_1(x, y, t_{n+1})$, $u_2(x, y, t_{n+1})$ can be computed over the domain Ω, as well as the phase-field variable $\varphi(x, y, t_{n+1})$. A brief overview is presented in Paragraph 4.3.2.1.

The FOM is used to generate data for a set of randomly generated loading paths. For each path and for each time step t_n, the data set contains the loading matrix $G(t_n)$ as well as the fields $u_1(x, y, t_n)$, $u_2(x, y, t_n)$ and $\varphi(x, y, t_n)$ which are averaged over each element of the FE mesh. As only Cartesian grids are used, this element-wise averaging results in a three-channel 2D picture for each time step. The data set is presented in details in Paragraph 4.3.2.2.

The objective is to use the FOM only to generate a data set for a limited number of loading paths, and then rely on deep learning to make predictions for a greater number of unseen loading paths. This approach will only be relevant if the deep learning approach is less computationally expensive than the FOM when making predictions, and yields results that compare well to those computed using the FOM. As stated previously, the chosen strategy consists in first reducing the dimensionality of the problem, in order to have only some unknowns to predict instead of three-channel 2D pictures. The history-dependence due to the damage model introduces an additional complexity that can also be treated using dimensionality reduction. This nonlinear data-driven ROM based on deep learning is presented in Paragraph 4.3.2.3.

4.3.2.1 Full order model

The FOM used in this work on deep learning is the one presented in Subsection 3.4, with the difference that Wu's phase-field cohesive zone model (Wu and Huang, 2020) is used instead of Miehe's model. The equilibrium model is the one in Equation (22), with the exception that boundary conditions do not rely on a decomposition of the displacement field into linear and periodic parts. This decomposition is replaced by the boundary conditions of Equation (40). Expressions of the Cauchy stress tensor with the nonlinear degradation function and the phase-field evolution equation based on a history variable can be found in Reference (Shinde et al., 2023).

Regarding the numerical method, an FE implementation of the phase-field cohesive zone model in *FEMS* is used. A time discretization with weak coupling between the displacement field and the phase-field is chosen. In a time step t_{n+1}, balance equations are first solved using $g(\varphi(x, y, t_n))$ in the constitutive model to compute $u(x, y, t_{n+1})$, history variable $\mathcal{H}(\sigma(u(x, y, t_{n+1})))$ is then updated and the phase-field evolution equation is solved to compute $\varphi(x, y, t_{n+1})$. This strategy typically requires a very small time step to obtain converged results.

4.3.2.2 Data set

The FOM based on a phase-field model solved with the FE method is designed for generating data for learning. This data set is composed of displacement fields and phase-fields. It is defined as a tensor:

$$X = (u_1^m(x_i, y_j, t_n), u_2^m(x_i, y_j, t_n), \varphi^m(x_i, y_j, t_n))_{m=1\ldots N_L, n=1\ldots N_T, i=1\ldots N_x, j=1\ldots N_y}, \tag{42}$$

with N_L the number of loading paths, N_T the number of time steps, and $N_x \times N_y$ the number of pixels. The data set X corresponds to different loading paths which are stored in the data set

$$S = (G_{xx}^m(t_n), G_{xy}^m(t_n), G_{yx}^m(t_n), G_{yy}^m(t_n))_{m=1\ldots N_L, n=1\ldots N_T}. \tag{43}$$

For any applied loading $S_{mn} \in \mathbb{R}^4$ at time step t_n, the displacements fields $u_1 \in \mathbb{R}^{N_x \times N_y}$, $u_2 \in \mathbb{R}^{N_x \times N_y}$, and the phase-field variable $\varphi \in \mathbb{R}^{N_x \times N_y}$ are grouped as a data point $X_{mn} \in \mathbb{R}^{N_x \times N_y \times 3}$, which is a three-channel picture. As described in Paragraph 4.3.2.1, the FOM predicts X_{mn+1} from S_{mn+1} and a history variable depending on $(X_{mp})_{p \leq n}$ by solving physics-based partial differential equations. The nonlinear data-driven ROM based on deep learning presented in Paragraph 4.3.2.3 is designed to make predictions following the same route as the FOM.

4.3.2.3 Data-driven reduced order model

To construct the nonlinear data-driven ROM, the dimensionality of each three-channel picture $X_{mn} \in \mathbb{R}^{N_x \times N_y \times 3}$ is first reduced using the encoder part of the autoencoder as per Paragraph 4.3.1. The result is the reduced state $H_{mn} \in \mathbb{R}^K$, $K \ll N_x \times N_y \times 3$. To do this, the autoencoder should be trained beforehand on the training samples of the data set X.

The next step is to design and train a deep learning model of which the inputs are the applied loading conditions $(S_{mn})_{n=2\ldots N_T} \in \mathbb{R}^{(N_T-1)\times 4}$, and the outputs are the reduced states $(H_{mn})_{n=2...N_T} \in \mathbb{R}^{(N_T-1)\times K}$. The initial reduced state H_{m1} may also be provided as input unless it is the same for all loading paths as in the present work. In this case, there is one input/output pair per loading path.

As shown in Fig. 43, the proposed deep learning model works in a recurrent manner. An ANN predicts H_{m2} from a combination of H_{m1} and S_{m2}. The same ANN with exactly the same weights then predicts H_{m3} from a combination of H_{m2} and S_{m3}, and so on for each time step. Consequently, the input layer of the ANN has $K+4$ neurons and the output layer K neurons. This ANN can hence compute reduced states for a new unseen loading path progressively, time step by time step.

The full displacements fields and phase-field variables can be reconstructed using the decoder part of the autoencoder. The combination of the iterative procedure of the ANN relying on the reduced space and the decoder forms a data-driven nonlinear ROM that works the same way as the FOM based on the FE method.

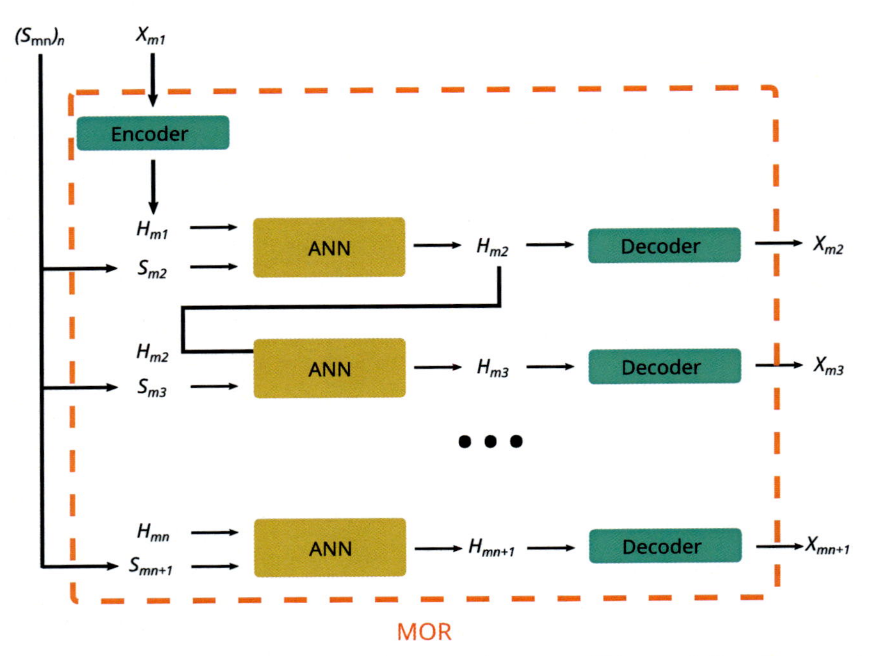

Fig. 43 Data-driven nonlinear ROM relying on the same ANN at each time increment for fracture modeling.

4.3.3 Results

In this part, the autoencoder and the ROM are applied to a fracture mechanics problem. Both neural networks have been implemented with *Keras* (Chollet et al., 2015), which is the high-level API of *TensorFlow 2* (Abadi et al., 2016). All computations were performed on a machine with an *Intel(R) Xenon(R) E5–2630 V4* processor with 10 CPUs (20 cores) and an *Nvidia Tesla K80* graphical processing unit (2496 cores). Acquiring and configuring workstations equipped with graphical processing units and enough memory and storage space is an additional work that is vital to research on deep learning.

The considered problem and the data sets are presented in Paragraph 4.3.3.1. The autoencoder's architecture is detailed in Paragraph 4.3.3.2 and the ROM's one in Paragraph 4.3.3.3. Results are then presented and analyzed in Paragraph 4.3.3.4 regarding compression, and Paragraph 4.3.3.5 regarding prediction.

4.3.3.1 Data sets

The problem's geometry is the squared notched plate $\Omega = [0,1] \times [0,1]\text{mm}^2$ illustrated in Fig. 44. Depending on boundary conditions, the crack may propagate in any direction. For non-proportional loading paths, the direction may even vary during loading.

The domain is composed of a brittle material of elastic properties $E = 210$ GPa and $\nu = 0.3$, and fracture properties $\sigma_c = 2445.42$ MPa and $G_c = 2.7\ \text{N mm}^1$. The regularization length of the phase-field model is set

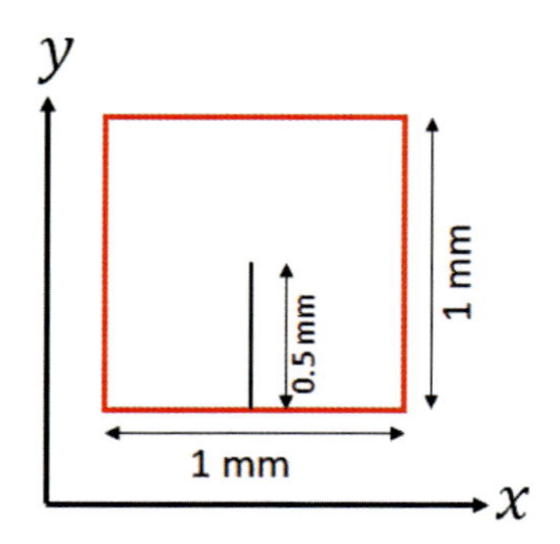

Fig. 44 Geometry for the notched test problem. Edges where displacements are imposed as per Equation (40) are shown in red. *Reprinted from Shinde, K., Itier, V., Mennesson, J., Vasiukov, D., & Shakoor, M. (2023). Dimensionality reduction through convolutional autoencoders for fracture patterns prediction. Applied Mathematical Modelling, 114, 94–113 with permission from Elsevier.*

to $l_c = 0.04$ mm and the element size to ≈ 0.016 mm in order to obtain a structured quadrangular FE mesh of 62×62 elements. Full integration with four integration points per element is used.

This mesh size and regularization length are voluntarily kept large enough to obtain pictures of $N_x \times N_y = 62 \times 62$ pixels and minimize the size of the database (which, as will be shown, will already be large enough with this resolution). For the same reason, the number of time steps is set to $N_T = 201$ (including the initial state), with the final time $T = 0.02$ s. The main objective of the present work is to show that the CNN and the nonlinear ROM can compute results that compare well to the FOM. The fact that the FOM could be more accurate is not limiting regarding that objective.

A database containing FOM results for 1024 randomly generated proportional loading paths is first built, with $||G^m(T)||_2^2 = T$, $m = 1...N_L$. Eight particular loading paths corresponding to

$$
\begin{aligned}
G^1(T) &= \begin{pmatrix} T & 0 \\ 0 & 0 \end{pmatrix}, \quad G^2(T) = \begin{pmatrix} 0 & T \\ 0 & 0 \end{pmatrix}, \\
& G^3(T) = \begin{pmatrix} 0 & 0 \\ T & 0 \end{pmatrix}, \\
G^4(T) &= \begin{pmatrix} 0 & 0 \\ 0 & T \end{pmatrix}, \quad G^5(T) = \begin{pmatrix} -T & 0 \\ 0 & 0 \end{pmatrix}, \\
& G^6(T) = \begin{pmatrix} 0 & -T \\ 0 & 0 \end{pmatrix}, \\
G^7(T) &= \begin{pmatrix} 0 & 0 \\ -T & 0 \end{pmatrix}, \quad G^8(T) = \begin{pmatrix} 0 & 0 \\ 0 & -T \end{pmatrix},
\end{aligned} \tag{44}
$$

are manually included in this database. It hence contains 205,824 images (corresponding to 1024 loading paths with 201 time steps) with 3 channels such that each image has a size of 62×62, *i.e.*, (1024,201,62,62,3). This database of nearly 18 GB is divided with respect to loading paths into a training and validation data set (175,071 samples corresponding to 871 loading paths), and a testing data set (30,753 samples corresponding to 153 loading paths). The eight particular loading paths are always included in the training data set. All data images are normalized by subtracting the mean image and dividing by the standard deviation image computed on the training and validation data set.

A second database is then built for further testing of the models. It contains 100 randomly generated non-proportional loading paths. The procedure to get a non-proportional loading path is to: generate a continuous curve of seven line segments of equal length $\frac{T}{7}$ in the space $(G_{xx}, G_{xy}, G_{yx}, G_{yy})$, but oriented randomly; compute a cubic B-spline going through the ends of these segments; interpolate the $G^m(t_n)$, $n = 1 \ldots 201$ matrices. This adds 20,100 images to be used as supplementary test data.

As illustration, proportional loading paths are shown in Figs. 45A and B, and non-proportional loading paths in Figs. 45C and D.

It is clearly visible that a wide range of directions are considered for proportional loading paths, but that the magnitude is always the same. For non-proportional loading paths, the visualization is more difficult as all paths are entangled, but their length is again approximately the same.

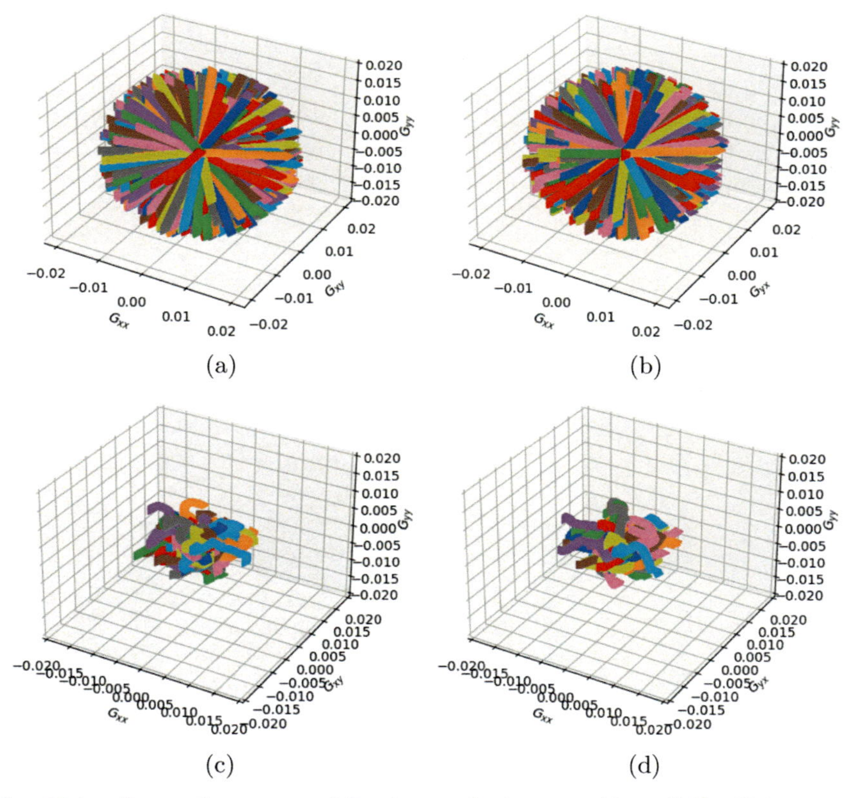

Fig. 45 Loading paths generated for the notched test problem: (A,B) 1024 proportional loading paths data set, (C,D) 100 non-proportional loading paths data set. *Reprinted from Shinde, K., Itier, V., Mennesson, J., Vasiukov, D., & Shakoor, M. (2023). Dimensionality reduction through convolutional autoencoders for fracture patterns prediction. Applied Mathematical Modelling, 114, 94–113 with permission from Elsevier.*

4.3.3.2 Autoencoder

The autoencoder used for compression is presented in Fig. 46A for the encoder and Fig. 46B for the decoder.

The encoder part of the CNN architecture takes an input with shape (62,62,3). It has two convolutional layers, each followed by a max-pooling layer. After the last max-pooling layer, a flattening layer formats the data in order to finish with one fully connected layer. This encoder compresses the input image to the encoding dimension of five.

The decoder part of the CNN architecture takes the compressed input with shape (5). It starts with a fully connected layer and then a reshaping layer to format the data into an image. This is followed by two deconvolutional layers (Conv2DTranspose) and a convolutional layer which reconstructs the original image back to its shape (62,62,3).

The activation functions used in the last dense layer of the encoder and the last convolutional layer of the decoder are linear. An *elu* activation function is used for the two convolutional layers of the encoder as well as those of the decoder. The mean squared error loss function is chosen to train the autoencoder.

Depending on applications (particularly in fluid mechanics), different autoencoder architectures have been proposed in the literature (Pant et al., 2021). The architecture of the autoencoder shown in Fig. 46 is very similar in its structure to the one which is considered in Reference Wen and Zhang (2018) for biomedical applications. The proposed architecture has been tuned according to the considered input image shape and content.

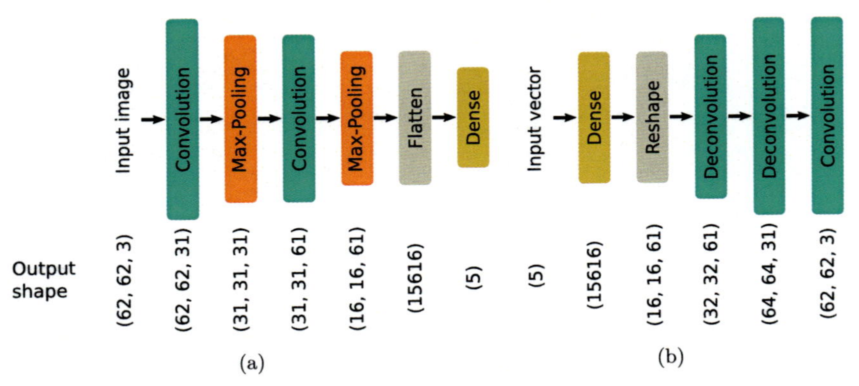

Fig. 46 Architecture of the convolutional autoencoder: (A) encoder, (B) decoder. *Reprinted from Shinde, K., Itier, V., Mennesson, J., Vasiukov, D., & Shakoor, M. (2023). Dimensionality reduction through convolutional autoencoders for fracture patterns prediction. Applied Mathematical Modelling, 114, 94–113 with permission from Elsevier.*

4.3.3.3 Data-driven reduced order model

As explained in Paragraph 4.3.2.3, the ROM is actually a recurrent network embedding an ANN that predicts the outputs progressively, time step by time step. The architecture of this ANN consists of dense layers. It takes an input with shape ($K + 4$). The input shape is a combination of the dimension of the latent space K and the number of components of the loading conditions at each time step (four). The input layer is followed by three hidden dense layers and one dense output layer. The three hidden layers contain 216, 324, and 540 neurons, respectively, and the *elu* activation function is used. The output layer uses a linear activation function to compute an output with shape (K).

Training the ROM actually means training the embedded ANN that is shown in Fig. 43. The mean squared error loss function is chosen for this training. To avoid the computational cost of repetitively decoding reduced states during training, the loss is computed directly on reduced states, which are pre-encoded for training and validation data sets.

4.3.3.4 Reconstruction of fracture patterns for unseen loading paths

In this part, the efficiency of the autoencoder for nonlinear data compression is evaluated. This comparison is done for different levels of compression depending on the encoding dimension. Reconstruction error as well as computation time are analyzed.

A mean RMSE over the 153 unseen proportional loading paths of 0.0331 is obtained using an encoding dimension of 5, 0.0234 using 10 and 0.0119 using 50. It is reminded that the RMSE is computed over normalized data. The very low error obtained using an encoding dimension of only five is hence particularly promising.

The mean value may however hide large errors for some samples. The RMSE for each sample is shown in Fig. 47. There are several samples for which the RMSE is greater than 0.05, and a few samples for which it is over 0.5. A systematic decrease of the RMSE can be obtained when using a larger encoding dimension. It is in fact possible to reduce the RMSE for most samples below 0.2 by using an encoding dimension of 50.

It is interesting to analyze the worst result, which is the sample leading to the highest peak in the middle of Fig. 47. This sample corresponds to the last time step of some loading path. All non-normalized fields for this particular sample are shown in Fig. 48. Boundary conditions for this sample correspond to shear loading, as the crack propagates along the diagonal in the FOM result. Due to the boundary conditions used to generate the data set, however, crack initiation and propagation also occurs along domain boundaries.

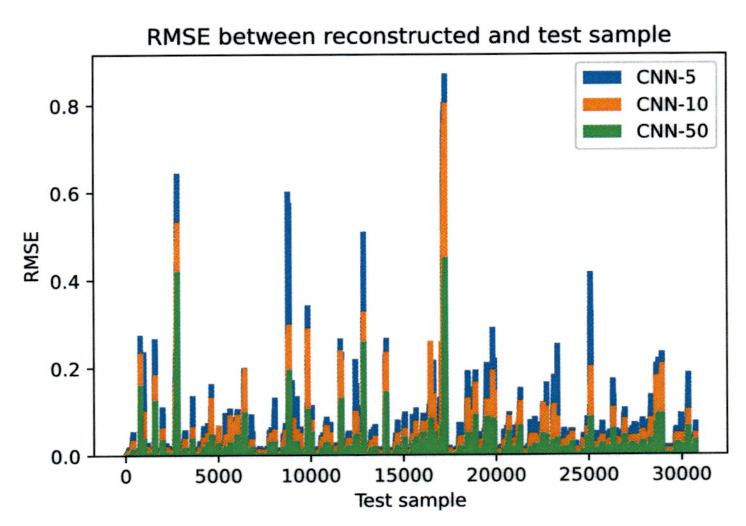

Fig. 47 RMSE between each reconstructed image (CNN) and the associated original (HFM) from the proportional loading paths data set using different latent space dimensions. *Reprinted from Shinde, K., Itier, V., Mennesson, J., Vasiukov, D., & Shakoor, M. (2023). Dimensionality reduction through convolutional autoencoders for fracture patterns prediction. Applied Mathematical Modelling, 114, 94–113 with permission from Elsevier.*

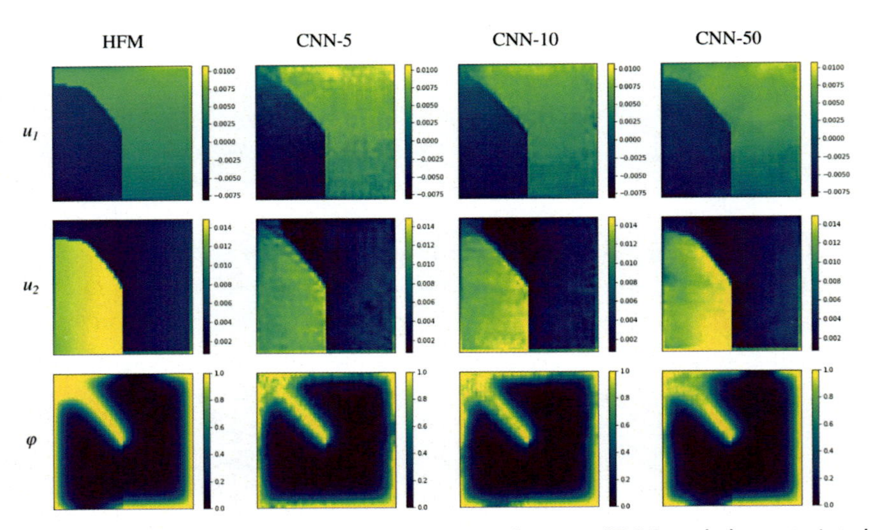

Fig. 48 Comparison between each reconstructed image (CNN) and the associated original (HFM) from the proportional loading paths data set using different latent space dimensions. Each line corresponds to a different field and each column to a different result. Displacements are in millimeters. *Reprinted from Shinde, K., Itier, V., Mennesson, J., Vasiukov, D., & Shakoor, M. (2023). Dimensionality reduction through convolutional autoencoders for fracture patterns prediction. Applied Mathematical Modelling, 114, 94–113 with permission from Elsevier.*

The autoencoder clearly has a difficulty in capturing this complex failure pattern, especially for the phase-field variable. This is easier to analyze by looking at Fig. 49, which shows the FOM result and the absolute error fields on non-normalized data with respect to the FOM result for the different encoding dimensions.

The difficulty comes from the phase-field variable and the propagation of the crack along domain boundaries. With an encoding dimension of five, the autoencoder over-estimates failure, as there are two regions with no crack along the boundary which are lost after reconstruction. These regions are recovered when using an encoding dimension of 10 or 50.

Since a crack along domain boundaries has no significant effect on the displacement field, the latter is accurately compressed by the autoencoder, even for an encoding dimension of five. The only errors that remain and that can be reduced using a larger encoding dimension are located near the crack and are due to the displacement jump.

Regarding computation time, it is mainly spent training the CNN. On the one hand, the training time is indeed close to 50 h using an encoding

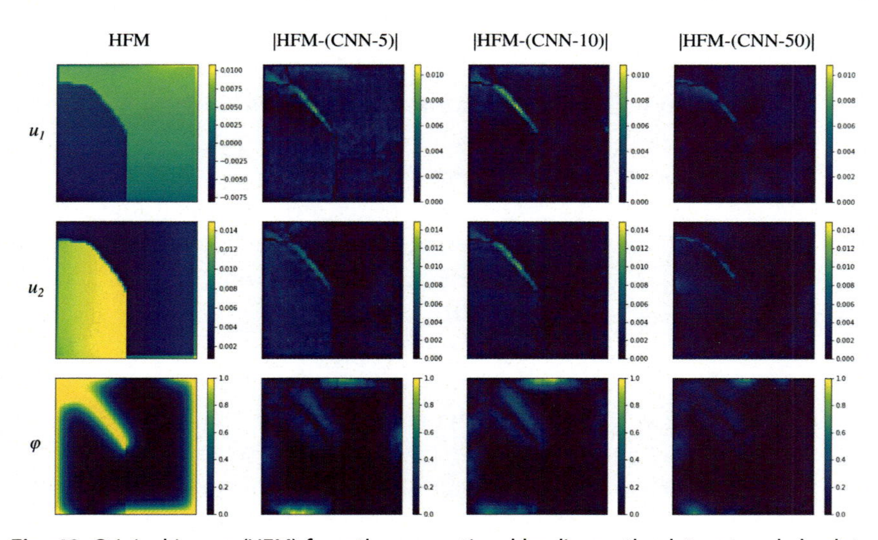

Fig. 49 Original image (HFM) from the proportional loading paths data set and absolute error between this original image and the reconstructed image (CNN) obtained using different latent space dimensions. Each line corresponds to a different field and each column to a different result. Displacements are in millimeters. *Reprinted from Shinde, K., Itier, V., Mennesson, J., Vasiukov, D., & Shakoor, M. (2023). Dimensionality reduction through convolutional autoencoders for fracture patterns prediction. Applied Mathematical Modelling, 114, 94–113 with permission from Elsevier.*

dimension of five, and 60 h using 50. On the other hand, computing FOM results for the 1024 proportional loading paths takes approximately 16 h (55 s per loading path) on a workstation with an Intel Xeon(R) W-2175 processor with 14 CPUs (28 cores). Generating data and training the autoencoder hence takes days for this problem, with no significant increase in the training time when using a larger encoding dimension.

The autoencoder is relatively fast once trained, as encoding and decoding the 201 samples for a whole loading path requires in average 1.3 s (6.5 ms per sample). This reconstruction time does not vary significantly when changing the encoding dimension.

To summarize, although the mean RMSE is already very low when using an encoding dimension of only five, large errors are obtained for some samples. These can be reduced by using a larger encoding dimension, with no significant increase in the training time or the reconstruction time. To build the ROM, an encoding dimension of 50 is chosen.

4.3.3.5 Prediction of fracture patterns for unseen loading paths

In the following, the ROM described in Fig. 43 is deployed with an encoding dimension of 50 as low dimensional space, thanks to the auto-encoder trained with the proportional loading paths data set. To train the ROM, only the training data set from the proportional loading paths database is used. This trained ROM is then employed to make predictions for unseen loading conditions of proportional as well as non-proportional loading paths.

A mean RMSE for the 153 unseen proportional loading paths of 0.0559 is obtained, which is of the same order as the reconstruction error for the chosen encoding dimension (0.0119). In this work, even though the autoencoder takes longer to train, its primary purpose is to produce a low dimensional space to construct the ROM. These results hence confirm that the autoencoder successfully eases the task of making predictions.

The RMSE distribution over all testing samples is shown in Fig. 50. The ROM almost systematically leads to larger errors as compared to the autoencoder. This is no surprising as making a prediction requires going through the ANN and then the decoder.

There are more samples for which the RMSE is larger than 0.5, and some samples even lead to an error larger than 1. This shows that the autoencoder is a promising tool for compressing highly nonlinear mechanical data, but that more research is needed regarding the ROM.

It is interesting to look again at the same sample as previously shown in Fig. 49. The prediction error is, indeed, at least twice larger than the

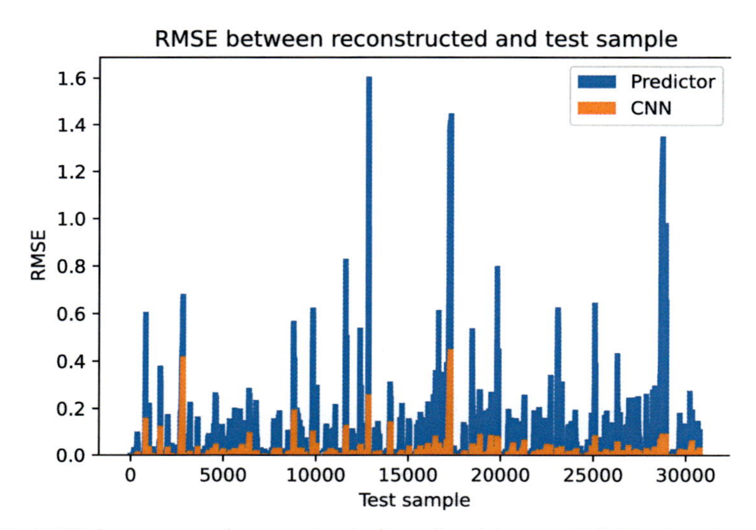

Fig. 50 RMSE between each reconstructed/predicted image (CNN/Predictor) and the associated original (HFM) from the proportional loading paths data set. *Reprinted from Shinde, K., Itier, V., Mennesson, J., Vasiukov, D., & Shakoor, M. (2023). Dimensionality reduction through convolutional autoencoders for fracture patterns prediction. Applied Mathematical Modelling, 114, 94–113 with permission from Elsevier.*

reconstruction error for this sample, and corresponds to the second highest peak in the middle of Fig. 50. Absolute errors using the ROM for this sample are shown in Fig. 51.

As opposed to what the large errors may indicate, the ROM accurately captures the propagation of the crack along the diagonal due to shear loading. The errors are again localized along the plate boundary for the phase-field variable, and are relatively low for the displacement field.

The capabilities of the autoencoder and the ROM to capture complex crack propagation patterns are also tested for 100 unseen non-proportional loading paths. A mean RMSE of 0.0124 is obtained for reconstruction and 0.0887 for prediction. This increase in the errors is expected as both the CNN and the ROM were trained only on the training data set of proportional loading paths. This increase is nevertheless not significant, which demonstrates the robustness and good generalization capabilities of the proposed approach.

Error distributions are shown in Fig. 52. For these non-proportional loading paths, there are many cases where no crack initiation occurs and which lead to errors very close to zero for both the autoencoder and the ROM. For the remaining cases, errors are of the same order as those found for proportional loading paths in Fig. 50.

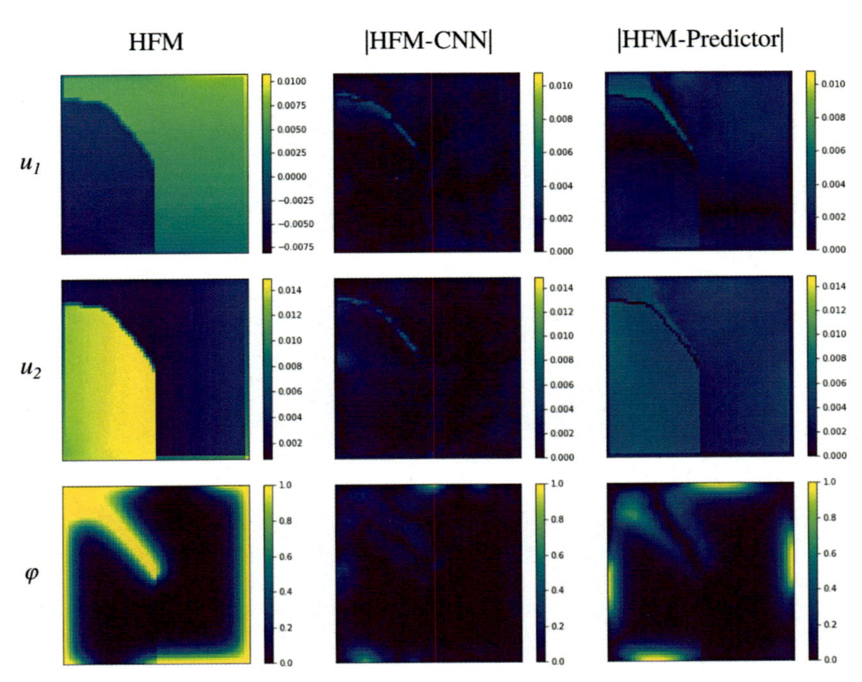

Fig. 51 Original image (HFM) from the proportional loading paths data set and absolute error between this original image and the reconstructed (CNN) and predicted (Predictor) images. Each line corresponds to a different field and each column to a different result. Displacements are in millimeters. *Reprinted from Shinde, K., Itier, V., Mennesson, J., Vasiukov, D., & Shakoor, M. (2023). Dimensionality reduction through convolutional autoencoders for fracture patterns prediction. Applied Mathematical Modelling, 114, 94–113 with permission from Elsevier.*

It is chosen again to analyze the result for the sample with largest reconstruction error. Absolute errors for this sample are shown in Fig. 53. On the one hand, this example confirms that the autoencoder compresses images very efficiently, even for unseen non-proportional loading paths. It is indeed important to point out that there is no sample with a curved crack in the training data set.

On the other hand, the ROM finds a straight crack along the wrong direction, which leads to very large errors. Improving the integration of loading history in the ROM seems necessary to improve results. This could be achieved by developing an autoencoder for the history variable of the phase-field model and introducing it in the ROM. This will be considered in a future work.

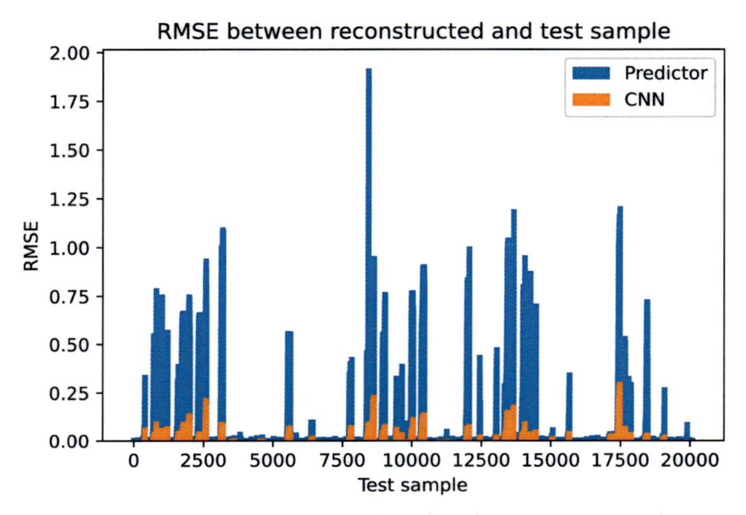

Fig. 52 RMSE between each reconstructed/predicted image (CNN/Predictor) and the associated original (HFM) from the non-proportional loading paths data set. *Reprinted from Shinde, K., Itier, V., Mennesson, J., Vasiukov, D., & Shakoor, M. (2023). Dimensionality reduction through convolutional autoencoders for fracture patterns prediction. Applied Mathematical Modelling, 114, 94–113 with permission from Elsevier.*

Regarding computation times, finally, the training time for the ROM is found to be negligible, as it is of only 40 min. Once trained, the time spent predicting reduced states and then decoding them is very low as it is close to 0.22 s per loading path. The proposed ROM is thus drastically cheaper than the FOM, with a computation time reduction by a factor of at least 250.

4.4 Conclusions

Two model order reduction methods have been presented in this section. The two methods are based on a three-step process. The first step consists in generating data with a FOM based on a conventional approach such as the FFT-based numerical method or the FE method. In the second step, the data is compressed. Finally, predictions are computed in the third step with an ROM where the number of unknowns to predict and hence the computation time are drastically reduced.

The first approach considered in this section is the SCA method. In this method, compression is achieved with K-means clustering and predictions are computed using a mean-field approach where the unknowns are assumed to be constant cluster-wise. The advantage of the SCA method over POD-based approaches is the simultaneous reduction of the number

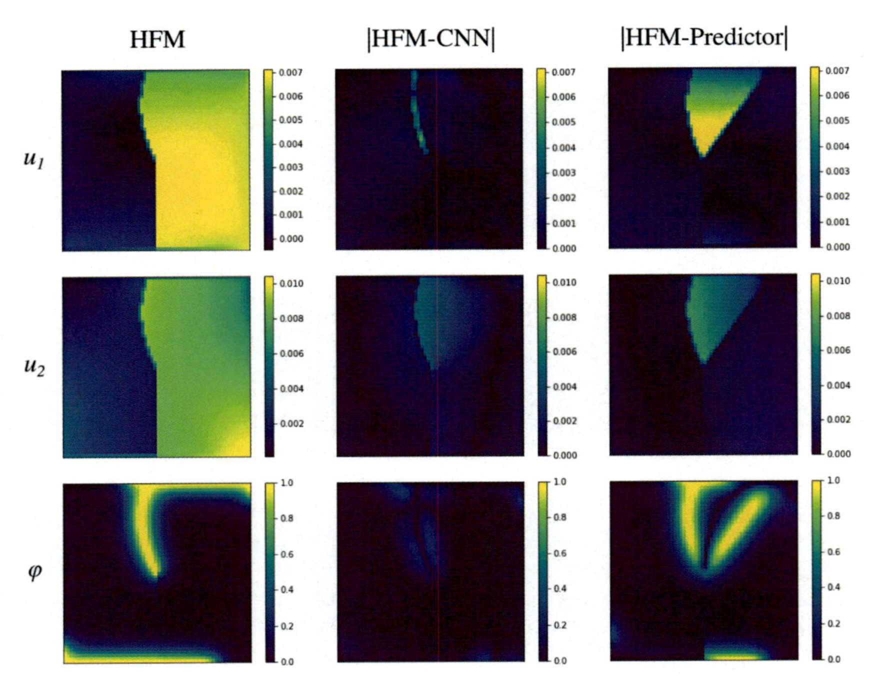

Fig. 53 Original image (HFM) from the non-proportional loading paths data set and absolute error between this original image and the reconstructed (CNN) and predicted (Predictor) images. Each line corresponds to a different field and each column to a different result. Displacements are in millimeters. *Reprinted from Shinde, K., Itier, V., Mennesson, J., Vasiukov, D., & Shakoor, M. (2023). Dimensionality reduction through convolutional autoencoders for fracture patterns prediction. Applied Mathematical Modelling, 114, 94–113 with permission from Elsevier.*

of unknowns and the number of integration points. Consequently, it does not require hyper-reduction. This advantage is also a weakness, as it limits the accuracy of ROM predictions at the local level.

Although there are applications where this approach is very interesting, the lack of accuracy at the local level is limiting. The SCA method can, therefore, be considered as a data-driven mean-field method that is quite reliable, but only at the global level.

To obtain reliable predictions at the local level, a second approach based on a nonlinear compression method has been presented. This method relies on two deep neural networks: an autoencoder for compression and an ROM integrating an ANN for prediction. This approach has been developed for crack initiation and propagation problems with proportional and non-proportional loading paths. Results show that the autoencoder compresses the data efficiently to a space of sufficiently small size, and that it generalizes quite well. The ROM, however, does not sufficiently integrate loading history in its architecture even though it has a recurrent form. It cannot, therefore, deal with complex loading paths.

5. Multiscale modeling

5.1 Introduction

Sections 2 and 3 have dealt with numerical modeling of highly nonlinear and localized physical phenomena. Despite an extensive research work on improving the numerical methods' efficiency, in particular using parallel computing, results have exclusively been obtained for small simulation domains containing a heterogeneity. In Section 4, it has been demonstrated that for large simulation campaigns with some redundancy, an acceleration by several orders of magnitude could be obtained thanks to compression methods and model order reduction, especially when only boundary conditions change between two calculations. This section focuses on the deployment of highly nonlinear models such as those developed in Sections 2 and 3 at the scale of relatively large structures, such as laboratory specimens, automotive parts, hydrogen tanks, wind turbine blades, *etc.*

The objective is to develop multiscale methods separating the discretization of the simulation domain from the discretization of the heterogeneity. The underlying assumption is that the characteristic scale of the heterogeneity is small compared to that of the simulation domain, the latter being called coarse scale in the following. The two main challenges to overcome are the development of the multiscale model linking the two scales, and its robust and efficient implementation. To reach this second objective, the Reduced Order Models (ROMs) developed in Section 4 have a part to play.

Multiscale modeling has been studied in the field of mechanics of materials for decades with the aim of producing more predictive behavior laws inspired from the microstructure (Geers et al., 2010; Matouš et al., 2017). Homogenization theory enables, for instance, to find the stress-strain response of a material from a Taylor expansion of the displacement field and an internal energy conservation principle known as Hill-Mandel's lemma (Geers et al., 2010). With this theory, it is even possible to make simplifying assumptions at the microscale so that an analytic solution can be derived and a behavior law that is directly exploitable at the macroscale can be obtained. Mori and Tanaka (1973) and (Gurson, 1977) models can be derived that way. For more complex models that aim at being closer to reality, one may turn to computational homogenization with the Finite Element (FE) method, the Fast Fourier Transform (FFT)-based numerical method or other computational methods (Matouš et al., 2017).

For transient and nonlinear phenomena, concurrent homogenization with an FE × FE (FE^2) scheme is particularly relevant (Feyel, 1999). This method consists in discretizing the simulation domain without the heterogeneity, and discretizing the latter at the scale of Representative Volume Elements (RVEs). These RVEs are small simulation domains that are discretized separately and attached at each integration point of the coarse scale FE model. At each time increment, the coarse scale problem should be solved simultaneously with an additional problem for each RVE. This approach is of great interest to the industry as it draws its reliability and predictive capability from the microscale, and it can be used to visualize and explain physical phenomena at different scales. Its application to industrial-scale problems is nevertheless limited, especially due to its great computational requirements.

Although they have been largely studied and exploited in structural mechanics, multiscale methods and especially FE^2 approaches have received little attention for fluid mechanics. The only remarkable exception is modeling of flows in porous or permeable media. In these flows, the heterogeneity is due to the presence of obstacles of a very small characteristic size compared to the size of the simulation domain. These flows are well understood and they can be modeled accurately even with obstacles of very small size and complex distribution and morphology as long as they are steady and laminar (Beliaev and Kozlov, 1996).

For viscous flows, indeed, governing equations at the fine scale are Stokes equations. Various theories have been proposed to relate the velocity and pressure fields between the two scales, with different choices of boundary conditions to employ at the fine scale. From these theories, Darcy's law (Beliaev and Kozlov, 1996) emerges for modeling the flow at the coarse scale. This law relies on the definition of a permeability tensor that can be computed analytically or numerically depending on the complexity of the distribution and morphology of the heterogeneity at the fine scale.

It is generally not possible to use Darcy's law for inertial flows. Extensions such as Forchheimer's law have been proposed but they are not based on homogenization theory (Skjetne & Auriault, 1999a,b). Recently, a theoretical framework known as Principle of Multiscale Virtual Power (PMVP) has been proposed for steady flows with obstacles and inertia (Blanco et al., 2017). Governing equations at the fine scale were steady-state Navier-Stokes equations, including the auto-advection term. This framework involved the solution of a nonlinear problem at the fine scale, and also for the coarse scale. A concurrent implementation with full coupling between the two scales, however, was not developed. This simplification was only possible because the unsteady case was not considered.

A work in relation with the industry on concurrent simulation of the failure of composite structures is presented in Subsection 5.2. It includes the development of an FE $\times$ ROM approach in order to accelerate multiscale simulations. A particular attention is given to model validation by relying on experimental tests.

More fundamental contributions on multiscale modeling of unsteady flows in domains containing small obstacles are presented in Subsection 5.3. The FE^2 implementation with full coupling is presented and compared with a straightforward approach where the heterogeneity is meshed directly at the scale of the simulation domain.

5.2 Fracture

As presented in Subsection 4.2, the Self-consistent Clustering Analysis (SCA) method has been proposed in 2016 (Liu et al., 2016) and has quickly been the object of numerous developments. Along those, there is the extension to finite strains, localized plastic strains and ductile damage modeling. Another important research track has been the integration of the SCA method in a concurrent homogenization scheme (Liu et al., 2018), and in particular in the context of a collaboration with the Ford Motor Company.

In this project, various polymer composite materials have been considered with the aim of replacing a structural automotive part previously made of steel. This effort included characterization and modeling. Characterization targeted both components (carbon fibers and epoxy matrix) and various composites including short fiber composites, unidirectional long fiber composites, and woven composites. The work presented in the following is about the unidirectional long fiber composite with epoxy matrix and carbon fibers, and has been published in Reference (Gao et al., 2020). The methodology consists in:

- Characterizing the mechanical behavior and properties of the matrix with pure epoxy specimens, as well as that of the fibers based on data provided by the manufacturer.

- Analyzing the composite microstructure in order to generate an RVE.
- Using the FE method to generate data by applying six orthogonal loading of an amplitude small enough to remain in the elastic range.
- Using the generated data to deploy the SCA method.
- Integrating the SCA method in a concurrent homogenization approach, which can provide the macroscopic response of the material at each integration point of an FE mesh of the structure.

The objective is to evaluate the reliability of this methodology in comparison with unseen experimental tests, *i.e.*, tests that have not been used to establish the multiscale model.

Materials and models used for the RVE are presented in Paragraph 5.2.1, as well as the multiscale method integrating the SCA method. A compelling example is presented in Paragraph 5.2.2.

5.2.1 Method

The multiscale model is described in the following for the example of Fig. 54, which is a tension specimen with all fibers oriented at 10 with respect to the tensile direction.

The structure in Fig. 54 is composed of layers of unidirectional composite. It is possible to vary fiber orientation in the thickness, although this is not done in this example. Each layer is discretized with 3D thick shell elements which enable to stretch elements up to an aspect ratio of ten without deteriorating predictions significantly. A single element in the thickness is used for each layer. Reduced integration is chosen, with only two integration points in the thickness of each element. At each integration point, a 3D RVE containing 93 fibers discretized with $600 \times 600 \times 100$ voxels is attached. This RVE has a very small dimension in the fiber direction thanks to periodic boundary conditions and the assumption that fibers are straight. Fiber volume fraction is close to 50%.

The FE $\times$ SCA concurrent simulation algorithm relies on information exchange between the coarse scale of the structure and the fine scale of the RVE. For the latter, the problem to solve is the equilibrium problem introduced in Subsection 3.2. It is solved with the SCA method based on the procedure detailed in Subsection 4.2, with the difference that in the following small strains are assumed. Fibers behavior is assumed linear elastic and transversely isotropic, with brittle fracture in the longitudinal direction modeled with a different strength under compression and tension. Matrix behavior is modeled with an elasto-plastic behavior law integrating a

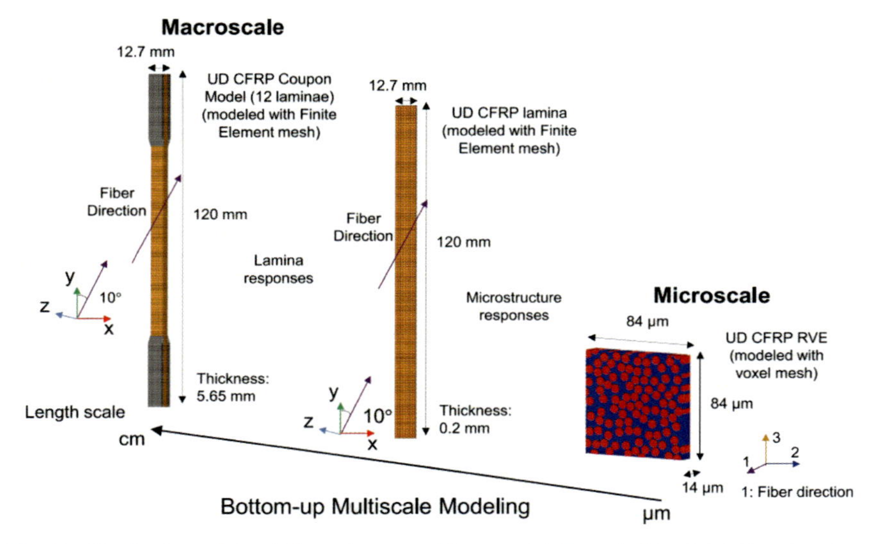

Fig. 54 Illustration of the different scales of the model. From left to right, the structure composed of 12 layers of a thickness of about 0.2 mm, each layer being composed of matrix and circular fibers of a diameter of 7 μm. *Reprinted from Gao, J., Shakoor, M., Domel, G., Merzkirch, M., Zhou, G., Zeng, D., Su, X., & Liu, W. K. (2020). Predictive multiscale modeling for unidirectional carbon fiber reinforced polymers. Composites Science and Technology, 186, 107922 with permission from Elsevier.*

damage model and a plasticity criterion where the hardening rule is different under compression and tension (Melro et al., 2013). Details on these behavior laws and the associated material properties can be found in Reference (Gao et al., 2020).

In the preparation step, six simulations are conducted with the FE method at RVE level by solving the equilibrium problem of Equation (22) with applied strains of an amplitude small enough to remain in the elastic range. Results from these simulations are used for K-means clustering and pre-calculation of the two parts of the interaction tensor for Algorithm (35). The structure FE model with its 3D thick shell elements is also prepared.

In the prediction step, although the problem is static, an explicit scheme is used at the coarse scale, with an artificial and automatic increase of the mass to limit the number of time steps. This requires to control for all results that the kinetic energy remains negligible.

At each time increment of the explicit solve, the macroscopic stress $\boldsymbol{\sigma}^M(\boldsymbol{x})$ is computed at each integration point $\boldsymbol{x}$ of each shell element of the structure from a known macroscopic strain $\boldsymbol{\varepsilon}^M(\boldsymbol{x})$. The advantage of the explicit scheme is that it avoids the computation of the tangent matrix for the macroscopic solve. To compute the macroscopic stress $\boldsymbol{\sigma}^M(\boldsymbol{x})$, a small strain version of Algorithm (35) is used. The fibers failure criterion and the matrix elasto-plasticity and damage behavior law are integrated in the algorithm.

Damage is not actually modeled at the microscopic scale within the partitions of the SCA method, but at the macroscopic scale. A macroscopic damage variable is introduced to soften the mechanical response computed for each RVE, and it is regularized with an integral non-local damage model. This is advantageous as it prevents any dependency on the discretization of the RVE, and also on the discretization of the structure. The drawback is that local phenomena such as debonding or crack initiation and propagation within the RVE are not actually modeled and it is not, therefore, possible to predict microstructure evolution. Details on this regularized damage model can be found in Reference (Liu et al., 2018).

A first analysis at the scale of the RVE has led to the conclusion that two partitions in fibers and eight in the matrix are enough to predict the macroscopic response of the composite accurately (Gao et al., 2020). This has been verified for applied macroscopic strains up to 2%. This choice of clustering is shown in Fig. 55.

5.2.2 Results

In this part, the FE × SCA approach is validated for an experimental tension test with fibers oriented at 10 with respect to the tensile direction. Details on this test and coupon manufacturing can be found in Reference (Gao et al., 2020). Comparison in terms of stress-strain response is

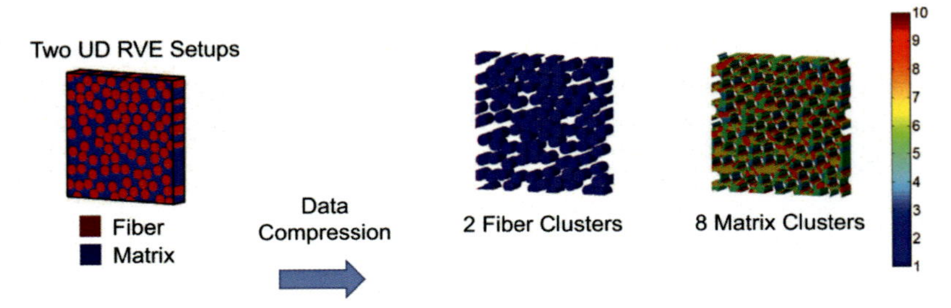

Fig. 55 K-means clustering of the RVE from data computed for six orthogonal loadings. *Reprinted from Gao, J., Shakoor, M., Domel, G., Merzkirch, M., Zhou, G., Zeng, D., Su, X., & Liu, W. K. (2020). Predictive multiscale modeling for unidirectional carbon fiber reinforced polymers. Composites Science and Technology, 186, 107922 with permission from Elsevier.*

presented in Fig. 56. A slight underestimation of material stiffness is observed in the elastic part. It could be caused by the absence of interphase in the RVE model. The concurrent approach still gives accurate results as the error on the maximum stress is of 2.32%.

For this test, digital image correlation measurements of displacement and strain fields are also available. Comparison between simulation and experiment for these fields is presented in Fig. 57. A shear band along the fiber direction is observed. This band is predicted quite well by the FE × SCA approach, as the error is of 3.95%, but there is a gap that could be due to a variation of the microstructure in the experiment.

Microstructure variation could also explain the difference regarding the final macroscopic crack position, which is shown in Fig. 58. The crack direction, however, is correctly predicted by the simulation. This proves that information is exchanged reliably between the structure modeled by the FE method and RVEs modeled by the SCA method, and that the influence of the microstructure on the macroscopic response is correctly captured.

This simulation, finally, required 540 h of computation time with the FE × SCA approach. Based on RVE simulations performed with the FE method in the clustering data generation step, it can be estimated that

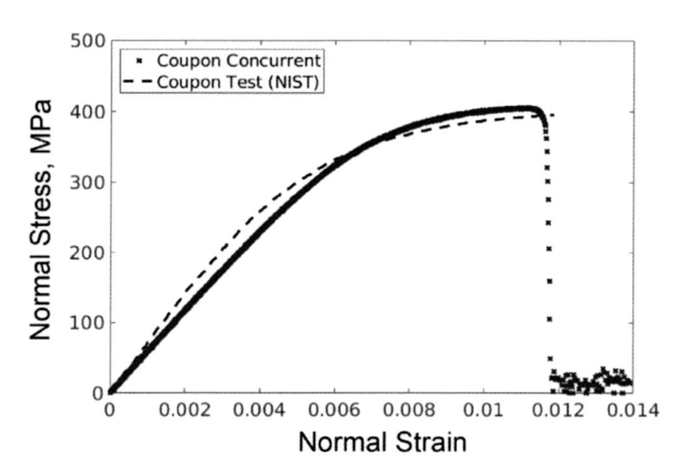

Fig. 56 Experimental (Coupon Test) and numerical (Coupon Concurrent) stress-strain curves for the tension test with fibers oriented at 10. *Reprinted from Gao, J., Shakoor, M., Domel, G., Merzkirch, M., Zhou, G., Zeng, D., Su, X., & Liu, W. K. (2020). Predictive multiscale modeling for unidirectional carbon fiber reinforced polymers. Composites Science and Technology, 186, 107922 with permission from Elsevier.*

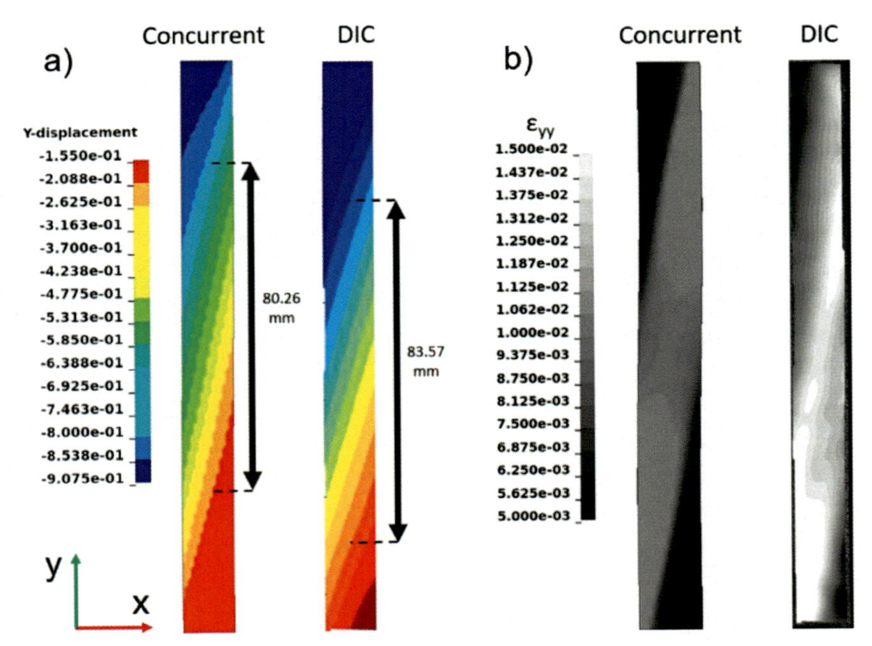

Fig. 57 Vertical direction (A) displacement and (B) strain fields comparing the digital image correlation measurements (DIC) with the simulation result (Concurrent). *Reprinted from Gao, J., Shakoor, M., Domel, G., Merzkirch, M., Zhou, G., Zeng, D., Su, X., & Liu, W. K. (2020). Predictive multiscale modeling for unidirectional carbon fiber reinforced polymers. Composites Science and Technology, 186, 107922 with permission from Elsevier.*

Fig. 58 Final macroscopic crack for (A) the simulation and (B) the experiment. *Reprinted from Gao, J., Shakoor, M., Domel, G., Merzkirch, M., Zhou, G., Zeng, D., Su, X., & Liu, W. K. (2020). Predictive multiscale modeling for unidirectional carbon fiber reinforced polymers. Composites Science and Technology, 186, 107922 with permission from Elsevier.*

the same structure simulation would have required at least 5000 times this computation time using an FE^2 approach.

Results for a dynamic three-point bending test including comparison with experiments can be found in Reference (Gao et al., 2020). Other tests have also been simulated in the context of the project with the Ford Motor Company, as well as other materials such as woven composites.

Applications to manufacturing have also been considered, especially regarding composites cutting processes (Shakoor et al., 2019b). Ductile damage simulations with void growth have been conducted as well (Shakoor et al., 2018b).

These different applications of the FE × SCA approach have shown that it can reduce computation times by several orders of magnitude as compared to an FE^2 approach. The SCA method smartly takes advantage of the redundancy of simulations where boundary conditions change but not the geometry. The intrinsic integration of periodic boundary conditions in the SCA method is also interesting for concurrent homogenization. This method is, however, limited regarding the prediction of local fields, especially for applications to manufacturing processes. For these applications, indeed, it is essential to predict the final state of the material, which may undergo local phenomena such as void formation.

5.3 Flows

This part summarizes a work published in Reference (Shakoor and Park, 2023) on homogenization for fluid mechanics problems in domains containing a great number of small obstacles. The first step is to extend the methodology presented in Reference (Blanco et al., 2017) to unsteady flows. This problem being nonlinear due to the auto-advection term in Navier-Stokes equations, the approach that is developed consists in an FE^2 scheme where the coarse scale problem and the fine scale problems are fully coupled during the whole simulation. In other words, it is not possible to pre-compute a permeability tensor or any other quantity that could circumvent the systematic solve of fine scale problems.

The multiscale model is presented in Paragraph 5.3.1 and its numerical implementation in Paragraph 5.3.2. An example of result is presented in Paragraph 5.3.3.

5.3.1 Multiscale model

The simulation domain is denoted $\Omega^M \subset \mathbb{R}^d$, with Dirichlet boundary conditions $\boldsymbol{v}_D^M$ on Γ_D^M and Neumann boundary conditions $\boldsymbol{t}_N^M$ on Γ_N^M. Regarding time, it is defined as $t \in [0, T]$, with $T \in \mathbb{R}^{+*}$. The variational equation to satisfy at the coarse scale is to find $\boldsymbol{v}^M(.,t) \in \mathcal{V}^M(t)$, $p^M \in L^2(\Omega^M)$, such that

$$
\begin{aligned}
&\int_{\Omega^M} (\boldsymbol{f}^M(\boldsymbol{x}, t).\ \delta\boldsymbol{v}^M(\boldsymbol{x}) + \boldsymbol{\sigma}^{M,dev}(\boldsymbol{x}, t)\colon \nabla_{\boldsymbol{x}}\delta\boldsymbol{v}^M(\boldsymbol{x}))\,\mathrm{d}\boldsymbol{x} \\
&\int_{\Omega^M} (-p^M(\boldsymbol{x})\nabla_{\boldsymbol{x}}.\ \delta\boldsymbol{v}^M(\boldsymbol{x}) - \delta p^M(\boldsymbol{x})\nabla_{\boldsymbol{x}}.\ \boldsymbol{v}^M(\boldsymbol{x}))\,\mathrm{d}\boldsymbol{x} \\
&\quad = \int_{\Gamma_N^M} \boldsymbol{t}_N^M(\boldsymbol{x}, t).\ \delta\boldsymbol{v}^M(\boldsymbol{x})\,\mathrm{d}\boldsymbol{x}, \\
&\qquad\qquad \forall\ \delta\boldsymbol{v}^M \in \mathcal{V}^M(t), \\
&\qquad\qquad \forall\ \delta p^M \in L^2(\Omega^M),
\end{aligned}
\tag{45}
$$

with functional spaces

$$
\begin{aligned}
\mathcal{V}^M(t) &= \{\boldsymbol{w} \in (H^1(\Omega^M))^d,\ \boldsymbol{w}(\boldsymbol{x}) \\
&\qquad = \boldsymbol{v}_D^M(\boldsymbol{x}, t),\ \forall\ (\boldsymbol{x}, t) \in \Gamma_D^M \times [0, T]\}, \\
H^1(\Omega^M) &= \{w \in L^2(\Omega^M),\ \nabla_{\boldsymbol{x}} w \in (L^2(\Omega^M))^d\}, \\
L^2(\Omega^M) &= \left\{w\colon \Omega^M \mapsto \mathbb{R},\ \int_{\Omega^M} w(\boldsymbol{x})^2\mathrm{d}\boldsymbol{x} < +\infty\right\}.
\end{aligned}
$$

This is standard and easier to understand if incompressible Newtonian flow is modeled directly at the coarse scale. Navier-Stokes equations then lead to:

$$
\begin{cases}
\boldsymbol{f}^M(\boldsymbol{x}, t) = \rho^M(\boldsymbol{x})\left(\frac{\partial \boldsymbol{v}^M}{\partial t}(\boldsymbol{x}, t) + \boldsymbol{v}^M(\boldsymbol{x}, t).\ \nabla_{\boldsymbol{x}}\boldsymbol{v}^M(\boldsymbol{x}, t)\right), \\
\boldsymbol{\sigma}^{M,dev}(\boldsymbol{x}, t) = 2\mu^M(\boldsymbol{x})\nabla_{\boldsymbol{x}}^{S,dev}\ \boldsymbol{v}^M(\boldsymbol{x}, t),
\end{cases}
$$

$$
\forall\ \boldsymbol{x} \in \Omega^M,\ \forall\ t \in [0, T],
$$

where the deviatoric part is

$$
\nabla^{S,dev}\boldsymbol{v}^M = \nabla^S\boldsymbol{v}^M - \frac{1}{d}\mathrm{tr}(\nabla^S\boldsymbol{v}^M)\boldsymbol{I},
$$

and the symmetric part

$$
\nabla^S\boldsymbol{v}^M = \frac{1}{2}(\nabla\boldsymbol{v}^M + \nabla^T\boldsymbol{v}^M).
$$

The problem with this direct approach is that the computational cost is going to blow up if there are obstacles of a small size compared to Ω^M, as element or cell size should be smaller than obstacle size. The alternative proposed herein is a multiscale approach, in which $\boldsymbol{f}^M(\boldsymbol{x}, t)$ and $\boldsymbol{\sigma}^{M,dev}(\boldsymbol{x}, t)$ are unknown functions of $\boldsymbol{v}^M(\boldsymbol{x}, t)$ and $\nabla_{\boldsymbol{x}}\boldsymbol{v}^M(\boldsymbol{x}, t)$, while p^M is still a Lagrange multiplier.

Consequently, a fine scale domain $\Omega^m = \Omega^m(\boldsymbol{x}) \subset \mathbb{R}^d$ is introduced at each point $\boldsymbol{x}$ of the coarse scale domain. Small obstacles are represented within this domain instead of the coarse scale domain. The boundary of those obstacles is $\Gamma_O^m = \Gamma_O^m(\boldsymbol{x}) \subset \partial\Omega^m(\boldsymbol{x})$, and Dirichlet boundary conditions $\boldsymbol{0}$ are imposed on Γ_O^m. The proposed multiscale approach is kinematic, which means averaging constraints are going to be imposed on the velocity field, while the expressions of $\boldsymbol{f}^M(\boldsymbol{x}, t)$ and $\boldsymbol{\sigma}^{M,dev}(\boldsymbol{x}, t)$ are going to be derived from a so-called PMVP (Blanco et al., 2016).

After different steps detailed in Reference (Shakoor and Park, 2023), it is possible to prove that the problem to solve for each fine scale domain is to find $(\boldsymbol{v}^m(.,t), p^m, \boldsymbol{\alpha}, \boldsymbol{\beta}) \in \mathcal{V}^m \times L^2(\Omega^m) \times \mathbb{R}^d \times \mathbb{R}^{d\times d}$, with $(\boldsymbol{v}^M(.,t), p^M) \in \mathcal{V}^M(t) \times L^2(\Omega^M)$ such that

$$\begin{aligned}
&\frac{1}{|\Omega^m|}\int_{\Omega^m}\begin{bmatrix} \rho^m(\boldsymbol{y})\left(\frac{\partial \boldsymbol{v}^m}{\partial t}(\boldsymbol{y}, t) + \boldsymbol{v}^m(\boldsymbol{y}, t).\ \nabla_{\boldsymbol{y}}\boldsymbol{v}^m(\boldsymbol{y}, t)\right). \\ \delta\boldsymbol{v}^m(\boldsymbol{y}) \\ +2\mu^m(\boldsymbol{y})\nabla_{\boldsymbol{y}}^{S,dev}\boldsymbol{v}^m(\boldsymbol{y}, t): \nabla_{\boldsymbol{y}}\delta\boldsymbol{v}^m(\boldsymbol{y}) \\ -p^m(\boldsymbol{y})\nabla_{\boldsymbol{y}}.\ \delta\boldsymbol{v}^m(\boldsymbol{y}) - \delta p^m(\boldsymbol{y})\nabla_{\boldsymbol{y}}.\ \boldsymbol{v}^m(\boldsymbol{y}, t) \end{bmatrix}\mathrm{d}\boldsymbol{y} \\
&- \delta\boldsymbol{\alpha}.\left(\frac{1}{|\Omega^m|}\int_{\Omega^m}\boldsymbol{v}^m(\boldsymbol{y}, t)\,\mathrm{d}\boldsymbol{y} - \boldsymbol{v}^M(\boldsymbol{x}, t)\right) \\
&- \delta\boldsymbol{\beta}: \left(\frac{1}{|\Omega^m|}\int_{\Omega^m}\nabla_{\boldsymbol{y}}\boldsymbol{v}^m(\boldsymbol{y}, t)\,\mathrm{d}\boldsymbol{y} - \nabla_{\boldsymbol{x}}\boldsymbol{v}^M(\boldsymbol{x}, t)\right) \\
&- \boldsymbol{\alpha}.\left(\frac{1}{|\Omega^m|}\int_{\Omega^m}\delta\boldsymbol{v}^m(\boldsymbol{y})\,\mathrm{d}\boldsymbol{y}\right) \\
&- \boldsymbol{\beta}: \left(\frac{1}{|\Omega^m|}\int_{\Omega^m}\nabla_{\boldsymbol{y}}\delta\boldsymbol{v}^m(\boldsymbol{y})\,\mathrm{d}\boldsymbol{y}\right) = 0, \\
&\forall\ (\delta\boldsymbol{v}^m, \delta p^m) \in \mathcal{V}^m \times L^2(\Omega^m), \\
&\forall\ (\delta\boldsymbol{\alpha}, \delta\boldsymbol{\beta}) \in \mathbb{R}^d \times \mathbb{R}^{d\times d},
\end{aligned} \tag{46}$$

with $\mathcal{V}^m = \{\boldsymbol{w} \in (H^1(\Omega^m))^d,\ \boldsymbol{w}(\boldsymbol{y}) = \boldsymbol{0},\ \forall\ \boldsymbol{y} \in \Gamma_O^m\}$. This variational form can directly be solved by the FE method. It features a first block corresponding to Navier-Stokes equations which are written at the fine scale with the incompressibility constraint on the velocity field and its variation. It also features constraints relating the velocity field and its gradient between the two scales, as well as their variations. These constraints are applied with Lagrange multipliers $\boldsymbol{\alpha}$ and $\boldsymbol{\beta}$, for which it can be shown that $\boldsymbol{f}^M = \boldsymbol{\alpha}$ and $\boldsymbol{\sigma}^{M,dev} = \boldsymbol{\beta} + p^M\boldsymbol{I}$. The definition of the functional space for the fine scale velocity field, moreover, includes boundary conditions for obstacles.

To summarize:

- The coarse scale problem in Equation (45) is similar to the well-known Navier-Stokes equations for unsteady incompressible Newtonian flows, except that it is written in terms of an unknown force per unit volume and an unknown deviatoric stress.
- Small obstacles are not represented directly within the coarse scale. Instead, they are embedded in fine scale domains, which are placed at each point of the coarse scale domain.
- For each fine scale domain, the boundary value problem in Equation (46) should be solved. Boundary conditions for this problem come from averaging constraints relating the fine scale velocity field and its gradient to their coarse scale counterparts.
- The force per unit volume and deviatoric stress for the coarse scale are computed from the solution of the fine scale problem.

5.3.2 Numerical method

The coarse scale FE mesh relies on the Taylor-Hood P2/P1 element. The variational formulation being mixed, this element adds the necessary ingredient to obtain discretization spaces that are compatible with the incompressibility constraint. Quadratic interpolation for the velocity and linear interpolation for the pressure leads to a non-negligible overhead regarding numerical integration, as six integration points are required per triangle in 2D, and 16 per tetrahedron in 3D. This is even more challenging as it means six fine scale domains per element in 2D, and 16 in 3D, as illustrated in Fig. 59.

To avoid increasing the complexity of this method, no stabilization and no turbulence model are used. The two-way coupling algorithm to simultaneously solve the coarse scale problem and all fine scale problems is effectively already quite complex. At instant t^{n+1}, this algorithm consists in:

1. Setup initial guesses $\boldsymbol{v}_0^M$, p_0^M for the coarse scale quasi-Newton algorithm from the last computed solutions
2. For each coarse scale quasi-Newton algorithm iteration $i \rightarrow i+1$
   (a) For each fine scale domain
      i. Setup initial guesses $\boldsymbol{v}_0^m$, p_0^m, $\boldsymbol{\alpha}_0$, $\boldsymbol{\beta}_0$ for the fine scale Newton-Raphson algorithm from the last computed solutions
      ii. For each fine scale Newton-Raphson algorithm iteration $j \rightarrow j+1$
         A. Approximate the fine scale problem and assemble and solve the associated linear system for iteration j
         B. Test for convergence and go back to 2(a)ii if necessary

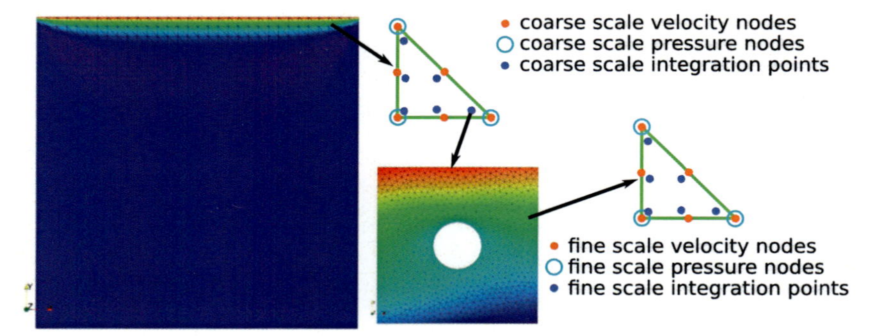

Fig. 59 FE^2 discretization with the Taylor-Hood P2/P1 pair showing the coarse scale mesh, one coarse scale element, one fine scale mesh, and one fine scale element. In this example, there are 2, 048 coarse scale elements, 2, 048 × 6 = 12, 288 coarse scale integration points and as many fine scale domains, 12, 288 × 1, 524 = 18, 726, 912 fine scale elements and 12, 288 × 3, 172 × 2 = 77, 955, 072 fine scale velocity degrees of freedom. Copyright (2022) Wiley. *Used with permission from Shakoor, M., & Park, C.H. (2023). Computational homogenization of unsteady flows with obstacles. International Journal for Numerical Methods in Fluids, 95(4), 499–527.*

iii. Get the converged $\boldsymbol{\alpha} = \boldsymbol{f}^M(\boldsymbol{x}, t^{n+1})$ and $\boldsymbol{\beta} = \boldsymbol{\sigma}^{M,dev}(\boldsymbol{x}, t^{n+1}) - p^M(\boldsymbol{x})\boldsymbol{I}$

(b) Assemble and solve the linear system associated to coarse scale iteration i

(c) Test for convergence and go back to 2 if necessary

(d) Get the converged coarse scale velocity-pressure solutions

In practice, advanced procedures such as line search or sub-stepping might be added to improve the convergence of this algorithm, especially regarding the solution of the fine scale problems. This algorithm has been implemented in *FEMS* with a distributed memory parallel paradigm. The fine scale solve is not implemented in parallel, as distributing the coarse scale mesh automatically distributes all fine scale problems.

5.3.3 Results

The multiscale approach has been compared for two examples to a single-scale approach where the heterogeneity is directly discretized within the coarse scale FE mesh in Reference (Shakoor and Park, 2023). The proposed method appears to be more accurate when the Reynolds number and/or the obstacle size decrease. This is interesting as this FE^2 approach also becomes more interesting from a computational cost point of view when the obstacle size decreases.

The example considered in the following is based on simulations in an L-shaped domain. It is the $1 \times 1\ m^2$ square split into a light phase and a dark phase as illustrated in Fig. 60A. Fine scale domain size is set to $l = 1$ mm and a circular obstacle of radius $R = 0.15$ mm is placed at the center of the fine scale domain in the light phase while the radius is changed to $R = 0.39$ mm in the dark phase. The obstacle area fraction, consequently, is of 7.1% in the light phase and 48% in the dark phase. The fine scale domain is systematically discretized with a mesh size of 0.04 mm.

The adapted mesh in Fig. 60A contains 3, 081 elements in the dark phase and 3, 149 in the light phase. For the dark phase, this means $3,081 \times 6 = 18,486$ coarse scale integration points and as many fine scale domains, $18,486 \times 844 = 15,602,184$ fine scale elements and $18,486 \times 1,852 \times 2 = 68,472,144$ fine scale degrees of freedom. For the light phase, there are $3,149 \times 6 = 18,894$ coarse scale integration points and as many fine scale domains, $18,894 \times 1,524 = 28,794,456$ fine scale elements and $18,894 \times 3,712 \times 2 = 119,863,536$ fine scale degrees of freedom. The total for the fine scale is, therefore, of 44, 396, 640 elements and 188, 335, 680 velocity degrees of freedom. It can be estimated that for this problem the number of elements in a single-scale mesh would narrow down to $\frac{844 \times 1 + 1,524 \times 3}{4} \times 1,000 \times 1,000 \approx 10^9$ and that the number of velocity

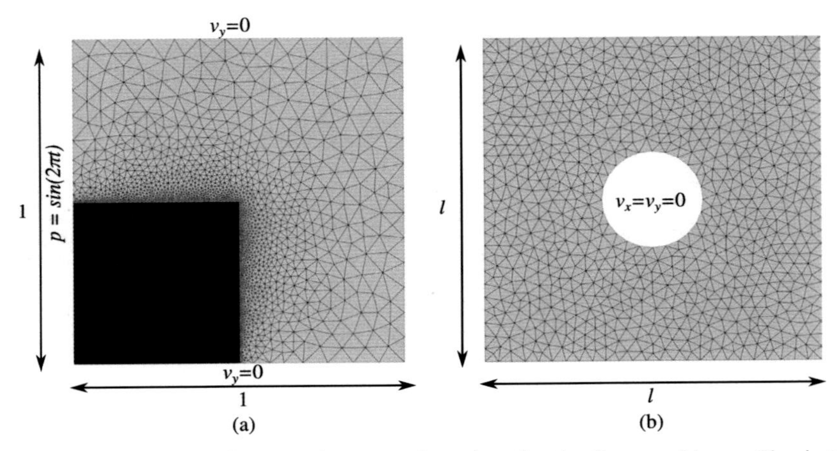

Fig. 60 Domain, boundary conditions and meshes for the flow problem with obstacles: (A) coarse scale domain, (B) fine scale domain with an obstacle. Lengths are in meters, velocities in meters par second and pressures in Pa. Copyright (2022) Wiley. *Used with permission from Shakoor, M., & Park, C. H. (2023). Computational homogenization of unsteady flows with obstacles. International Journal for Numerical Methods in Fluids, 95(4), 499–527.*

degrees of freedom would be of the same order. There is hence a reduction by a ratio of at least ten of the computational complexity with the multiscale method.

The traction vector is defined as $\boldsymbol{t}_N^M(\boldsymbol{x}, t) = (\sin(2\pi t), 0)$ to obtain a fluctuating pressure on the left side of the domain, while top and bottom sides are defined as non-penetrating walls. The fluid mass density is set to $1\,\text{KG}\,\text{M}^3$, gravity is neglected and the fluid dynamic viscosity is set to $1\,\mu\,\text{Pa s}$. The problem is solved with a time step of $\Delta t = 0.03$ s for 100 time increments.

The results are shown in Fig. 61 for the velocity field in the horizontal direction. The velocity transition at the coarse scale between the two phases is quite sharp.

Fine scale results at different locations at $t = 2.40$ s are shown in Fig. 62. The comparison with Fig. 61D shows that the different flow orientations

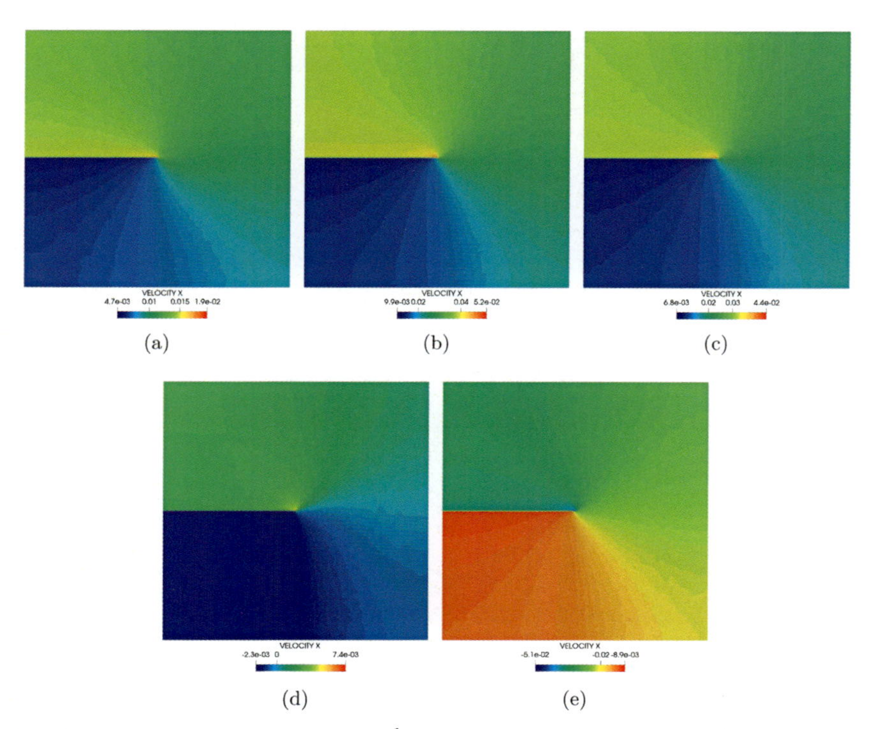

Fig. 61 Velocity field evolution $\boldsymbol{v}_x$ (mm^1) at: (A) $t = 2.10$ s, (B) $t = 2.25$ s, (C) $t = 2.40$ s, (D) $t = 2.55$ s, (E) $t = 2.70$ s. Copyright (2022) Wiley. *Used with permission from Shakoor, M., & Park, C. H. (2023). Computational homogenization of unsteady flows with obstacles. International Journal for Numerical Methods in Fluids, 95(4), 499–527.*

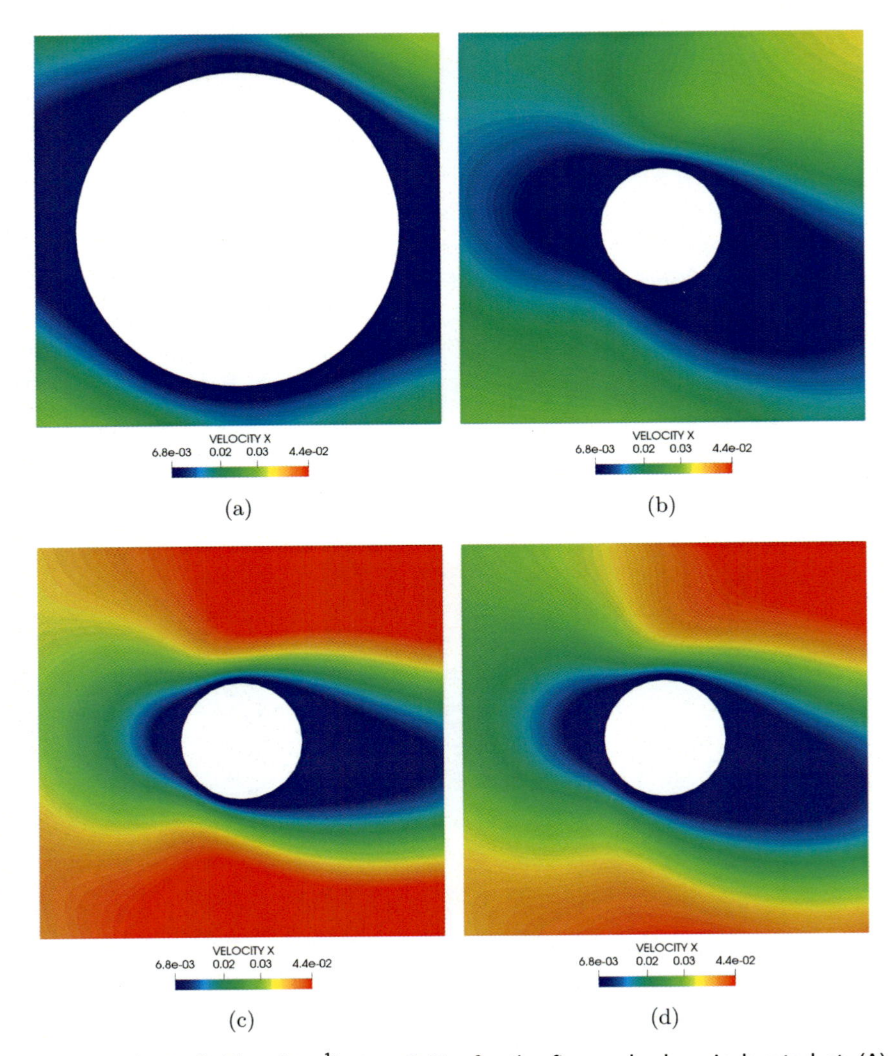

Fig. 62 Velocity field $\boldsymbol{v}_x$ (m s[1]) at $t = 2.40$ s for the fine scale domain located at: (A) $x = 0.45$ m, $y = 0.45$ m, (B) $x = 0.55$ m, $y = 0.45$ m, (C) $x = 0.45$ m, $y = 0.55$ m, (D) $x = 0.55$ m, $y = 0.55$ m. Copyright (2022) Wiley. *Used with permission from Reference (Shakoor and Park, 2023).*

are correctly transferred from the coarse scale to the fine scale. As shown in Fig. 62C, in the boundary layer on the light phase side, very large velocities can be observed at both scales. This demonstrates the capabilities of the multiscale method to predict the flow simultaneously at both scales.

5.4 Conclusions

Contributions to multiscale modeling have been presented in this section for the problems related to flows with obstacles and fracture mechanics from Sections 2 and 3. In both cases, the physical phenomena at hand being highly nonlinear and transient, a concurrent FE^2 homogenization scheme has been developed. The scale of the simulation domain is hence separated from the scale of the heterogeneity, which is discretized in a separate fictitious domain. Discretizing the heterogeneity directly at the scale of the simulation domain, indeed, reveals to be impossible for instance for composite structures or injection molds containing millions of fibers.

For fracture mechanics, the presented approach is an FE $\times$ SCA scheme. It integrates the SCA method presented in Subsection 4.2. This method has been extended for modeling the behavior of a unidirectional long fiber composite material with an elasto-plastic matrix and damage criteria in both matrix and fibers. Damage has been regularized at both the fine and coarse scales to avoid any mesh dependency. Comparison with experimental results demonstrates the predictive capability of the method at the macroscale, both locally and globally. At the microscale, the SCA method has the drawback that it cannot predict the final microstructure of the material. It is nevertheless very advantageous regarding computational complexity, with a reduction ratio estimated at more than 5000 compared to an FE^2 approach.

For fluid mechanics, the nonlinear and unsteady case has not been studied extensively in the literature, especially regarding FE^2 approaches. The first step has hence been to develop a PMVP to relate the coarse scale of the domain to the fine scale of the obstacles. The FE^2 implementation with discretization spaces compatible with the incompressibility constraint has then been described and deployed for domains containing 1,000,000 obstacles. It is estimated that the complexity of this computation compared to a single-scale approach is reduced by a ratio of at least 10. This ratio increases as obstacle size decreases.

6. Conclusions

Research activities on numerical modeling of heterogeneous materials and domains have been presented. The applications that have often been targeted in this work are related to manufacturing and service life of long fiber composite structures. These structures can have a large size of several meters but contain fibers of a diameter of a few micrometers. The ambition and objective of the research activities conducted in this work is, therefore, to push towards more predictive models for highly nonlinear phenomena at a very fine scale.

The phenomena considered in this work have included steady and unsteady flows featuring one or two phases, and which occur during composites manufacturing. They have also included fracture under static loading conditions, which can occur during the service life of composite structures.

For the various phenomena that have been modeled, this work's contributions have mainly focused on numerical methods. Both liquid/gas interfaces in flows and cracks in structures, hence, have been discretized and modeled using advanced numerical methods. These methods have consisted in Finite Element (FE) mesh generation and adaption, Level-Set (LS) functions and phase-fields. The numerical method based on the Fast Fourier Transform (FFT) has also been introduced to eliminate meshing difficulties.

Accelerating computations has revealed to be vital, first through parallel computing and then through model order reduction. The Reduced Order Models (ROMs) developed in this work, in particular, show that it is possible to take advantage of the partial redundancy of some simulations in order to reduce their cost by several orders of magnitude. This is especially interesting for integration in an FE^2-type multiscale scheme.

This work's main contributions are:

- A method for FE mesh generation with explicit discretization of interfaces from LS functions, including LS functions computed from images.
- An LS method with quadratic FE interpolation and anisotropic mesh adaption for two-phase flows modeling. A balanced surface tension model thanks to Residual-Based Variational MultiScale (RBVMS) stabilization. A study on the choice of coupling (weak or strong) and setting (Eulerian or Lagrangian) for modeling two-phase flows depending on the capillary number.
- An analysis of the parasitic oscillations present in fields computed with the FFT-based numerical method, and their causes. An alternative composite voxels method based on LS functions in order to ease the implementation for geometries provided as images.
- An investigation of a phase-field model for heterogeneous materials implemented using the FFT-based numerical method revealing abnormal diffusion of damage. The elimination of this diffusion with a revisited approach.

- The extension of a model order reduction approach, the Self-consistent Clustering Analysis (SCA) method to various material behaviors, especially in finite strains.
- The development of a ROM relying on an autoencoder for nonlinear compression and an Artificial Neural Network (ANN) for prediction.
- The deployment of an FE $\times$ ROM-type multiscale method relying on SCA for modeling the fracture of a composite structure, and its validation using experimental measurements.
- The development of a multiscale method for modeling inertial flows in heterogeneous media containing a large number of small obstacles. The validation of this method and its FE^2 implementation using results obtained with a single-scale method.

Future work should focus on extending the approaches presented in this work to more complex and realistic problems. This should include multiscale modeling of two-phase flows with an FE $\times$ ROM approach and applications of this approach to real composites manufacturing processes. Another research track should be improving the nonlinear ROM for fracture and integrating it in an FE $\times$ ROM approach to predict both the static and fatigue failure of composite structures and develop digital twins.

References

Abadi, M., Barham, P., Chen, J., Chen, Z., Davis, A., Dean, J., ... Zheng, X. (2016). Tensorflow: A system for large-scale machine learning. In: *12th USENIX Symposium on Operating Systems Design and Implementation, OSDI'16* (USA), pp. 265–283, USENIX Association.

Abgrall, R., Beaugendre, H., & Dobrzynski, C. (2014). An immersed boundary method using unstructured anisotropic mesh adaptation combined with level-sets and penalization techniques. *Journal of Computational Physics, 257*, 83–101.

Ahmed, S. E., San, O., Rasheed, A., & Iliescu, T. (2021). Nonlinear proper orthogonal decomposition for convection-dominated flows. *Physics of Fluids, 33*(12), 121702.

Arsigny, V., Fillard, P., Pennec, X., & Ayache, N. (2006). Log-Euclidean metrics for fast and simple calculus on diffusion tensors. *Magnetic Resonance in Medicine, 56*(2), 411–421.

Bažant, Z. P., & Jirásek, M. (2002). Nonlocal integral formulations of plasticity and damage: Survey of progress. *Journal of Engineering Mechanics, 128*(11), 1119–1149.

Bazilevs, Y., Calo, V., Cottrell, J., Hughes, T., Reali, A., & Scovazzi, G. (2007). Variational multiscale residual-based turbulence modeling for large eddy simulation of incompressible flows. *Computer Methods in Applied Mechanics and Engineering, 197*(1–4), 173–201.

Beliaev, A. Y., & Kozlov, S. M. (1996). Darcy equation for random porous media. *Communications on Pure and Applied Mathematics, 49*(1), 1–34.

Bessa, M., Bostanabad, R., Liu, Z., Hu, A., Apley, D. W., Brinson, C., ... Liu, W. K. (2017). A framework for data-driven analysis of materials under uncertainty: Countering the curse of dimensionality. *Computer Methods in Applied Mechanics and Engineering, 320*, 633–667.

Besson, J. (2010). Continuum models of ductile fracture: A review. *International Journal of Damage Mechanics, 19*(1), 3–52.

Blanco, P. J., Sánchez, P. J., de Souza Neto, E. A., & Feijóo, R. A. (2016). Variational foundations and generalized unified theory of RVE-based multiscale models. *Archives of Computational Methods in Engineering, 23*(2), 191–253.
Blanco, P. J., Clausse, A., & Feijóo, R. A. (2017). Homogenization of the Navier-Stokes equations by means of the multi-scale virtual power principle. *Computer Methods in Applied Mechanics and Engineering, 315*, 760–779.
Bodaghi, M., Gnaba, I., Legrand, X., Soulat, D., Wang, P., Deléglise-Lagardère, M., & Park, C. H. (2021). In-plane permeability changes of plain weave glass fabric induced by tufting. *Advanced Composite Materials, 30*(5), 478–494.
Bourdin, B., Francfort, G. A., & Marigo, J. (2008). The variational approach to fracture. *Journal of Elasticity, 91*, 5–148.
Brackbill, J., Kothe, D., & Zemach, C. (1992). A continuum method for modeling surface tension. *Journal of Computational Physics, 100*(2), 335–354.
Bui, C., Dapogny, C., & Frey, P. (2012). An accurate anisotropic adaptation method for solving the level set advection equation. *International Journal for Numerical Methods in Fluids, 70*(7), 899–922.
Chaturantabut, S., & Sorensen, D. C. (2010). Nonlinear model reduction via discrete empirical interpolation. *SIAM Journal on Scientific Computing, 32*(5), 2737–2764.
Chen, Y., Vasiukov, D., Gélébart, L., & Park, C. H. (2019). A FFT solver for variational phase-field modeling of brittle fracture. *Computer Methods in Applied Mechanics and Engineering, 349*, 167–190.
Chinesta, F., Leygue, A., Bordeu, F., Aguado, J. V., Cueto, E., Gonzalez, D., ... Ammar, A., Huerta, A. (2013). PGD-based computational vademecum for efficient design, optimization and control. *Archives of Computational Methods in Engineering, 20*(1), 31–59.
Chollet, F. et al. (2015). Keras. ⟨https://github.com/fchollet/keras⟩.
Coulaud, O., & Loseille, A. (2016). Very high order anisotropic metric-based mesh adaptation in 3D. *Procedia Engineering, 163*, 353–365.
Dapogny, C., Dobrzynski, C., & Frey, P. (2014). Three-dimensional adaptive domain remeshing, implicit domain meshing, and applications to free and moving boundary problems. *Journal of Computational Physics, 262*, 358–378.
Denner, F., Evrard, F., Serfaty, R., & van Wachem, B. G. (2017). Artificial viscosity model to mitigate numerical artefacts at fluid interfaces with surface tension. *Computers & Fluids, 143*, 59–72.
Di Pietro, D. A., Lo Forte, S., & Parolini, N. (2006). Mass preserving finite element implementations of the level set method. *Applied Numerical Mathematics, 56*(9), 1179–1195.
Dobrzynski, C., & Frey, P. (2008). Anisotropic delaunay mesh adaptation forunsteady simulations. In R. V. Garimella (Ed.). *Proceedings of the 17th international meshing roundtable* (pp. 177–194). Berlin, Heidelberg: Springer.
Doitrand, A., Fagiano, C., Irisarri, F. X., & Hirsekorn, M. (2015). Comparison between voxel and consistent meso-scale models of woven composites. *Composites Part A: Applied Science and Manufacturing, 73*, 143–154.
Dowell, E., & Hall, K. (2001). Modeling of fluid-structure interaction. *Annual Review of Fluid Mechanics, 33*, 445–490.
Eloh, K. S., Jacques, A., & Berbenni, S. (2019). Development of a new consistent discrete green operator for FFT-based methods to solve heterogeneous problems with eigenstrains. *International Journal of Plasticity, 116*, 1–23.
Enright, D., Losasso, F., & Fedkiw, R. (2005). A fast and accurate semi-Lagrangian particle level set method. *Computers & Structures, 83*(6–7), 479–490.
Feyel, F. (1999). Multiscale FE2 elastoviscoplastic analysis of composite structures. *Computational Materials Science, 16*(1–4), 344–354.

Francois, M. M., Cummins, S. J., Dendy, E. D., Kothe, D. B., Sicilian, J. M., & Williams, M. W. (2006). A balanced-force algorithm for continuous and sharp interfacial surface tension models within a volume tracking framework. *Journal of Computational Physics, 213*(1), 141–173.

Frigo, M., & Johnson, S. (2005). The design and implementation of FFTW3. *Proceedings of the IEEE, 93*(2), 216–231.

Fukami, K., Hasegawa, K., Nakamura, T., Morimoto, M., & Fukagata, K. (2021). Model order reduction with neural networks: Application to laminar and turbulent flows. *SN Computer Science, 2*, 467.

Gao, J., Shakoor, M., Jinnai, H., Kadowaki, H., Seta, E., & Liu, W. K. (2019). An inverse modeling approach for predicting filled rubber performance. *Computer Methods in Applied Mechanics and Engineering, 357*, 112567.

Gao, J., Shakoor, M., Domel, G., Merzkirch, M., Zhou, G., Zeng, D., ... Liu, W. K. (2020). Predictive multiscale modeling for unidirectional carbon fiber reinforced polymers. *Composites Science and Technology, 186*, 107922.

Geers, M., Kouznetsova, V., & Brekelmans, W. (2010). Multi-scale computational homogenization: Trends and challenges. *Journal of Computational and Applied Mathematics, 234*(7), 2175–2182.

Geuzaine, C., & Remacle, J.-F. (2009). Gmsh: A 3-D finite element mesh generator with built-in pre- and post-processing facilities. *International Journal for Numerical Methods in Engineering, 79*(11), 1309–1331.

Goury, O., Amsallem, D., Bordas, S. P. A., Liu, W. K., & Kerfriden, P. (2016). Automatised selection of load paths to construct reduced-order models in computational damage micromechanics: From dissipation-driven random selection to Bayesian optimization. *Computational Mechanics, 58*(2), 213–234.

Gruau, C., & Coupez, T. (2005). 3D tetrahedral, unstructured and anisotropic mesh generation with adaptation to natural and multidomain metric. *Computer Methods in Applied Mechanics and Engineering, 194*(48–49), 4951–4976.

Gurson, A. L. (1977). Continuum theory of ductile rupture by void nucleation and growth: Part I—yield criteria and flow rules for porous ductile media. *Journal of Engineering Materials and Technology, 99*(1), 2–15.

Hirt, C., & Nichols, B. (1981). Volume of fluid (VOF) method for the dynamics of free boundaries. *Journal of Computational Physics, 39*(1), 201–225.

Hitchcock, F. L. (1927). The expression of a tensor or a polyadic as a sum of products. *Journal of Mathematics and Physics, 6*(1–4), 164–189.

Hou, G., Wang, J., & Layton, A. (2012). Numerical methods for fluid-structure interaction - A review. *Communications in Computational Physics, 12*(2), 337–377.

Hu, Z.-H., & Song, Y.-L. (2009). Dimensionality reduction and reconstruction of data based on autoencoder network. *Journal of Electronics & Information Technology, 31*(5), 1189–1192.

Huang, W. (2005). Metric tensors for anisotropic mesh generation. *Journal of Computational Physics, 204*(2), 633–665.

Hysing, S., Turek, S., Kuzmin, D., Parolini, N., Burman, E., Ganesan, S., & Tobiska, L. (2009). Quantitative benchmark computations of two-dimensional bubble dynamics. *International Journal for Numerical Methods in Fluids, 60*(11), 1259–1288.

Jeulin, D. (2020). Towards crack paths simulations in media with a random fracture energy. *International Journal of Solids and Structures, 184*, 279–286.

Kabel, M., Böhlke, T., & Schneider, M. (2014). Efficient fixed point and Newton-Krylov solvers for FFT-based homogenization of elasticity at large deformations. *Computational Mechanics, 54*(6), 1497–1514.

Kabel, M., Merkert, D., & Schneider, M. (2015). Use of composite voxels in FFT-based homogenization. *Computer Methods in Applied Mechanics and Engineering, 294*, 168–188.

Kabel, M., Fliegener, S., & Schneider, M. (2016). Mixed boundary conditions for FFT-based homogenization at finite strains. *Computational Mechanics, 57*(2), 193–210.

Kafka, O. L., Yu, C., Shakoor, M., Liu, Z., Wagner, G. J., & Liu, W. K. (2018). Data-driven mechanistic modeling of influence of microstructure on high-cycle fatigue life of nickel titanium. *JOM, 70*(7), 1154–1158.

M. Lagardère, et al., Skate v2. ⟨https://youtu.be/QWRjgJEI1ao⟩, 2019.

Laug, P., & Borouchaki, H. (2013). Construction d'un champ continu de métriques. *Comptes Rendus Mathematique, 351*(15–16), 639–644.

Leuschner, M., & Fritzen, F. (2018). Fourier-Accelerated Nodal Solvers (FANS) for homogenization problems. *Computational Mechanics, 62*, 359–392.

Liang, Y., Lee, H., Lim, S., Lin, W., Lee, K., & Wu, C. (2002). Proper orthogonal decomposition and its applications—Part I: Theory. *Journal of Sound and Vibration, 252*(3), 527–544.

Lieu, T., Farhat, C., & Lesoinne, M. (2006). Reduced-order fluid/structure modeling of a complete aircraft configuration. *Computer Methods in Applied Mechanics and Engineering, 195*(41–43), 5730–5742.

Liu, Z., Bessa, M., & Liu, W. K. (2016). Self-consistent clustering analysis: An efficient multi-scale scheme for inelastic heterogeneous materials. *Computer Methods in Applied Mechanics and Engineering, 306*, 319–341.

Liu, Y., Straumit, I., Vasiukov, D., Lomov, S. V., & Panier, S. (2017). Prediction of linear and non-linear behavior of 3D woven composite using mesoscopic voxel models reconstructed from X-ray micro-tomography. *Composite Structures, 179*, 568–579.

Liu, Z., Fleming, M., & Liu, W. K. (2018). Microstructural material database for self-consistent clustering analysis of elastoplastic strain softening materials. *Computer Methods in Applied Mechanics and Engineering, 330*, 547–577.

Loseille, A., & Alauzet, F. (2011a). Continuous mesh framework part I: Well-posed continuous interpolation error. *SIAM J. Numer. Anal. 49*(1), 38–60.

Loseille, A., & Alauzet, F. (2011b). Continuous mesh framework part II: Validations and applications. *SIAM Journal on Numerical Analysis, 49*(1), 61–86.

Ma, X., Shakoor, M., Vasiukov, D., Lomov, S. V., & Park, C. H. (2021). Numerical artifacts of Fast Fourier Transform solvers for elastic problems of multi-phase materials: Their causes and reduction methods. *Computational Mechanics, 67*, 1667–1683.

Ma, X., Chen, Y., Shakoor, M., Vasiukov, D., Lomov, S. V., & Park, C. H. (2023). Simplified and complete phase-field fracture formulations for heterogeneous materials and their solution using a Fast Fourier Transform based numerical method. *Engineering Fracture Mechanics, 279*, 109049.

Ma, X. (2022). The elastic and damage modeling of heterogeneous materials based on the fast Fourier transform. PhD thesis, IMT Nord Europe.

Matouš, K., Geers, M. G., Kouznetsova, V. G., & Gillman, A. (2017). A review of predictive nonlinear theories for multiscale modeling of heterogeneous materials. *Journal of Computational Physics, 330*, 192–220.

Maurer, C., Qi, R., & Raghavan, V. (2003). A linear time algorithm for computing exact Euclidean distance transforms of binary images in arbitrary dimensions. *IEEE Transactions on Pattern Analysis and Machine Intelligence, 25*(2), 265–270.

May, D., Syerko, E., Schmidt, T., Binetruy, C., Silva, L. R. D., Lomov, S., & Advani, S. (2021). *Benchmarking virtual permeability predictions of real fibrous microstructure. American Society for Composites 2021*. Destech Publications, Inc.,.

Mehdikhani, M., Gorbatikh, L., Verpoest, I., & Lomov, S. V. (2019). Voids in fiber-reinforced polymer composites: A review on their formation, characteristics, and effects on mechanical performance. *Journal of Composite Materials, 53*(12), 1579–1669.

Melro, A., Camanho, P., Andrade Pires, F., & Pinho, S. (2013). Micromechanical analysis of polymer composites reinforced by unidirectional fibres: Part I - Constitutive modelling. *International Journal of Solids and Structures, 50*(11–12), 1897–1905.

Michaud, V. (2016). A review of non-saturated resin flow in liquid composite moulding processes. *Transport in Porous Media, 115*(3), 581–601.
Michel, J., & Suquet, P. (2003). Nonuniform transformation field analysis. *International Journal of Solids and Structures, 40*(25), 6937–6955.
Miehe, C., Welschinger, F., & Hofacker, M. (2010a). Thermodynamically consistent phase-field models of fracture: Variational principles and multi-field FE implementations. *International Journal for Numerical Methods in Engineering, 83*(10), 1273–1311.
Miehe, C., Hofacker, M., & Welschinger, F. (2010b). A phase field model for rate-independent crack propagation: Robust algorithmic implementation based on operator splits. *Computer Methods in Applied Mechanics and Engineering, 199*(45–48), 2765–2778.
Mori, T., & Tanaka, K. (1973). Average stress in matrix and average elastic energy of materials with misfitting inclusions. *Acta Metallurgica, 21*(5), 571–574.
Moulinec, H., & Suquet, P. (1994). A fast numerical method for computing the linear and nonlinear mechanical properties of composites. *Comptes rendus de l'Académie des sciences. Série II, Mécanique, physique, chimie, astronomie, 318*(11), 1417–1423.
Moulinec, H., & Suquet, P. (1998). A numerical method for computing the overall response of nonlinear composites with complex microstructure. *Computer Methods in Applied Mechanics and Engineering, 157*(1–2), 69–94.
Murata, T., Fukami, K., & Fukagata, K. (2020). Nonlinear mode decomposition with convolutional neural networks for fluid dynamics. *Journal of Fluid Mechanics, 882*, A13.
Nikishkov, Y., Airoldi, L., & Makeev, A. (2013). Measurement of voids in composites by X-ray computed tomography. *Composites Science and Technology, 89*, 89–97.
Osher, S., & Sethian, J. A. (1988). Fronts propagating with curvature-dependent speed: Algorithms based on Hamilton-Jacobi formulations. *Journal of Computational Physics, 79*(1), 12–49.
Pant, P., Doshi, R., Bahl, P., & Barati Farimani, A. (2021). Deep learning for reduced order modelling and efficient temporal evolution of fluid simulations. *Physics of Fluids, 33*(10), 107101.
Park, C. H., Lebel, A., Saouab, A., Bréard, J., & Lee, W. I. (2011). Modeling and simulation of voids and saturation in liquid composite molding processes. *Composites: Part A, 42*, 658–668.
Parvathaneni, K. K. (2020). Characterization and multiscale modeling of textile reinforced composite materials considering manufacturing defects. PhD thesis, IMT Nord Europe.
Peerlings, R. H. J., De Borst, R., Brekelmans, W. A. M., & De Vree, J. H. P. (1996). Gradient enhanced damage for quasi-brittle materials. *International Journal for Numerical Methods in Engineering, 39*(19), 3391–3403.
Plaut, E. (2018). From principal subspaces to principal components with linear autoencoders, arXiv.
Pochet, F., Hillewaert, K., Geuzaine, P., Remacle, J.-F., & Marchandise, É. (2013). A 3D strongly coupled implicit discontinuous Galerkin level set-based method for modeling two-phase flows. *Computers & Fluids, 87*, 144–155.
Quan, D.-L., Toulorge, T., Marchandise, E., Remacle, J.-F., & Bricteux, G. (2014). Anisotropic mesh adaptation with optimal convergence for finite elements using embedded geometries. *Computer Methods in Applied Mechanics and Engineering, 268*, 65–81.
Rasthofer, U., Henke, F., Wall, W., & Gravemeier, V. (2011). An extended residual-based variational multiscale method for two-phase flow including surface tension. *Computer Methods in Applied Mechanics and Engineering, 200*(21–22), 1866–1876.
Ryckelynck, D. (2005). A priori hyperreduction method: An adaptive approach. *Journal of Computational Physics, 202*(1), 346–366.
Safi, M. A., Prasianakis, N., & Turek, S. (2017). Benchmark computations for 3D two-phase flows: A coupled lattice Boltzmann-level set study. *Computers and Mathematics with Applications, 73*(3), 520–536.

Schindelin, J., Arganda-Carreras, I., Frise, E., Kaynig, V., Longair, M., Pietzsch, T., ... Cardona, A. (2012). Fiji: An open-source platform for biological-image analysis. *Nature Methods, 9*(7), 676–682.

Schneider, M., Merkert, D., & Kabel, M. (2017). FFT-based homogenization for microstructures discretized by linear hexahedral elements. *International Journal for Numerical Methods in Engineering, 109*, 1461–1489.

Sethian, J. A., & Vladimirsky, A. (2000). Fast methods for the Eikonal and related Hamilton- Jacobi equations on unstructured meshes. *Proceedings of the National Academy of Sciences of the United States of America, 97*(11), 5699–5703.

Shakoor, M. & Delbeke, L. (2021). Topological optimization of triangulations on graphics processing units: A bad idea? In: *World congress on computational mechanics and European congress on computational methods in applied sciences and engineering (WCCM-ECCOMAS)*, (Virtual Congress).

Shakoor, M. & Park, C. H. (2019). Simulation de la migration de bulles dans le milieu fibreux pendant l'imprégnation de composites. In: *Journées scientifiques du GdR Mise en œuvre des composites et propriétés induites (GDR Week - MIC)*, (Tarbes (France).

Shakoor, M., & Park, C. H. (2021a). A higher-order finite element method with unstructured anisotropic mesh adaption for two phase flows with surface tension. *Computers & Fluids, 230*, 105154.

Shakoor, M. & Park, C. H. (2021b). Modélisation numérique de la migration de porosités pendant la mise en œuvre des composites. In: *Journées Nationales sur les Composites (JNC)*, (Conférence virtuelle).

Shakoor, M. & Park, C. H. (2021c). Adaptive higher-order finite element modeling of multiphase flow. In: *World congress on computational mechanics and European congress on computational methods in applied sciences and engineering (WCCM-ECCOMAS)*, (Virtual Congress).

Shakoor, M., & Park, C. H. (2023). Computational homogenization of unsteady flows with obstacles. *International Journal for Numerical Methods in Fluids, 95*(4), 499–527.

Shakoor, M., Scholtes, B., Bouchard, P.-O., & Bernacki, M. (2015a). An efficient and parallel level set reinitialization method - Application to micromechanics and microstructural evolutions. *Applied Mathematical Modelling, 39*(23–24), 7291–7302.

Shakoor, M., Bernacki, M., & Bouchard, P.-O. (2015b). A new body-fitted immersed volume method for the modeling of ductile fracture at the microscale: Analysis of void clusters and stress state effects on coalescence. *Engineering Fracture Mechanics, 147*, 398–417.

Shakoor, M., Bouchard, P.-O., & Bernacki, M. (2017a). An adaptive level-set method with enhanced volume conservation for simulations in multiphase domains. *International Journal for Numerical Methods in Engineering, 109*(4), 555–576.

Shakoor, M., Buljac, A., Neggers, J., Hild, F., Morgeneyer, T. F., Helfen, L., ... Bouchard, P.-O. (2017b). On the choice of boundary conditions for micromechanical simulations based on 3D imaging. *International Journal of Solids and Structures, 112*, 83–96.

Shakoor, M., Bernacki, M., & Bouchard, P.-O. (2018a). Ductile fracture of a metal matrix composite studied using 3D numerical modeling of void nucleation and coalescence. *Engineering Fracture Mechanics, 189*, 110–132.

Shakoor, M., Yu, C., Kafka, O. L., & Liu, W. K. (2018b). A multiscale computational homogenization theory with data-driven model reduction for the prediction of ductile damage. In: *World congress on computational mechanics (WCCM)*, (New York, NY, USA).

Shakoor, M., Kafka, O. L., Yu, C., & Liu, W. K. (2019a). Data science for finite strain mechanical science of ductile materials. *Computational Mechanics, 64*(1), 33–45.

Shakoor, M., Gao, J., Liu, Z., & Liu, W. K. (2019b). A data-driven multiscale theory for modeling damage and fracture of composite materials. 129 of Lecture Notes in Computational Science and Engineering In M. Griebel, & M. Schweitzer (Eds.). *Meshfree Methods for Partial Differential Equations IX. IWMMPDE 2017* (pp. 135–148). Cham: Springer 129 of Lecture Notes in Computational Science and Engineering.

Shakoor, M., TrejoNavas, V. M., PinoMuñoz, D., Bernacki, M., & Bouchard, P.-O. (2019c). Computational methods for ductile fracture modeling at the microscale. *Archives of Computational Methods in Engineering, 26*(4), 1153–1192.

Shakoor, M. (2021). FEMS - A mechanics-oriented finite element modeling software. *Computer Physics Communications, 260*, 107729.

Shakoor, M. (2022). FEMS – finite element modeling software. ⟨https://hal.science/hal-03781711⟩.

Shinde, K., Itier, V., Mennesson, J., Vasiukov, D., & Shakoor, M. (2023). Dimensionality reduction through convolutional autoencoders for fracture patterns prediction. *Applied Mathematical Modelling, 114*, 94–113.

Simo, J. (1992). Algorithms for static and dynamic multiplicative plasticity that preserve the classical return mapping schemes of the infinitesimal theory. *Computer Methods in Applied Mechanics and Engineering, 99*(1), 61–112.

Skjetne, E., & Auriault, J. L. (1999a). New insights on steady, non-linear flow in porous media, European. *Journal of Mechanics - B/Fluids, 18*(1), 131–145.

Skjetne, E., & Auriault, J. L. (1999b). High-velocity laminar and turbulent flow in porous media. *Transport in Porous Media, 36*(2), 131–147.

Sussman, M., Smereka, P., & Osher, S. (1994). A level set approach for computing solutions to incompressible two-phase flow. *Journal of Computational Physics, 114*(1), 146–159.

Sussman, M., Fatemi, E., Smereka, P., & Osher, S. (1998). An improved level set method for incompressible two-phase flows. *Computers & Fluids, 27*(5–6), 663–680.

Sussman, M. (2005). A parallelized, adaptive algorithm for multiphase flows in general geometries. *Computers & Structures, 83*(6–7), 435–444.

Syerko, E., Schmidt, T., May, D., Binetruy, C., Advani, S. G., Lomov, S., ... Vorobyev, R. (2023). Benchmark exercise on image-based permeability determination of engineering textiles: Microscale predictions. *Composites Part A: Applied Science and Manufacturing, 167*, 107397.

Vernet, N., Ruiz, E., Advani, S., Alms, J., Aubert, M., Barburski, M., ... Ziegmann, G. (2014). Experimental determination of the permeability of engineering textiles: Benchmark II. *Composites Part A: Applied Science and Manufacturing, 61*, 172–184.

Wang, Y., Yao, H., & Zhao, S. (2016). Auto-encoder based dimensionality reduction. *Neurocomputing, 184*, 232–242.

Wen, T., & Zhang, Z. (2018). Deep convolution neural network and autoencoders-based unsupervised feature learning of eeg signals. *IEEE Access, 6*, 25399–25410.

Willot, F. (2015). Fourier-based schemes for computing the mechanical response of composites with accurate local fields. *Comptes Rendus Mécanique, 343*(3), 232–245.

Wu, J.-Y., & Huang, Y. (2020). Comprehensive implementations of phase-field damage models in Abaqus. *Theoretical and Applied Fracture Mechanics, 106*(December 2019), 102440.

Yvonnet, J., & He, Q. C. (2007). The reduced model multiscale method (R3M) for the non-linear homogenization of hyperelastic media at finite strains. *Journal of Computational Physics, 223*(1), 341–368.

Zhang, Z., & Naga, A. (2005). A new finite element gradient recovery method: Superconvergence property. *SIAM Journal on Scientific Computing, 26*(4), 1192–1213.

Zhao, J. X., Coupez, T., Decencière, E., Jeulin, D., Cárdenas-Peña, D., & Silva, L. (2016). Direct multiphase mesh generation from 3D images using anisotropic mesh adaptation and a redistancing equation. *Computer Methods in Applied Mechanics and Engineering, 309*, 288–306.

Zheng, X., Lowengrub, J., Anderson, A., & Cristini, V. (2005). Adaptive unstructured volume remeshing - II: Application to two- and three-dimensional level-set simulations of multiphase flow. *Journal of Computational Physics, 208*(2), 626–650.

Zienkiewicz, O. C., & Zhu, J. Z. (1987). A simple error estimator and adaptive procedure for practical engineerng analysis. *International Journal for Numerical Methods in Engineering, 24*(2), 337–357.

Printed in the United States
by Baker & Taylor Publisher Services